A Contemporary Study of Iterative Methods

A Contemporary Study of Iterative Methods

Convergence, Dynamics and Applications

Á. Alberto Magreñán
Universidad Internacional de La Rioja (UNIR),
Escuela Superior de Ingeniería y Tecnología,
26002 Logroño, La Rioja, Spain

Ioannis K. Argyros
Cameron University,
Department of Mathematical Sciences,
Lawton, OK 73505, USA

ACADEMIC PRESS
An imprint of Elsevier

Academic Press is an imprint of Elsevier
125 London Wall, London EC2Y 5AS, United Kingdom
525 B Street, Suite 1800, San Diego, CA 92101-4495, United States
50 Hampshire Street, 5th Floor, Cambridge, MA 02139, United States
The Boulevard, Langford Lane, Kidlington, Oxford OX5 1GB, United Kingdom

Notices

Knowledge and best practice in this field are constantly changing. As new research and experience
broaden our understanding, changes in research methods, professional practices, or medical treatment
may become necessary.

Practitioners and researchers must always rely on their own experience and knowledge in evaluating and
using any information, methods, compounds, or experiments described herein. In using such information
or methods they should be mindful of their own safety and the safety of others, including parties for
whom they have a professional responsibility.

To the fullest extent of the law, neither the Publisher nor the authors, contributors, or editors, assume any
liability for any injury and/or damage to persons or property as a matter of products liability, negligence
or otherwise, or from any use or operation of any methods, products, instructions, or ideas contained in
the material herein.

Library of Congress Cataloging-in-Publication Data
A catalog record for this book is available from the Library of Congress

British Library Cataloguing-in-Publication Data
A catalogue record for this book is available from the British Library

ISBN: 978-0-12-809214-9

For information on all Academic Press publications
visit our website at https://www.elsevier.com/books-and-journals

Working together
to grow libraries in
developing countries

www.elsevier.com • www.bookaid.org

Publisher: Candice Janco
Acquisition Editor: Graham Nisbet
Editorial Project Manager: Susan Ikeda
Production Project Manager: Swapna Srinivasan
Designer: Matthew Limbert

Typeset by VTeX

Dedication

To my parents Alberto and Mercedes, my grandmother Ascensión,
my beloved Lara.

Á. Alberto Magreñán

To my wife Diana, my children Christopher, Gus, Michael and Stacey,
my parents: Anastasia and Constantinos.

Ioannis Argyros

Contents

List of Figures

Chapter 1

The majorization method in the Kantorovich theory

1.1 INTRODUCTION

Let $\mathcal{X}$, $\mathcal{Y}$ be Banach spaces and let $\mathcal{D} \subseteq \mathcal{X}$ be a closed and convex subset of $\mathcal{X}$. In the present chapter, we are concerned with the problem of approximating a locally unique solution $x^\star$ of

$$F(x) = 0, \tag{1.1.1}$$

where $F : \mathcal{D} \longrightarrow \mathcal{Y}$ is a Fréchet-differentiable operator.

By means of using mathematical modeling as it can be seen in [10], [13], and [18], many problems in Applied Sciences can be brought in the form of equation (1.1.1).

Finding a solution of equation (1.1.1) in a closed form is not usually easy, and that is why we use iterative methods.

The most well-known and studied one is Newton's method, which is defined as

$$x_{n+1} = x_n - F'(x_n)^{-1} F(x_n) \quad \text{for each} \quad n = 0, 1, 2, \ldots, \tag{1.1.2}$$

where $x_0 \in \mathcal{D}$ is an initial point and $F'(x)$ denotes the Fréchet-derivative of F at the point $x \in \mathcal{D}$. Newton's method requires the inversion of linear operator $F'(x)$, as well as the computation of the function F at $x = x_n$ ($n \in \mathbb{N}$) at each step. If the starting point is close enough to the solution, Newton's method converges quadratically [10], [31]. The inversion of $F'(x)$ ($x \in \mathcal{D}$) at each step may be too expensive or unavailable. That is why the modified Newton's method

$$y_{n+1} = y_n - F'(y_0)^{-1} F(y_n) \quad \text{for each} \quad n = 0, 1, 2, \ldots \quad (y_0 = x_0 \in \mathcal{D}) \tag{1.1.3}$$

can be used in this case instead of Newton's method. However, the convergence of this modification is only linear as it is stated in [6], [9], [10], and [31].

The study about convergence of iterative methods is usually centered on two types: semilocal and local convergence analysis. The semilocal convergence matter is, based on the information around an initial point, to give criteria ensuring the convergence of the iterative method; while the local one is, based on the information around a solution, to find estimates of the radii of convergence balls. There exist several studies about the convergence of Newton's method, see

A Contemporary Study of Iterative Methods. DOI: 10.1016/B978-0-12-809214-9.00001-2

[10], [13], [18], [31], [34], and the ones presented by the authors of this chapter [1–3,19–24,30,32,33].

Concerning the semilocal convergence of Newton's method or Newton-like methods, one of the most important results is the celebrated Kantorovich theorem for solving nonlinear equations. This theorem provides a simple and transparent convergence criterion for operators with bounded second derivatives F'' or Lipschitz continuous first derivatives. The second type analysis for numerical methods is the local convergence. Traub and Woźniakowsi [46], Rheinboldt [44], [45], Rall [43], Argyros [10], and other authors gave estimates of the radii of local convergence balls when the Fréchet-derivatives are Lipschitz continuous around a solution.

Concerning the semilocal convergence of both methods, the Lipschitz-type condition

$$\| F'(x_0)^{-1} (F'(x) - F'(y)) \| \leq \omega(\| x - y \|) \quad \text{for each} \quad x, y \in \mathcal{D} \quad (1.1.4)$$

has been used [4], [5], [10], [25], [26–29], [35–42], [47–49], where ω is a strictly increasing and continuous function with

$$\omega(0) = 0. \quad (1.1.5)$$

If $\omega(t) = L t$ for $t \geq 0$, we obtain the Lipschitz case, whereas if $\omega(t) = L t^{\mu}$ for $t \geq 0$ and fixed $\mu \in [0, 1)$, we obtain the Hölder case. Sufficient convergence criteria in the references, as well as error bounds on the distances $\| x_{n+1} - x_n \|$, $\| x_n - x^* \|$ for each $n = 0, 1, 2, \ldots$, have been established using the majorizing sequence $\{u_n\}$ given by

$$u_0 = 0, \quad u_1 = \eta > 0,$$

$$u_{n+2} = u_{n+1} + \frac{\int_0^1 \omega(\theta (u_{n+1} - u_n)) \, d\theta (u_{n+1} - u_n)}{1 - \omega(u_{n+1})} \quad (1.1.6)$$

$$\text{for each} \quad n = 0, 1, 2, \ldots$$

Using (1.1.6) (see [10], [4], [5]) it is easy that

$$u_{n+2} \leq u_{n+1} + \frac{\chi(u_{n+1})}{1 - \omega(u_{n+1})} \quad \text{for each} \quad n = 0, 1, 2, \ldots$$

where

$$\chi(t) = \eta - t + \int_0^1 \overline{\omega}(r) \, dr \quad \text{and} \quad \overline{\omega}(t) = \sup_{t_1 + t_2 = t} (\omega(t_1) + \omega(t_2)).$$

Under the same or weaker convergence criteria, we provided a convergence analysis [6], [10], [16], [18] with the following advantages over the earlier stated works:

- tighter error bounds on the distances $\| x_{n+1} - x_n \|, \| x_n - x^\star \|$ for each $n = 0, 1, 2, \ldots$,
- and an at least as precise information on the location of the solution $x^\star$.

We obtain these advantages by means of introducing the center Lipschitz-condition

$$\| F'(x_0)^{-1} (F'(x) - F'(x_0)) \| \le \omega_0(\| x - x_0 \|) \quad \text{for each} \quad x \in \mathcal{D} \quad (1.1.7)$$

where ω_0 is a strictly increasing and continuous function with the same property as (1.1.5). Condition (1.1.7) follows from (1.1.4), and

$$\omega_0(t) \le \omega(t) \quad \text{for each} \quad t \ge 0 \tag{1.1.8}$$

holds in general. Note also that $\dfrac{\omega(t)}{\omega_0(t)}$ $(t \ge 0)$ can be arbitrarily large [6–18]. Using (1.1.4) it can be shown that

$$\| F'(x)^{-1} F'(x_0) \| \le \frac{1}{1 - \omega(\| x - x_0 \|)} \tag{1.1.9}$$

for each x in a certain subset $\mathcal{D}_0$ of $\mathcal{D}$. This estimate leads to majorizing sequence $\{u_n\}$ [10], [4], [5]. Now, using the less expensive and more precise (1.1.7), we obtain that

$$\| F'(x)^{-1} F'(x_0) \| \le \frac{1}{1 - \omega_0(\| x - x_0 \|)} \quad \text{for each} \quad x \in \mathcal{D}. \tag{1.1.10}$$

Inequality (1.1.10) is tighter than (1.1.9) if strict inequality holds in (1.1.8). This way we used majorizing sequence $\{t_n\}$ given by

$$t_0 = 0, \quad t_1 = \eta,$$

$$t_{n+2} = t_{n+1} + \frac{\int_0^1 \omega(\theta (t_{n+1} - t_n)) \, d\theta \, (t_{n+1} - t_n)}{1 - \omega_0(t_{n+1})} \quad \text{for each} \quad n = 0, 1, 2, \ldots$$

$$\tag{1.1.11}$$

A simple inductive argument shows

$$t_n \le u_n, \tag{1.1.12}$$

$$t_{n+1} - t_n \le u_{n+1} - u_n, \tag{1.1.13}$$

and

$$t^\star = \lim_{n \to \infty} t_n \le u^\star = \lim_{n \to \infty} u_n. \tag{1.1.14}$$

In the case of modified Newton's method [6], [9], [10], [31], the majorizing sequences used are

$$\bar{u}_0 = 0, \quad \bar{u}_1 = \eta,$$

$$\bar{u}_{n+2} = \bar{u}_{n+1} + \int_0^1 \omega(\bar{u}_n + \theta\,(\bar{u}_{n+1} - \bar{u}_n))\,d\theta\,(\bar{u}_{n+1} - \bar{u}_n) \qquad (1.1.15)$$

for each $n = 0, 1, 2, \ldots$

Using (1.1.15) we obtain

$$\bar{u}_{n+2} \le \bar{u}_{n+1} + \chi(\bar{u}_{n+1}) \quad \text{for each} \quad n = 0, 1, 2, \ldots$$

However, we used tighter majorizing sequences $\{\bar{t}_n\}$ given by

$$\bar{t}_0 = 0, \quad \bar{t}_1 = \eta,$$

$$\bar{t}_{n+2} = \bar{t}_{n+1} + \int_0^1 \omega_0(\bar{t}_n + \theta\,(\bar{t}_{n+1} - \bar{t}_n))\,d\theta\,(\bar{t}_{n+1} - \bar{t}_n) \qquad (1.1.16)$$

for each $n = 0, 1, 2, \ldots$

Moreover, we obtain

$$\bar{t}_{n+2} \le \bar{t}_{n+1} + \chi_0(t) \quad \text{for each} \quad n = 0, 1, 2, \ldots,$$

where

$$\chi_0(t) = \eta - t + \int_0^1 \bar{\omega}_0(r)\,dr \quad \text{and} \quad \bar{\omega}_0(t) = \sup_{t_1 + t_2 = t}\,(\omega_0(t_1) + \omega_0(t_2)).$$

It follows from (1.1.7) that there exist functions ω_{-2} and ω_{-1} with the same properties as ω_0 such that

$$\| F'(x_0)^{-1}\,(F'(x_0 + \theta\,(F'(x_0)^{-1}\,F(x_0))) - F'(x_0)) \|$$
$$\le \omega_{-1}(\theta \parallel F'(x_0)^{-1}\,F(x_0) \parallel) \quad \text{for each} \quad \theta \in [0, 1] \qquad (1.1.17)$$

and

$$\| F'(x_0)^{-1}\,(F'(x_0 - F'(x_0)^{-1}\,F(x_0)) - F'(x_0)) \| \le \omega_{-2}(\parallel F'(x_0)^{-1}\,F(x_0) \parallel). \qquad (1.1.18)$$

Function ω_{-2} can certainly be chosen to be a constant function. For example,

$$\omega_{-2}(t) = L_{-2} > 0 \quad \text{for each} \quad t \ge 0.$$

We have that

$$\omega_{-2}(t) \le \omega_{-1}(t) \le \omega_0(t) \quad \text{for each} \quad t \ge 0.$$

Note that (1.1.7), (1.1.17), and (1.1.18) are not additional to (1.1.4) hypotheses, since in practice finding function ω requires also finding functions ω_{-2}, ω_{-1}, and ω_0. Furthermore, the verification of (1.1.17) and (1.1.18) requires only computations at the initial data.

In the present chapter, we still use (1.1.4) and (1.1.7). However, we use an even tighter than $\{t_n\}$ majorizing sequence $\{s_n\}$ given by

$$s_0 = 0, \quad s_1 = \eta, \quad s_2 = s_1 + \frac{\int_0^1 \omega_{-1}(\theta\,(s_1 - s_0))\,d\theta\,(s_1 - s_0)}{1 - \omega_{-2}(s_1)},$$

$$s_{n+2} = s_{n+1} + \frac{\int_0^1 \omega(\theta\,(s_{n+1} - s_n))\,d\theta\,(s_{n+1} - s_n)}{1 - \omega_0(s_{n+1})}$$

$$\text{for each} \quad n = 1, 2, 3, \ldots$$

(1.1.19)

Under the same or weaker convergence criteria (than those for $\{t_n\}$), we get

$$s_n \leq t_n,$$ (1.1.20)

$$s_{n+1} - s_n \leq t_{n+1} - t_n,$$ (1.1.21)

and

$$s^\star = \lim_{n \to \infty} s_n \leq t^\star.$$ (1.1.22)

Our results for the semilocal convergence extend those in [17] when restricted to the Lipschitz case.

Related to the local convergence matter, the condition

$$\| F'(x^\star)^{-1}\,(F'(x) - F'(y)) \| \leq v(\| x - y \|) \quad \text{for each} \quad x, y \in \mathcal{D} \quad (1.1.23)$$

has been used [10], [18], [31], [34], [43], where function v is as ω. The radius of convergence has been enlarged and error bounds on the distances $\| x_n - x^\star \|$ $(n \geq 0)$ have been found to be tighter [10], [18] using (1.1.23) in combination with

$$\| F'(x^\star)^{-1}\,(F'(x) - F'(x^\star)) \| \leq v_\star(\| x - x^\star \|) \quad \text{for each} \quad x \in \mathcal{D}, \quad (1.1.24)$$

where the function $v_\star$ is as ω_0. In this case we obtain

$$\| F'(x)^{-1}\,F'(x^\star) \| \leq \frac{1}{1 - v(\| x - x^\star \|)} \quad \text{for each} \quad x \in \mathcal{D}, \quad (1.1.25)$$

used by different authors (see [6], [10], [16], [18])

$$\| F'(x)^{-1}\,F'(x^\star) \| \leq \frac{1}{1 - v_\star(\| x - x^\star \|)} \quad \text{for each} \quad x \in \mathcal{D}, \quad (1.1.26)$$

which is tighter than (1.1.25). We shall also use the conditions

$$\| F'(x^\star)^{-1} (F'(x^\star + \theta (x_0 - x^\star)) - F'(x_0)) \|$$
$$\leq v^0((1 - \theta) \| x_0 - x^\star \|) \quad \text{for each} \quad \theta \in [0, 1], \tag{1.1.27}$$

$$\| F'(x^\star)^{-1} (F'(x_0) - F'(x^\star)) \| \leq v_\star^0(\| x_0 - x^\star \|) \quad \text{for each} \quad \theta \in [0, 1], \tag{1.1.28}$$

and

$$\| F'(x^\star)^{-1} (F'(x^\star + \theta (x - x^\star)) - F'(x_0)) \|$$
$$\leq v_1(\| x^\star + \theta (x - x^\star) - x_0 \|) \quad \text{for each} \quad x \in \mathcal{D}, \tag{1.1.29}$$

where v^0, $v_\star^0$, and v_1 are functions with the same properties as the other v functions. In the present chapter we show that the new error bounds can be tighter than the old ones [10], [17], [18].

The chapter is organized as follows. Section 1.2 contains the semilocal convergence of Newton's methods. The local convergence is examined in Section 1.3. Finally, some applications are shown in Section 1.4 while some numerical examples are given in Section 1.5.

1.2 SEMILOCAL CONVERGENCE

First of all, we need to introduce some results for the convergence of majorizing sequences for Newton's method. Let s_1, s_2 be given by (1.1.19). It is convenient for us to define a sequence of functions g_n on $[0, 1)$ for each $n = 1, 2, \ldots$ by

$$
\begin{aligned}
g_n(t) &= \int_0^1 \left(\omega(\theta (s_2 - s_1) t^{k+1}) - \omega(\theta (s_2 - s_1) s^\star) \right) dt \\
&\quad + t \left(\omega_0(s_1 + \frac{1 - t^{k+2}}{1 - t} (s_2 - s_1)) - \omega_0(s_1 + \frac{1 - t^{k+1}}{1 - t} (s_2 - s_1)) \right).
\end{aligned}
\tag{1.2.1}
$$

Denote by $\mathcal{C}(\omega_0, \omega, \alpha)$ the class of strictly increasing and continuous functions on the interval $[0, +\infty)$ with $\omega(0) = \omega_0(0) = 0$ and $\alpha \in [0, 1)$ such that

$$g_n(\alpha) = 0 \quad \text{for each} \quad n = 1, 2, \ldots \tag{1.2.2}$$

Note that $\mathcal{C}(\omega_0, \omega, \alpha)$ is nonempty. Moreover, let us define

$$\omega(t) = L t \quad \text{and} \quad \omega_0(t) = L_0 t \tag{1.2.3}$$

for some positive constants L_0 and L. Then, we have by (1.2.1) and (1.2.3) that

$$g_n(t) = \frac{1}{2} (2 L_0 t^2 + L t - L) t^n (s_2 - s_1). \tag{1.2.4}$$

Set

$$\alpha = \frac{2L}{L + \sqrt{L^2 + 8LL_0}}.$$

(1.2.5)

Note that $\alpha \in (0, 1)$. Then (1.2.2) holds for α given by (1.2.5).

Lemma 1.2.1. *Set*

$$\alpha_n = \frac{\int_0^1 \omega(t(s_{n+1} - s_n)) \, dt}{1 - \omega_0(s_{n+1})} \qquad for\ each \quad n = 1, 2, \dots$$

(1.2.6)

Furthermore, suppose that functions ω_0 and ω are in the class $\mathcal{C}(\omega_0, \omega, \alpha)$,

$$\omega_0(\eta) < 1,$$

(1.2.7)

and there exists $\alpha_1 \in (0, 1)$ such that

$$0 < \alpha_1 \le \alpha \le 1 - \frac{s_2 - s_1}{\omega_0^{-1}(1) - \eta}.$$

(1.2.8)

Then, sequence $\{s_n\}$ given by (1.1.19) is well defined, strictly increasing, bounded from above by

$$s^{**} = \left(1 + \frac{\int_0^1 \omega(\theta \eta) \, d\theta}{(1 - \alpha)(1 - \omega_0(\eta))}\right) \eta$$

(1.2.9)

and converges to its unique least upper bound s^* which satisfies

$$0 \le s^* \le s^{**}.$$

(1.2.10)

Moreover, the following estimates hold:

$$0 < s_{n+2} - s_{n+1} \le \frac{\int_0^1 \omega(\theta \eta) \eta \, d\theta}{1 - \omega_0(\eta)} \alpha^n \qquad for\ each \quad n = 1, 2, \dots$$

(1.2.11)

Proof. By means of using mathematical induction, we will prove that

$$0 < \alpha_k \le \alpha \quad for\ each \quad k = 1, 2, \dots$$

(1.2.12)

Estimate (1.2.12) holds for $k = 1$ by (1.2.8). Then we have by (1.1.19) and (1.2.12) (for $k = 1$) that

$$
\begin{aligned}
0 < s_3 - s_2 \le \alpha(s_2 - s_1) \implies & \ s_3 \le s_2 + \alpha(s_2 - s_1) \\
\implies & \ s_3 \le s_2 + (1 + \alpha)(s_2 - s_1) - (s_2 - s_1) \\
\implies & \ s_3 \le s_1 + \frac{1 - \alpha^2}{1 - \alpha}(s_2 - s_1) < s^{**}.
\end{aligned}
$$

(1.2.13)

Assume that (1.2.12) holds for all natural integer $n \leq k$. Then we get by (1.1.19) and (1.2.12) that

$$0 < s_{k+2} - s_{k+1} \leq \alpha^k (s_2 - s_1) \tag{1.2.14}$$

and

$$s_{k+2} \leq s_1 + \frac{1 - \alpha^{k+1}}{1 - \alpha} (s_2 - s_1) < s^{**}. \tag{1.2.15}$$

Evidently, estimate (1.2.12) is true if k is replaced by $k+1$, provided that

$$\int_0^1 \omega(\theta (s_{n+1} - s_n)) d\theta + \alpha \omega_0(s_{k+2}) - \alpha \leq 0$$

or

$$\int_0^1 \omega(\theta (s_2 - s_1) \alpha^k) d\theta + \alpha \omega_0(s_1 + \frac{1 - \alpha^{k+1}}{1 - \alpha} (s_2 - s_1)) - \alpha \leq 0. \tag{1.2.16}$$

Moreover, estimate (1.2.16) motivates us to define recurrent functions f_k on $[0, 1)$ by

$$f_k(t) = \int_0^1 \omega(\theta (s_2 - s_1) t^k) d\theta + t \omega_0(s_1 + \frac{1 - t^{k+1}}{1 - t} (s_2 - s_1)) - t. \tag{1.2.17}$$

We need to find a relationship between two consecutive functions f_k. We get

$$f_{k+1}(t) = f_k(t) + g_k(t), \tag{1.2.18}$$

where g_k is given by (1.2.1). Then estimate (1.2.16) is certainly true if

$$f_k(\alpha) \leq 0 \quad \text{for each} \quad k = 1, 2, \ldots \tag{1.2.19}$$

In view of (1.2.2) and (1.2.18), we obtain

$$f_{k+1}(\alpha) = f_k(\alpha) \quad \text{for each} \quad k = 1, 2, \ldots \tag{1.2.20}$$

Define function f_∞ on $[0, 1)$ by

$$f_\infty(t) = \lim_{k \to \infty} f_k(t). \tag{1.2.21}$$

Then we get by (1.2.19)–(1.2.21) that (1.2.19) is satisfied, provided that

$$f_\infty(\alpha) \leq 0. \tag{1.2.22}$$

Using (1.2.17) and (1.2.21) we obtain that

$$f_\infty(\alpha) = \alpha \left(\omega_0(s_1 + \frac{s_2 - s_1}{1 - \alpha}) - 1 \right). \tag{1.2.23}$$

That is, (1.2.22) is true by (1.2.8) and (1.2.23). Consequently, induction for (1.2.11) is complete. Hence, sequence $\{s_n\}$ is increasing, bounded from above by $s^{\star\star}$ given by (1.2.9), and as such that it converges to its unique least upper bound $s^{\star}$ which satisfies (1.2.10). □

Next, we have the following useful extension of Lemma 1.2.1.

Lemma 1.2.2. *Suppose there exists a minimal natural integer $N > 1$ and $\alpha \in (0, 1)$ such that*

$$s_1 < s_2 < \cdots < s_{N+1} < \omega_0^{-1}(1)$$

and

$$0 < \alpha_N \le \alpha \le 1 - \frac{s_{N+1} - s_N}{\omega_0^{-1}(1) - s_N}.$$

Then, the sequence $\{s_n\}$ given by (1.1.19) is well defined, strictly increasing, bounded from above by

$$s_N^{\star\star} = s_{N-1} + \frac{s_N - s_{N-1}}{1 - \alpha},$$

and converges to its unique least upper bound $s_N^{\star}$ which satisfies $0 \le s_N^{\star} \le s_N^{\star\star}$. Moreover, the following estimates hold:

$$0 < s_{N+n} - s_{N+n-1} \le (s_{N+1} - s_N)\alpha^n \quad for\ each \quad n = 1, 2, \ldots$$

Remark 1.2.3. *The conclusions of Lemma 1.2.1 hold if $g_n(\alpha) \ge 0$ for each $n = 1, 2, \ldots$ holds instead of (1.2.2). Another possibility is given by the set of conditions*

$$g_n(\alpha) \le 0 \quad for\ each \quad n = 1, 2, \ldots \quad and \quad f_1(\alpha) \le 0.$$

In this case we have again by (1.2.18) that (1.2.19) and (1.2.22) hold. Finally, the conclusions of Lemma 1.2.2 hold if

$$g_n(\alpha) \ge 0 \quad for\ each \quad n = N, N+1, \ldots$$

or

$$g_n(\alpha) \le 0 \quad for\ each \quad n = N, N+1, \ldots \quad and \quad f_N(\alpha) \le 0.$$

Next, we present results for computing upper bounds on the limit point $t^{\star}$ using sequence $\{t_n\}$.

Lemma 1.2.4. *Let $\lambda \in [0, \omega_0^{-1}(1)]$. Let f, ω_0 be differentiable functions on $[0, \omega_0^{-1}(1)]$. Suppose f has a zero in $[\lambda, \omega_0^{-1}(1)]$. Denote by ϱ the smallest zero of function f in $[\lambda, \omega_0^{-1}(1)]$. Define functions r and g on $[0, \omega_0^{-1}(1))$ by*

$$r(t) = \frac{f(t)}{1 - \omega_0(t)} \quad and \quad g(t) = t + r(t).$$

Moreover, suppose

$$g'(t) > 0 \quad for\ each \quad t \in [\lambda, \varrho].$$

Then, function g is strictly increasing and bounded above by ϱ.

Proof. Function r is well defined on $[\lambda, \varrho]$ with the possible exception when $\varrho = \omega_0^{-1}(1)$; but the L'Hospital theorem implies that f admits a continuous extension on the interval $[\lambda, \varrho]$. The function g is strictly increasing, since $g'(t) > 0$ on $[\lambda, \varrho]$. Therefore, we have for each $t \in [\lambda, \varrho]$ that

$$g(t) = t + r(t) \leq \varrho + r(\varrho) = \varrho. \qquad \square$$

Now, we have the following lemma.

Lemma 1.2.5. *Suppose that hypotheses of Lemma 1.2.4 hold. Define functions ϕ and ψ on $\mathcal{I} := [\lambda, t] \times [s, \varrho]$ for each $t \in [\lambda, \varrho]$ by*

$$\phi(s, t) = t + \frac{\int_0^1 \omega(\theta\,(t - s))\,d\theta\,(t - s)}{1 - \omega_0(t)}$$

and

$$\psi(s, t) = \begin{cases} \int_0^1 \omega(\theta\,(t - s))\,d\theta\,(t - s) - f(t) & if \quad t \neq \varrho, \\ 0 & if \quad t = \varrho. \end{cases}$$

Moreover, suppose that

$$\psi(s, t) \leq 0 \quad for\ each \quad (s, t) \in \mathcal{I}.$$

As a consequence, the following assertion holds:

$$\phi(s, t) \leq g(t) \quad for\ each \quad (s, t) \in \mathcal{I}.$$

Proof. The result follows immediately from the definition of functions g, ϕ, ψ, and the hypothesis of the lemma. $\qquad \square$

Lemma 1.2.6. *Let $N = 0, 1, 2, \ldots$ be fixed. Under the hypotheses of Lemma 1.2.5 with $\lambda = t_N$, further suppose that*

$$t_1 \leq t_2 \leq \cdots \leq t_N \leq t_{N+1} \leq \varrho$$

and

$$f(t_{N+1}) \geq 0.$$

Then, sequence $\{t_n\}$ generated by (1.1.11) is nondecreasing, bounded by ϱ, and converges to its unique least upper bound $t^\star$ which satisfies $t^\star \in [t_N, \varrho]$.

Proof. We can write

$$t_{n+1} = \phi(t_{n-1}, t_n).$$

Then, we obtain that

$$t_{N+2} = \phi(t_N, t_{N+1}) \le g(t_{N+1}) \le \varrho. \qquad \square$$

Next, we have the following remark.

Remark 1.2.7. *(a) Hypotheses of Lemma 1.2.6 are satisfied in the Kantorovich case [10], [31], i.e., if ω is given by (1.2.3) and $\omega_0 = \omega$ provided that*

$$h_\star = L\eta \le \frac{1}{2},$$

$$f(t) = \frac{L}{2}t^2 - t + \eta,$$

and

$$\varrho = \frac{1 - \sqrt{1 - 2h_\star}}{L}.$$

(b) Cartesian product $\mathcal{I}$ can be replaced by the more practical $\mathcal{J} = [\lambda, \varrho]^2$ in Lemma 1.2.6.

From now on, let $U(x, r)$ and $\overline{U}(x, r)$ stand, respectively, for the open and closed ball in $\mathcal{X}$ with center x and radius $r > 0$. Let also $\mathcal{L}(\mathcal{Y}, \mathcal{X})$ stand for the space of bounded linear operators from $\mathcal{Y}$ into $\mathcal{X}$. For $x_0 \in D$, let us denote by $\mathcal{K}(x_0, F, \omega_0, \omega)$ the class of triplets (F, ω_0, ω) such that operators F satisfy (1.1.4) and (1.1.7), functions ω, ω_0 are in $\mathcal{C}(\omega_0, \omega)$ and $\omega(0) = \omega_0(0)$. We also denote by $\mathcal{K}(x_0, F, \omega_0)$ the class of pairs (F, ω_0) such that $(F, \omega_0, \omega_0) \in \mathcal{K}(x_0, F, \omega_0, \omega_0)$.

We present respectively the following semilocal results for Newton's method and modified Newton's method. The proofs can be found from the corresponding ones in [6], [18] by simply replacing the hypotheses given on the convergence of $\{t_n\}$, $\{\bar{t}_n\}$ by the corresponding ones for $\{s_n\}$, $\{\bar{s}_n\}$ given in Lemma 1.2.1, Lemma 1.2.2, or Lemma 1.2.6.

Theorem 1.2.8. *Suppose the triplet $(F, \omega_0, \omega) \in \mathcal{K}(x_0, F, \omega_0, \omega)$; hypotheses of Lemma 1.2.1, Lemma 1.2.2, or Lemma 1.2.6 hold;*

$$\| F'(x_0)^{-1} F(x_0) \| \le \eta,$$

and $U(x_0, s^\star) \subseteq D$ if Lemma 1.2.1 or Lemma 1.2.2 applies (or $U(x_0, t^\star) \subseteq D$ if Lemma 1.2.6 applies). Then, sequence $\{x_n\}$ generated by Newton's method is well defined, remains in $\overline{U}(x_0, t^\star)$ for all $n \ge 0$, and converges to a solution $x^\star \in \overline{U}(x_0, t^\star)$ of equation $F(x) = 0$. Moreover, the following estimates hold:

(a) *If Lemma 1.2.1 or Lemma 1.2.2 applies, we have that*

$$\| x_{n+1} - x_n \| \le s_{n+1} - s_n \quad and \quad \| x_n - x^* \| \le s^* - s_n;$$

(b) *If Lemma 1.2.6 applies, we obtain*

$$\| x_{n+1} - x_n \| \le t_{n+1} - t_n \quad and \quad \| x_n - x^* \| \le t^* - t_n \le \varrho - t_n.$$

Furthermore, if there exists $R \ge t^*$ such that $\overline{U}(x_0, R) \subseteq D$ and $\int_0^1 \omega_0((1 - \theta)R + t^*)\, d\theta < 1$, then the solution x^* of equation $F(x) = 0$ is unique in $U(x_0, R)$.

Theorem 1.2.9. *Suppose the pair $(F, \omega_0) \in \mathcal{K}(x_0, F, \omega_0)$; hypotheses of Lemma 1.2.4 hold with $r(t) = f(t)$; $\phi(t, s) = t + \int_0^1 \omega(s + \theta (t - s))\, d\theta\, (t - s)$ and $U(x_0, \overline{t^*}) \subseteq D$, where $\overline{t}_n$ is given by (1.1.16) and $\overline{t^*} = \lim\limits_{n \to \infty} \overline{t}_n$. Then the following assertions hold*

(a) *Sequence $\{\overline{t}_n\}$ is strictly increasing and converges to $\overline{t^*} \in [0, \varrho]$.*

(b) *Modified Newton's method $\{y_n\}$ generated by (1.1.3) is well defined, remains in $\overline{U}(x_0, \overline{t^*})$ for all $n \ge 0$, and converges to a solution $x^* \in \overline{U}(x_0, \overline{t^*})$ of equation $F(x) = 0$. Moreover, the following estimates hold:*

$$\| y_{n+1} - y_n \| \le \overline{t}_{n+1} - \overline{t}_n \quad and \quad \| y_n - x^* \| \le \overline{t^*} - \overline{t}_n.$$

Furthermore, if there exists $R \ge \overline{t^*}$ such that $\overline{U}(x_0, R) \subseteq D$ and $\int_0^1 \omega_0((1 - \theta)R + \overline{t^*})\, d\theta < 1$, then the solution x^* of equation $F(x) = 0$ is unique in $U(x_0, R)$.

1.3   LOCAL CONVERGENCE

In this section, we present the local convergence results for Newton's method and modified Newton's method.

It follows from (1.1.23) that there exists a function v_0 like v such that

$$\| F'(x^*)^{-1} (F'(x) - F'(x_0)) \| \le v_0(\| x - x_0 \|) \quad \text{for each} \quad x \in D. \quad (1.3.1)$$

Let $x^* \in D$ be such that $F(x^*) = 0$ and $F'(x^*)^{-1} \in \mathcal{L}(\mathcal{Y}, \mathcal{X})$. The proofs can be essentially found in [6], where, however, (1.3.1) was not used.

Theorem 1.3.1. *Suppose that (1.1.23), (1.1.24), and (1.3.1) hold. Denote by R^* the smallest positive zero of function*

$$q(t) = \int_0^1 v((1 - \theta)t)\, d\theta + v_*(t) - 1 = 0.$$

Suppose also that

$$\overline{U}(x_0, R^\star) \subseteq \mathcal{D}.$$

Then sequence $\{x_n\}$ generated by Newton's method is well defined, remains in $\overline{U}(x_0, R^\star)$ for all $n \geq 0$, and converges to $x^\star$ provided that $x_0 \in \overline{U}(x^\star, R^\star)$. Moreover, the following estimates hold:

$$\| x_{n+1} - x^\star \| \leq e_n \| x_n - x^\star \|,$$

where

$$e_n = \frac{\int_0^1 \overline{v}((1-\theta) \| x_n - x^\star \|)\, d\theta}{1 - \overline{v}_\star(\| x_n - x^\star \|)},$$

$$\overline{v} = \begin{cases} v^0 & \text{if } n = 0, \\ v & \text{if } n > 0, \end{cases} \quad \text{and} \quad \overline{v}_\star = \begin{cases} v_\star^0 & \text{if } n = 0, \\ v_\star & \text{if } n > 0. \end{cases}$$

Theorem 1.3.2. *Suppose that (1.1.24), (1.1.29), and (1.3.1) hold. Denote by $R_1^\star$ the smallest positive zero of function*

$$q_1(t) = \int_0^1 v_1((1+\theta)t)\, d\theta + v_\star(t) - 1 = 0.$$

Suppose also that

$$\overline{U}(x_0, R_1^\star) \subseteq \mathcal{D}.$$

Then sequence $\{y_n\}$ generated by modified Newton's method is well defined, remains in $\overline{U}(x_0, R_1^\star)$ for all $n \geq 0$, and converges to $x^\star$ provided that $x_0 \in \overline{U}(x^\star, R_1^\star)$. Moreover, the following estimates hold:

$$\| y_{n+1} - x^\star \| \leq e_n^1 \| y_n - x^\star \|,$$

where

$$e_n^1 = \frac{\int_0^1 \overline{v}^1(\| x^\star - x_0 \| + \theta \| y_n - x^\star \|)\, d\theta}{1 - \overline{v}_\star(\| y_n - x^\star \|)}, \quad \overline{v}^1 = \begin{cases} v^0 & \text{if } n = 0, \\ v_1 & \text{if } n > 0, \end{cases}$$

and $\overline{v}_\star$ is given in Theorem 1.3.1.

Remark 1.3.3. *If $v_0 = v = v_\star$, the results reduce to earlier ones [4], [47–49]. Moreover, if $v_0 = v$ and $v \neq v_\star$, the results reduce to those in [6], [9], [10], [13], [18]. Otherwise, these results provide tighter bounds on the distances. The results obtained in this chapter can be extended immediately with some slight modifications in the case when the ω or v functions are simply nondecreasing, nonnegative, with $\lim_{t \to 0} \omega(t) = \lim_{t \to 0} v(t) = 0$ (see [4], [10], [13], [18]). Indeed,* *notice that hypothesis (1.2.8) in Lemma 1.2.1 will be replaced by the following assumption:*

   There exists $\alpha \in (0, 1)$ such that

$$0 < \alpha_1 \leq \alpha \quad and \quad \omega_0(\eta + \frac{s_2 - s_1}{1 - \alpha}) < 1; \tag{1.3.2}$$

whereas hypotheses of Lemma 1.2.2 will be

$$s_1 \leq s_2 \leq \cdots \leq s_{N+1}, \quad \omega_0(s_{N+1}) < 1,$$

and there exists $\alpha \in (0, 1)$ such that

$$0 < \alpha_N \leq \alpha \quad and \quad \omega_0(s_N + \frac{s_{N+1} - s_N}{1 - \alpha}) < 1.$$

In this case, sequence $\{s_n\}$ will be nondecreasing. With the above changes the results of the semilocal case hold in this more general setting.

1.4   APPLICATIONS

Application 1.4.1. *(a) Let us consider the interesting Lipschitz case for Newton's method. Define*

$$\omega_{-2}(t) = L_{-2}\,t, \quad \omega_{-1}(t) = L_{-1}\,t$$

for some positive constants L_{-2} and L_{-1} with

$$L_{-2} \leq L_{-1} \leq L_0 \leq L. \tag{1.4.1}$$

The conditions of Lemma 1.2.2 will hold for $N = 1$ if

$$\eta + \frac{L_{-2}\,\eta^2}{2(1 - L_{-1}\eta)} < \frac{1}{L_0} \tag{1.4.2}$$

and

$$\alpha_1 = \frac{\dfrac{L}{2}\dfrac{L_{-2}\,\eta^2}{2(1 - L_{-1}\eta)}}{1 - L_0\left(\eta + \dfrac{L_{-2}\,\eta^2}{2(1 - L_{-1}\eta)}\right)} \leq \alpha \leq \alpha_2 = 1 - \frac{L_0\dfrac{L_{-2}\,\eta^2}{2(1 - L_{-1}\eta)}}{1 - L_0\eta}. \tag{1.4.3}$$

In order for us to solve the system of inequalities (1.4.2) and (1.4.3), it is convenient to define quadratic polynomials p_1 and p_2 by

$$p_1(t) = \beta_1\,t^2 + \beta_2\,t + \beta_3$$

and

$$p_2(t) = \gamma_1 t^2 + \gamma_2 t + \gamma_3,$$

where

$$\beta_1 = L,$$

$$L_{-2} + 2\alpha L_0 (2 L_{-1} - L_{-2}),$$

$$\beta_2 = 4\alpha (L_{-1} - L_0),$$

$$\beta_3 = -4\alpha,$$

$$\gamma_1 = L_0 (L_{-2} - 2 (1 - \alpha) L_{-1}),$$

$$\gamma_2 = 2 (1 - \alpha) (L_0 + L_{-1}),$$

and

$$\gamma_3 = -2 (1 - \alpha).$$

In view of (1.2.5), (1.4.1), (1.4.2), and (1.4.3), polynomial p_1 has a positive root

$$\lambda_1 = \frac{-\beta_2 + \sqrt{\beta_2^2 - 4\beta_1 \beta_3}}{2\beta_1}$$

and a negative root.
Inequalities (1.4.2) and (1.4.3) are satisfied if

$$p_1(\eta) \leq 0 \tag{1.4.4}$$

and

$$p_2(\eta) \leq 0. \tag{1.4.5}$$

Therefore, (1.4.4) is satisfied if

$$\eta \leq \lambda_1. \tag{1.4.6}$$

If $\gamma_1 < 0$ and $\Delta_2 := \gamma_2^2 - 4\gamma_1 \gamma_3 \leq 0$, (1.4.5) always holds. If $\gamma_1 < 0$ and $\Delta_2 > 0$, then p_2 has two positive roots. The smaller of the two is denoted by λ_2 and given by

$$\lambda_2 = \frac{-\gamma_2 - \sqrt{\Delta_2}}{2\gamma_1}. \tag{1.4.7}$$

If $\gamma_1 > 0$, polynomial p_2 has a positive root λ_3 given by

$$\lambda_3 = \frac{-\gamma_2 + \sqrt{\Delta_2}}{2\gamma_1} \tag{1.4.8}$$

and a negative root.
Let us define

$$\frac{L_4^{-1}}{2} = \begin{cases} \lambda_1 & \text{if} \quad \gamma_1 < 0 \quad \text{and} \quad \Delta_2 \le 0, \\ \min\{\lambda_1, \lambda_2\} & \text{if} \quad \gamma_1 < 0 \quad \text{and} \quad \Delta_2 > 0, \\ \min\{\lambda_1, \lambda_3\} & \text{if} \quad \gamma_1 > 0. \end{cases} \tag{1.4.9}$$

Summarizing we conclude that (1.4.2) and (1.4.3) are satisfied if

$$h_4 = L_4\,\eta \le \frac{1}{2}. \tag{1.4.10}$$

In the special case when $L_{-2} = L_{-1} = L_0$, elementary computations show $L_3 = L_4$. That is, (1.4.10) reduces to

$$h_3 = L_3\,\eta \le \frac{1}{2},$$

where

$$L_3 = \frac{1}{8}\left((L_0\,L)^{1/2} + 4\,L_0 + (8\,L_0^2 + L_0\,L)^{1/2}\right),$$

which is the weaker than the "h" conditions given in [17]. The rest of the "h" conditions are

$$h_\star = L\,\eta \le \frac{1}{2},$$

$$h_1 = L_1\,\eta \le \frac{1}{2},$$

$$h_2 = L_2\,\eta \le \frac{1}{2},$$

where

$$L_1 = \frac{L_0 + L}{2} \quad \text{and} \quad L_2 = \frac{1}{8}\left(L + 4\,L_0 + (L^2 + 8\,L_0\,L)^{1/2}\right).$$

Note that

$$h_\star \le \frac{1}{2} \implies h_1 \le \frac{1}{2} \implies h_2 \le \frac{1}{2} \implies h_3 \le \frac{1}{2} \implies h_4 \le \frac{1}{2},$$

but not necessarily vice versa, unless $L_{-2} = L_{-1} = L_0 = L$.

In view of the definition of β_1, condition h_4 can be improved if L can be replaced by $L_\star$ such that $0 < L_\star < L$. This requires the verification of condition

$$\| F'(x_0)^{-1} (F'(x_1 + \theta (x_2 - x_1)) - F'(x_1)) \| \le L_\star \theta \| x_2 - x_1 \|$$
for each $\theta \in [0, 1]$,

where x_1 and x_2 are given by Newton's iterations. The computation of $L_\star$ is possible and requires computations at the initial data. Note also that iterate s_3 will be given by

$$s_3 = s_2 + \frac{L_\star (s_2 - s_1)^2}{2 (1 - L_0 s_2)}$$

instead of

$$s_3 = s_2 + \frac{L (s_2 - s_1)^2}{2 (1 - L_0 s_2)}.$$

Moreover, condition h_4 is then replaced by the at least as weak

$$h_5 = L_5 \, \eta \le \frac{1}{2},$$

where L_5 is defined as L_4 with $L_\star$ replacing L in the definition of β_1.
(b) *Another extension of our results in the case where functions ω_0 and ω are defined by (1.2.3) is given as follows. Let $N = 1, 2, \ldots$ and $R \in (0, 1/L_0)$. Assume $x_1, x_2, \ldots, x_N$ can be computed by Newton's iterations, $F'(x_N)^{-1} \in \mathcal{L}(\mathcal{Y}, \mathcal{X})$ such that $\| F'(x_N)^{-1} F(x_N) \| \le R - \| x_N - x_0 \|$. Set $\mathcal{D} = \overline{U}(x_0, R)$ and $\mathcal{D}_N = \overline{U}(x_N, R - \| x_N - x_0 \|)$. Then, for all $x \in \mathcal{D}_N$, we have that*

$$\| F'(x_N)^{-1} (F'(x) - F'(y)) \|$$
$$\le \| F'(x_N)^{-1} F'(x_0) \| \, \| F'(x_0)^{-1} (F'(x) - F'(y)) \|$$
$$\le \frac{L}{1 - L_0 \| x_N - x_0 \|} \| x - y \|$$

and

$$\| F'(x_N)^{-1} F(x_N) \| \le \frac{L}{2} \frac{L}{1 - L_0 \| x_N - x_0 \|} \| x_N - x_{N-1} \|^2.$$

Set

$$L^N = L_0^N = \frac{L}{1 - L_0 \| x_N - x_0 \|} \quad and \quad \eta^N = \frac{L L_N}{2} \| x_N - x_{N-1} \|^2.$$

Then the Kantorovich hypothesis becomes

$$h_\star^N = L^N \eta^N \leq \frac{1}{2}.$$

Clearly, the most interesting case is when $N = 1$. In this case the Kantorovich condition becomes

$$h_\star^1 = L_6 \eta \leq \frac{1}{2},$$

where

$$L_6 = \frac{L_0 + L\sqrt{L}}{2}.$$

Note also that L_6 can be smaller than L.

1.5 EXAMPLES

First, we provide two examples where $v_\star < v$.

Example 1.5.1. *Let $\mathcal{X} = \mathcal{Y} = C[0, 1]$, the space of continuous functions defined on $[0, 1]$, be equipped with the max norm and $\mathcal{D} = \overline{U}(0, 1)$. Define function F on $\mathcal{D}$ by*

$$F(h)(x) = h(x) - 5 \int_0^1 x \theta \, h(\theta)^3 \, d\theta. \qquad (1.5.1)$$

Then we have

$$F'(h[u])(x) = u(x) - 15 \int_0^1 x \theta \, h(\theta)^2 \, u(\theta) \, d\theta \quad \text{for all } u \in \mathcal{D}.$$

Using (1.5.1) we see that hypotheses of Theorem 1.3.1 hold for $x^\star(x) = 0$, where

$$x \in [0, 1],$$

$$v(t) = 15\,t,$$

and

$$v_\star(t) = 7.5\,t.$$

So we can ensure the convergence by Theorem 1.3.1.

Example 1.5.2. *Let $\mathcal{X} = \mathcal{Y} = \mathbb{R}$. Define function F on $\mathcal{D} = [-1, 1]$ by*

$$F(x) = e^x - 1. \qquad (1.5.2)$$

Then, using (1.5.2) for $x^* = 0$, we get that

$$F(x^*) = 0$$

and

$$F'(x^*) = e^0 = 1.$$

Moreover, hypotheses of Theorem 1.3.1 hold for

$$v(t) = e\,t$$

and

$$v_\star(t) = (e - 1)\,t.$$

Note that $v_\star < v$. So we can ensure the convergence by Theorem 1.3.1.

Example 1.5.3. Let $\mathcal{X} = \mathcal{Y} = C[0, 1]$, equipped with the max-norm. Let $\theta \in [0, 1]$ be a given parameter. Consider the "cubic" integral equation

$$u(s) = u^3(s) + \lambda u(s) \int_0^1 \mathcal{G}(s, t)\, u(t)\, dt + y(s) - \theta. \qquad (1.5.3)$$

Nonlinear integral equations of the form (1.5.3) are considered Chandrasekhar-type equations [10], [13], [18], and they arise in the theories of radiative transfer, neutron transport, and in the kinetic theory of gasses. Here the kernel $\mathcal{G}(s, t)$ is a continuous function of two variables $(s, t) \in [0, 1] \times [0, 1]$ satisfying

(i) $0 < \mathcal{G}(s, t) < 1$,
(ii) $\mathcal{G}(s, t) + \mathcal{G}(t, s) = 1$.

The parameter λ is a real number called the "albedo" for scattering; $y(s)$ is a given continuous function defined on $[0, 1]$, and $x(s)$ is the unknown function sought in $C[0, 1]$. For simplicity, we choose

$$u_0(s) = y(s) = 1$$

and

$$\mathcal{G}(s, t) = \frac{s}{s + t}$$

for all $(s, t) \in [0, 1] \times [0, 1]$ $(s + t \neq 0)$.

Let $\mathcal{D} = U(u_0, 1 - \theta)$ and define the operator F on $\mathcal{D}$ by

$$F(x)(s) = x^3(s) - x(s) + \lambda x(s) \int_0^1 \mathcal{G}(s, t)\, x(t)\, dt + y(s) - \theta \qquad (1.5.4)$$

for all $s \in [0, 1]$.

TABLE 1.1 Different values of the parameters involved.

λ	θ	h_*	h_1	h_2	h_3
0.97548	0.954585	0.4895734208	0.4851994045	0.4837355633	0.4815518345
0.8457858	0.999987	0.4177974405	0.4177963046	0.4177959260	0.4177953579
0.3245894	0.815456854	0.5156159025	0.4967293568	0.4903278739	0.4809439506
0.3569994	0.8198589998	0.5204140737	0.5018519741	0.4955632842	0.4863389899
0.3789994	0.8198589998	0.5281518448	0.5093892893	0.5030331107	0.4937089648
0.458785	0.5489756	1.033941504	0.9590659445	0.9332478337	0.8962891928

Then every zero of F satisfies equation (1.5.3). Therefore, the operator F' sat-
isfies conditions of Theorem 1.2.8, with

$$\eta = \frac{|\lambda| \ln 2 + 1 - \theta}{2 \left(1 + |\lambda| \ln 2\right)},$$

$$\omega(t) := L t = \left(\frac{|\lambda| \ln 2 + 3 (2 - \theta)}{1 + |\lambda| \ln 2}\right) t,$$

and

$$\omega_0(t) := L_0 t = \left(\frac{2 |\lambda| \ln 2 + 3 (3 - \theta)}{2 \left(1 + |\lambda| \ln 2\right)}\right) t.$$

It follows from our main results that if one condition in Application 1.4.1 holds,
then problem (1.5.3) has a unique solution near u_0. This assumption is weaker
than the one given before using the Newton–Kantorovich hypothesis. Note also
that $L_0 < L$ for all $\theta \in [0, 1]$ and $\omega_0 < \omega$. Next we pick some values of λ and
θ such that all hypotheses are satisfied, so we can compare the "h" conditions
(see Table 1.1).

Example 1.5.4. *Let*

$$\mathcal{X} = \mathcal{Y} = \mathbb{R},$$

$$x_0 = 1,$$

$$\mathcal{D} = [\xi, 2 - \xi],$$

$$\xi \in [0, 0.5).$$

Define function F on $\mathcal{D}$ by

$$F(x) = x^3 - \xi. \tag{1.5.5}$$

(a) *Using (1.1.4), (1.1.7), we get that*

$$\eta = \frac{1}{3} (1 - \xi), \quad \omega_0(t) := L_0 t = (3 - \xi) t, \quad and$$
$$\omega(t) := L t = 2 (2 - \xi) t.$$

TABLE 1.2 Different values of the parameters involved.

ξ	$x^\star$	$h_\star$	h_1	h_2	h_3
0.486967	0.6978302086	0.5174905727	0.4736234295	0.4584042632	0.4368027442
0.5245685	0.7242710128	0.4676444075	0.4299718890	0.4169293786	0.3983631448
0.452658	0.6727986326	0.5646168433	0.5146862992	0.4973315343	0.4727617854
0.435247	0.6597325216	0.5891326340	0.5359749755	0.5174817371	0.4913332192
0.425947	0.6526461522	0.6023932312	0.5474704233	0.5283539600	0.5013421729
0.7548589	0.8688261621	0.2034901726	0.1934744795	0.1900595014	0.1850943832

Using Application 1.4.1(a), we have that

$$h_\star = \frac{2}{3}(1-\xi)(2-\xi) > 0.5 \quad \text{for all} \quad \xi \in (0,0.5).$$

Hence, there is no guarantee that Newton's method (1.1.2) starting at $x_0 = 1$ converges to $x^\star$. However, one can easily see that if, for example, $\xi = 0.49$, Newton's method (1.1.2) converges to $x^\star = \sqrt[3]{0.49}$.

(b) *Consider our "h" conditions given in Application 1.4.1. Then we obtain that*

$$h_1 = \frac{1}{6}(7-3\xi)(1-\xi) \le 0.5 \quad \text{for all} \quad \xi \in [0.4648162415, 0.5),$$

$$h_2 = \frac{1}{12}(8-3\xi+(5\xi^2-24\xi+28)^{1/2})(1-\xi) \le 0.5$$

$$\text{for all} \quad \xi \in [0.450339002, 0.5)$$

and

$$h_3 = \frac{1}{24}(1-\xi)(12-4\xi+(84-58\xi+10\xi^2)^{1/2}$$
$$+ (12-10\xi+2\xi^2)^{1/2}) \le 0.5 \quad \text{for all} \quad \xi \in [0.4271907643, 0.5).$$

Next we pick some values of ξ such that all hypotheses are satisfied, so we can compare the "h" conditions. Now, we can compare our "h" conditions (see Table 1.2).

(c) *Consider the case $\xi = 0.7548589$ where our "h" conditions given in Application 1.4.1 are satisfied. Let*

$$L_{-1} = 2$$

and

$$L_{-2} = \frac{\xi+5}{3} = 1.918286300.$$

Hence, we have that

$$\alpha = 0.5173648648,$$

$$\alpha_1 = 0.01192474572,$$

and

$$\alpha_2 = 0.7542728992.$$

Then conditions (1.4.1)–(1.4.3) hold. We also have that

$$\beta_1 = 9.613132975,$$

$$\beta_2 = -0.5073095684,$$

$$\beta_3 = -2.069459459,$$

$$\gamma_1 = -0.02751250012,$$

$$\gamma_2 = 4.097708498,$$

$$\gamma_3 = -0.965270270,$$

$$\lambda_1 = 0.4911124649,$$

$$\Delta_2 = 16.68498694,$$

and

$$\lambda_2 = 148.7039440.$$

Since $\gamma_1 < 0$, $\Delta_2 > 0$ and using (1.4.9), we get in turn that

$$\frac{1}{2\,L_4} = \min\{\lambda_1, \lambda_2\} = 0.4911124649.$$

Then, we deduce that condition (1.4.10) holds, provided that

$$L_4 = 1.018096741$$

and

$$h_4 = 0.08319245167 < 0.5.$$

If we consider $L_\star = 1.836572600$ in Application 1.4.1, we get that

$$h_5 = 0.07234026489 < h_4.$$

Finally, we pick the same values of ξ as in Table 1.2, so we can compare the h_4 and h_5 conditions (see Table 1.3).

TABLE 1.3 Different values of the parameters involved.

ξ	h_4	h_5
0.486967	0.1767312629	0.1377052696
0.5245685	0.1634740591	0.1290677624
0.452658	0.1888584478	0.1454742725
0.435247	0.1950234005	0.1493795529
0.425947	0.1983192281	0.1514558980
0.7548589	0.08319245167	0.07234026489

Example 1.5.5. *Let $\mathcal{X}$ and $\mathcal{Y}$ as in Example 1.5.3. Consider the following nonlinear boundary value problem [10]:*

$$\begin{cases} u'' = -u^3 - \gamma \, u^2, \\ u(0) = 0, \\ u(1) = 1. \end{cases}$$

It is well known that this problem can be formulated as the integral equation

$$u(s) = s + \int_0^1 \mathcal{Q}(s,t) \, (u^3(t) + \gamma \, u^2(t)) \, dt \qquad (1.5.6)$$

where $\mathcal{Q}$ is the Green function given by

$$\mathcal{Q}(s,t) = \begin{cases} t \, (1-s), & t \le s, \\ s \, (1-t), & s < t. \end{cases}$$

Then problem (1.5.6) is in the form (1.1.1), where $F : \mathcal{D} \longrightarrow \mathcal{Y}$ is defined as

$$[F(x)]\,(s) = x(s) - s - \int_0^1 \mathcal{Q}(s,t) \, (x^3(t) + \gamma \, x^2(t)) \, dt.$$

Set $u_0(s) = s$ and $\mathcal{D} = U(u_0, R_0)$. It is easy to verify that $U(u_0, R_0) \subset U(0, R_0 + 1)$ since $\| u_0 \| = 1$. If $2 \gamma < 5$, the operator F' satisfies conditions of Theorem 1.2.8 with

$$\eta = \frac{1 + \gamma}{5 - 2 \gamma},$$

$$\omega(t) := L\,t = \frac{\gamma + 6 \, R_0 + 3}{4} \, t,$$

and

$$\omega_0(t) := L_0\,t = \frac{2 \, \gamma + 3 \, R_0 + 6}{8} \, t.$$

Note that $\omega_0 < \omega$. Next we pick some values of γ and R_0 such that all hypotheses are satisfied, so we can compare the "h" conditions (see Table 1.4).

TABLE 1.4 Different values of the parameters involved.

γ	R_0	$h_\star$	h_1	h_2	h_3
0.00025	1	0.4501700201	3376306412	0.2946446274	0.2413108547
0.25	0.986587	0.6367723612	0.4826181423	0.4240511567	0.3508368298
0.358979	0.986587	0.7361726023	0.5600481163	0.4932612622	0.4095478068
0.358979	1.5698564	1.013838328	0.7335891949	0.6245310288	0.4927174588
0.341378	1.7698764	1.084400750	0.7750792917	0.6539239239	0.5088183074

REFERENCES

[1] S. Amat, S. Busquier, C. Bermúdez, Á.A. Magreñán, Expanding the applicability of a third order Newton-type method free of bilinear operators, Algorithms 8 (3) (2015) 669–679.

[2] S. Amat, S. Busquier, C. Bermúdez, Á.A. Magreñán, On the election of the damped parameter of a two-step relaxed Newton-type method, Nonlinear Dynam. 84 (1) (2016) 9–18.

[3] S. Amat, Á.A. Magreñán, N. Romero, On a two-step relaxed Newton-type method, Appl. Comput. Math. 219 (24) (2013) 11341–11347.

[4] J. Appell, E. De Pascale, J.V. Lysenko, P.P. Zabrejko, New results on Newton–Kantorovich approximations with applications to nonlinear integral equations, Numer. Funct. Anal. Optim. 18 (1997) 1–17.

[5] J. Appell, E. De Pascale, P.P. Zabrejko, On the application of the Newton–Kantorovich method to nonlinear integral equations of Uryson type, Numer. Funct. Anal. Optim. 12 (1991) 271–283.

[6] I.K. Argyros, A unifying local-semilocal convergence analysis and applications for two-point Newton-like methods in Banach space, J. Math. Anal. Appl. 298 (2004) 374–397.

[7] I.K. Argyros, On the Newton–Kantorovich hypothesis for solving equations, J. Comput. Appl. Math. 169 (2004) 315–332.

[8] I.K. Argyros, Concerning the "terra incognita" between convergence regions of two Newton methods, Nonlinear Anal. 62 (2005) 179–194.

[9] I.K. Argyros, Approximating solutions of equations using Newton's method with a modified Newton's method iterate as a starting point, Rev. Anal. Numér. Théor. Approx. 36 (2) (2007) 123–138.

[10] I.K. Argyros, Computational theory of iterative methods, in: C.K. Chui, L. Wuytack (Eds.), Stud. Comput. Math., vol. 15, Elsevier Publ. Co., New York, USA, 2007.

[11] I.K. Argyros, On a class of Newton-like methods for solving nonlinear equations, J. Comput. Appl. Math. 228 (2009) 115–122.

[12] I.K. Argyros, A semilocal convergence analysis for directional Newton methods, Math. Comp. 80 (2011) 327–343.

[13] I.K. Argyros, S. Hilout, Efficient Methods for Solving Equations and Variational Inequalities, Polimetrica Publisher, Milano, Italy, 2009.

[14] I.K. Argyros, S. Hilout, Enclosing roots of polynomial equations and their applications to iterative processes, Surveys Math. Appl. 4 (2009) 119–132.

[15] I.K. Argyros, S. Hilout, Extending the Newton–Kantorovich hypothesis for solving equations, J. Comput. Appl. Math. 234 (2010) 2993–3006.

[16] I.K. Argyros, S. Hilout, Convergence domains under Zabrejko–Zincenko conditions using recurrent functions, Appl. Math. (Warsaw) 38 (2011) 193–209.

[17] I.K. Argyros, S. Hilout, Weaker conditions for the convergence of Newton's method, J. Complexity 28 (3) (June 2012) 364–387.

[18] I.K. Argyros, S. Hilout, M.A. Tabatabai, Mathematical Modelling With Applications in Biosciences and Engineering, Nova Publishers, New York, 2011.

[19] I.K. Argyros, J.A. Ezquerro, M.A. Hernández-Verón, Á.A. Magreñán, Convergence of Newton's method under Vertgeim conditions: new extensions using restricted convergence domains, J. Math. Chem. 55 (7) (2017) 1392–1406.

[20] I.K. Argyros, S. González, Á.A. Magreñán, Majorizing sequences for Newton's method under centred conditions for the derivative, Int. J. Comput. Math. 91 (12) (2014) 2568–2583.

[21] I.K. Argyros, J.M. Gutiérrez, Á.A. Magreñán, N. Romero, Convergence of the relaxed Newton's method, J. Korean Math. Soc. 51 (1) (2014) 137–162.

[22] I.K. Argyros, Á.A. Magreñán, Extended convergence results for the Newton–Kantorovich iteration, J. Comput. Appl. Math. 286 (2015) 54–67.

[23] I.K. Argyros, Á.A. Magreñán, L. Orcos, J.A. Sicilia, Local convergence of a relaxed two-step Newton like method with applications, J. Math. Chem. 55 (7) (2017) 1427–1442, https://www.scopus.com/inward/record.uri?eid=2-.

[24] I.K. Argyros, Á.A. Magreñán, J.A. Sicilia, Improving the domain of parameters for Newton's method with applications, J. Comput. Appl. Math. 318 (2017) 124–135.

[25] X. Chen, T. Yamamoto, Convergence domains of certain iterative methods for solving nonlinear equations, Numer. Funct. Anal. Optim. 10 (1989) 37–48.

[26] J.A. Ezquerro, J.M. Gutiérrez, M.A. Hernández, N. Romero, M.J. Rubio, The Newton method: from Newton to Kantorovich, Gac. R. Soc. Mat. Esp. 13 (1) (2010) 53–76 (in Spanish).

[27] J.A. Ezquerro, M.A. Hernández, On the R-order of convergence of Newton's method under mild differentiability conditions, J. Comput. Appl. Math. 197 (1) (2006) 53–61.

[28] J.A. Ezquerro, M.A. Hernández, An improvement of the region of accessibility of Chebyshev's method from Newton's method, Math. Comp. 78 (267) (2009) 1613–1627.

[29] J.A. Ezquerro, M.A. Hernández, N. Romero, Newton-type methods of high order and domains of semilocal and global convergence, Appl. Math. Comput. 214 (1) (2009) 142–154.

[30] J.M. Gutiérrez, Á.A. Magreñán, N. Romero, On the semilocal convergence of Newton–Kantorovich method under center-Lipschitz conditions, Appl. Comput. Math. 221 (2013) 79–88.

[31] L.V. Kantorovich, G.P. Akilov, Functional Analysis, Pergamon Press, Oxford, 1982.

[32] Á.A. Magreñán, I.K. Argyros, Optimizing the applicability of a theorem by F. Potra for Newton-like methods, Appl. Comput. Math. 242 (2014) 612–623.

[33] Á.A. Magreñán, I.K. Argyros, J.A. Sicilia, New improved convergence analysis for Newton-like methods with applications, J. Math. Chem. 55 (7) (2017) 1505–1520.

[34] L.M. Ortega, W.C. Rheinboldt, Iterative Solution of Nonlinear Equations in Several Variables, Academic Press, New York, 1970.

[35] F.A. Potra, The rate of convergence of a modified Newton's process, Apl. Mat. 26 (1) (1981) 13–17. With a loose Russian summary.

[36] F.A. Potra, An error analysis for the secant method, Numer. Math. 38 (1981/1982) 427–445.

[37] F.A. Potra, On the convergence of a class of Newton-like methods, in: Iterative Solution of Nonlinear Systems of Equations, Oberwolfach, 1982, in: Lecture Notes in Math., vol. 953, Springer, Berlin, New York, 1982, pp. 125–137.

[38] F.A. Potra, Sharp error bounds for a class of Newton-like methods, Libertas Math. 5 (1985) 71–84.

[39] F.A. Potra, V. Pták, Sharp error bounds for Newton's process, Numer. Math. 34 (1) (1980) 63–72.

[40] F.A. Potra, V. Pták, Nondiscrete Induction and Iterative Processes, Res. Notes Math., vol. 103, Pitman (Advanced Publishing Program), Boston, MA, 1984.

[41] P.D. Proinov, General local convergence theory for a class of iterative processes and its applications to Newton's method, J. Complexity 25 (2009) 38–62.

[42] P.D. Proinov, New general convergence theory for iterative processes and its applications to Newton–Kantorovich type theorems, J. Complexity 26 (2010) 3–42.

[43] L.B. Rall, Computational Solution of Nonlinear Operator Equations, Wiley, New York, 1969.

[44] W.C. Rheinboldt, A unified convergence theory for a class of iterative processes, SIAM J. Numer. Anal. 5 (1968) 42–63.

[45] W.C. Rheinboldt, An adaptive continuation process for solving systems of nonlinear equations, in: Banach Ctr. Publ., vol. 3, Polish Academy of Science, 1977, pp. 129–142.

[46] J.F. Traub, H. Woźniakowsi, Convergence and complexity of Newton iteration for operator equations, J. Assoc. Comput. Mach. 26 (1979) 250–258.

[47] T. Yamamoto, A convergence theorem for Newton-like methods in Banach spaces, Numer. Math. 51 (1987) 545–557.

[48] P.P. Zabrejko, D.F. Nguen, The majorant method in the theory of Newton–Kantorovich approximations and the Pták error estimates, Numer. Funct. Anal. Optim. 9 (1987) 671–684.

[49] A.I. Zinčenko, Some approximate methods of solving equations with non-differentiable operators, Dopovidi Akad. Nauk Ukraïn. RSR (1963) 156–161 (in Ukrainian).

Chapter 2

Directional Newton methods

2.1 INTRODUCTION

In this chapter we are concerned with the problem of approximating a zero $x^\star$ of a differentiable function F defined on a convex subset $\mathcal{D}$ of $\mathbb{R}^n$, where $n \in \mathbb{N}$, with values in $\mathbb{R}$.

Many problems from applied sciences can be solved by means of finding the solutions of an equation like

$$F(x) = 0.$$

More specifically, when it comes to computer graphics, we often need to compute and display the intersection $\mathcal{C} = \mathcal{A} \cap \mathcal{B}$ of two surfaces $\mathcal{A}$ and $\mathcal{B}$ in $\mathbb{R}^3$ [5], [6]. If the two surfaces are explicitly given by

$$\mathcal{A} = \{(u, v, w)^T : w = F_1(u, v)\}$$

and

$$\mathcal{B} = \{(u, v, w)^T : w = F_2(u, v)\},$$

then the solution $x^\star = (u^\star, v^\star, w^\star)^T \in \mathcal{C}$ must satisfy the nonlinear equation

$$F_1(u^\star, v^\star) = F_2(u^\star, v^\star)$$

and

$$w^\star = F_1(u^\star, v^\star).$$

Hence, we must solve a nonlinear equation in two variables $x = (u, v)^T$ of the form

$$F(x) = F_1(x) - F_2(x) = 0.$$

The marching method can be used to compute the intersection $\mathcal{C}$. In this method, we first need to compute a starting point $x_0 = (u_0, v_0, w_0)^T \in \mathcal{C}$, and then compute the succeeding intersection points by successive updating.

In mathematical programming [8], for an equality-constraint optimization problem, e.g.,

$$\min \ \psi(x)$$
$$\text{s.t.} \quad F(x) = 0$$

A Contemporary Study of Iterative Methods. DOI: 10.1016/B978-0-12-809214-9.00002-4

where ψ, $F\colon \mathcal{D} \subseteq \mathbb{R}^n \longrightarrow \mathbb{R}$ are nonlinear functions, we need a feasible point to start a numerical algorithm. That is, we must compute a solution of equation $F(x) = 0$.

In the case of a system of nonlinear equations $G(x) = 0$, with $G : \mathcal{D} \subseteq \mathbb{R}^n \longrightarrow \mathbb{R}^n$, we may solve instead

$$\| G(x) \|_2 = 0,$$

if the zero of function G is isolated or locally isolated and if the rounding error is neglected [3], [7], [9], [10], [11].

We use the directional Newton method DNM [5] given by

$$x_{k+1} = x_k - \frac{F(x_k)}{\nabla F(x_k) \cdot d_k} \, d_k \quad (k \geq 0)$$

to generate a sequence $\{x_k\}$ converging to $x^\star$.

First, we must explain how DNM is conceived. We start with an initial guess $x_0 \in U_0$, where F is differentiable and a direction vector d_0. Then we restrict F on the line $\mathcal{A} = \{x_0 + \theta\, d_0, \ \theta \in \mathbb{R}\}$, where it is a function of one variable $f(\theta) = F(x_0 + \theta\, d_0)$.

Set $\theta_0 = 0$ to obtain the Newton iteration for f, that is, the next point:

$$v_1 = -\frac{f(0)}{f'(0)}.$$

The corresponding iteration for F is

$$x_1 = x_0 - \frac{F(x_0)}{\nabla F(x_0) \cdot d_0} \, d_0.$$

Note that $f(0) = F(x_0)$ and $f'(0)$ is the directional derivative

$$f'(0) = F'(x_0, d_0) = \nabla F(x_0) \cdot d_0.$$

By repeating this process, we arrive at DNM.

If $n = 1$, DNM reduces to the classical Newton method [1–3], [7].

An elegant semilocal convergence analysis for the DNM was provided by Levin and Ben–Israel in the well-known work [5].

The quadratic convergence of the method was established for directions d_k sufficiently close to the gradients $\nabla F(x_k)$, and under standard Newton–Kantorovich-type hypotheses [1–3], [7].

In this chapter, we are motivated by the paper [5], and optimization considerations. By introducing the new idea of restricted convergence domains, we find more precise location where the iterates lie than in [1,5]. This way, we provide a semilocal convergence analysis with the following advantages over the work in [1,5]:

- Weaker hypotheses;
- Larger convergence domain for DNM;
- Finer error bounds on the distances $\| x_{k+1} - x_k \|$, $\| x_k - x^\star \|$ $(k \geq 0)$;
- At least as precise information on the location of the zero $x^\star$.

Throughout the chapter, we use the Euclidean inner product, the corresponding norm $\| x \|$, and the corresponding matrix norm $\| A \|$, except in the cases where the ∞-norm is used for vectors and matrices, denoted by $\| x \|_\infty$, and $\| A \|_\infty$, respectively.

2.2 SEMILOCAL CONVERGENCE ANALYSIS

First of all, we need the following lemma on majorizing sequences for DNM. The proof can be found in [3].

Lemma 2.2.1. *Suppose that there exist constants $L_0 > 0$, $L > 0$, and $\eta \geq 0$ such that*

$$q_0 := \overline{L}\,\eta \leq \frac{1}{2} \tag{2.2.1}$$

where

$$\overline{L} = \frac{1}{8}\left(4\,L_0 + \sqrt{L_0\,L} + \sqrt{8\,L_0^2 + L_0\,L}\right). \tag{2.2.2}$$

Then sequence $\{t_k\}$ $(k \geq 0)$ given by

$$t_0 = 0, \quad t_1 = \eta, \quad t_2 = t_1 + \frac{L_0\,(t_1 - t_0)^2}{2\,(1 - L_0\,t_1)}, \quad t_{k+1} = t_k + \frac{L\,(t_k - t_{k-1})^2}{2\,(1 - L_0\,t_k)} \tag{2.2.3}$$
$(k \geq 2)$

is well defined, nondecreasing, bounded above by $t^{\star\star}$, and converges to its unique least upper bound $t^\star \in [0, t^{\star\star}]$, where

$$t^{\star\star} = \left[1 + \frac{L_0\eta}{2(1 - \delta)(1 - L_0\eta)}\right]\eta, \tag{2.2.4}$$

and

$$\delta = \frac{2\,L}{L + \sqrt{L^2 + 8\,L_0\,L}}, \quad L_0, L \neq 0. \tag{2.2.5}$$

Here, $\angle$ denotes the angle between two vectors u and v, given by

$$\angle(u, v) = \arccos \frac{u \cdot v}{\| u \| \cdot \| v \|}, \quad u \neq 0, \quad v \neq 0.$$

Throughout the chapter we will respectively denote by $B(x, \rho)$ and $\overline{B}(x, \rho)$ the open and closed ball in $\mathbb{R}^n$ with center $x \in \mathbb{R}^n$ and radius $\rho > 0$.

The main semilocal result for DNM is given by:

Theorem 2.2.2. *Let $F : \mathcal{D} \subseteq \mathbb{R}^n \longrightarrow \mathbb{R}$ be a differentiable function. Assume that*
(i) there exists a point $x_0 \in \mathcal{D}$ such that

$$F(x_0) \neq 0, \qquad \nabla F(x_0) \neq 0.$$

Let $d_0 \in \mathbb{R}^n$ be such that $\| d_0 \| = 1$, and set

$$h_0 = -\frac{F(x_0)}{\nabla F(x_0) \cdot d_0} \, d_0,$$

$$x_1 = x_0 + h_0.$$

(ii) For $F \in \mathcal{C}^2[\mathcal{D}]$, there exist constants $M_0 > 0$ and $M > 0$ such that

$$\| \nabla F(x) - \nabla F(x_0) \| \leq M_0 \; \| x - x_0 \|, \qquad x \in \mathcal{D}, \tag{2.2.6}$$

$$\sup_{x \in \mathcal{D}_0} \; \| F''(x) \| = M, \mathcal{D}_0 = \mathcal{D} \cap B(x_0, 1/M_0), \tag{2.2.7}$$

$$p_0 = |F(x_0)| \, \overline{M} \, |\nabla F(x_0) \cdot d_0|^{-2} \leq \frac{1}{2}, \tag{2.2.8}$$

and

$$B_0 := B(x_0, t^\star) \subseteq \mathcal{D}, \tag{2.2.9}$$

where

$$\overline{M} = \frac{1}{8}\left(4 \, M_0 + \sqrt{M_0 M} + \sqrt{8 \, M_0^2 + M_0 \, M} \right). \tag{2.2.10}$$

(iii) Sequence $\{x_k\}$ $(k \geq 0)$ given by

$$x_{k+1} = x_k + h_k, \tag{2.2.11}$$

where

$$h_k = -\frac{F(x_k)}{\nabla F(x_k) \cdot d_k} \, d_k, \tag{2.2.12}$$

satisfies

$$\angle(d_k, \nabla F(x_k)) \leq \angle(d_0, \nabla F(x_0)) \qquad k \geq 0, \tag{2.2.13}$$

where each $d_k \in \mathbb{R}^n$ is such that $\| d_k \| = 1$.
    Then sequence $\{x_k\}_{k \geq 0}$ remains in B_0 and converges to a zero $x^\star \in B_0$ of function F. Moreover, $\nabla F(x^\star) \neq 0$, unless $\| x^\star - x_0 \| = t^\star$. Furthermore, the following estimates hold for all $k \geq 0$:

$$\| x_{k+1} - x_k \| \leq t_{k+1} - t_k \tag{2.2.14}$$

and

$$\| x_k - x^\star \| \le t^\star - t_k,$$ (2.2.15)

where iteration $\{t_k\}$ is given by (2.2.3) for

$$L_0 = |\nabla F(x_0)|^{-1} M_0, \quad L = |\nabla F(x_0) \cdot d_0|^{-1} M,$$
$$\eta = |\nabla F(x_0) \cdot d_0|^{-1} |F(x_0)|.$$

Note that condition (2.2.13) is equivalent to

$$\frac{|\nabla F(x_k) \cdot d_k|}{\| \nabla F(x_k) \|} \ge \frac{|\nabla F(x_0) \cdot d_0|}{\| \nabla F(x_0) \|}, \quad k \ge 0.$$ (2.2.16)

Proof. The proof is similar to the corresponding one in [1]. The iterates $\{x_n\}$ are shown to be in D_0 which is a more precise location than D used in [1,5], since $D_0 \subseteq D$, leading to the advantages as already mentioned in the introduction of this study. We shall show this using mathematical induction on $k \ge 0$:

$$\| x_{k+1} - x_k \| \le t_{k+1} - t_k,$$ (2.2.17)

and

$$\overline{B}(x_{k+1}, t^\star - t_{k+1}) \subseteq \overline{B}(x_k, t^\star - t_k).$$ (2.2.18)

For every $z \in \overline{B}(x_1, t^\star - t_1)$, we have

$$\begin{aligned}\| z - x_0 \| &\le \| z - x_1 \| + \| x_1 - x_0 \| \\ &\le t^\star - t_1 + t_1 - t_0 = t^\star - t_0,\end{aligned}$$

which shows that $z \in \overline{B}(x_0, t^\star - t_0)$. Since also

$$\| x_1 - x_0 \| = \| h_0 \| \le \eta = t_1 - t_0,$$

estimates (2.2.17) and (2.2.18) hold for $k = 0$.

Assume (2.2.17) and (2.2.18) hold for all $i \le k$. Then we get

$$\begin{aligned}\| x_{k+1} - x_0 \| &\le \| x_{k+1} - x_k \| + \| x_k - x_{k-1} \| + \cdots + \| x_1 - x_0 \| \\ &\le (t_{k+1} - t_k) + (t_k - t_{k-1}) + \cdots + (t_1 - t_0) = t_{k+1}\end{aligned}$$ (2.2.19)

and

$$\| x_k + t (x_{k+1} - x_k) - x_0 \| \le t_k + t (t_{k+1} - t_k) \le t^\star, \quad t \in [0, 1].$$

Using condition (2.2.6) for $x = x_k$, we obtain

$$
\begin{aligned}
\| \nabla F(x_k) \| \;\geq\;& \| \nabla F(x_0) \| - \| \nabla F(x_k) - \nabla F(x_0) \| \\
\geq\;& \| \nabla F(x_0) \| - M_0 \, \| x_k - x_0 \| \\
\geq\;& \| \nabla F(x_0) \| - M_0 \, (t_k - t_0) \\
\geq\;& \| \nabla F(x_0) \| - M_0 \, t_k > 0 \qquad \text{(by (2.2.18) and Lemma 2.2.1).}
\end{aligned}
$$

$$(2.2.20)$$

We have the identity

$$
\begin{aligned}
\int_{x_{k-1}}^{x_k} (x_k - x) \, F''(x) \, dx \;=\;& -(x_k - x_{k-1}) \, \nabla F(x_{k-1}) + F(x_k) - F(x_{k-1}) \\
=\;& -h_{k-1} \, \nabla F(x_{k-1}) + F(x_k) - F(x_{k-1}) \\
=\;& \frac{F(x_{k-1})}{(F(x_{k-1}) \cdot d_{k-1})} \, (d_{k-1} \cdot \nabla F(x_{k-1})) \\
& + F(x_k) - F(x_{k-1}) \\
=\;& F(x_k).
\end{aligned}
$$

$$(2.2.21)$$

We prefer the integration to be from 0 to 1. That is why we introduce a change of variable given by $x = x_{k-1} + t \, h_{k-1}, t \in [0, 1]$. We can write

$$
x_k - x = x_k - x_{k-1} - t \, h_{k-1} = h_{k-1} - t \, h_{k-1} = (1-t) \, h_{k-1}, \qquad dx = h_{k-1} \, dt.
$$

Then (2.2.21) can be rewritten as

$$
F(x_k) = \int_0^1 (1-t) \, h_{k-1} \, F''(x_{k-1} + \theta \, h_{k-1}) \, h_{k-1} \, d\theta.
$$

$$(2.2.22)$$

Using (2.2.7), (2.2.11)–(2.2.16), we obtain

$$
\begin{aligned}
\| x_{k+1} - x_k \| = \| h_k \| \;=\;& \frac{|F(x_k)|}{|\nabla F(x_k) \cdot d_k|} \\[2mm]
\leq\;& \frac{\left| \int_0^1 (1-t) \, h_{k-1} \, F''(x_0 + \theta \, h_{k-1}) \, h_{k-1} \, d\theta \right|}{|\nabla F(x_k) \cdot d_k|} \\[2mm]
\leq\;& \frac{M \, \| h_{k-1} \|^2}{2 \, |\nabla F(x_k) \cdot d_k|} \\[2mm]
\leq\;& \frac{M \, \| h_{k-1} \|^2}{2 \, \| \nabla F(x_k) \|} \, \frac{\| \nabla F(x_0) \|}{|\nabla F(x_0) \cdot d_0|} \\[2mm]
\leq\;& \frac{M \, \| h_{k-1} \|^2 \, \| \nabla F(x_0) \|}{2 \, (\| \nabla F(x_0) \| - M_0 \, t_k) \, |\nabla F(x_0) \cdot d_0|}
\end{aligned}
$$

$$\leq \frac{M \parallel h_{k-1} \parallel^2}{2 \left(1 - \parallel \nabla F(x_0) \parallel^{-1} M_0 \, t_k\right) \mid \nabla F(x_0) \cdot d_0 \mid}$$

$$\leq \frac{M \, (t_k - t_{k-1})^2}{2 \left(1 - \mid \nabla F(x_0) \cdot d_0 \mid^{-1} M_0 \, t_k\right) \mid \nabla F(x_0) \cdot d_0 \mid}$$

$$= t_{k+1} - t_k,$$

which shows (2.2.17) for all $k \geq 0$. Then, for every $w \in \overline{B}(x_{k+2}, t^\star - t_{k+2})$, we get

$$\begin{aligned} \parallel w - x_{k+1} \parallel &\leq \parallel w - x_{k+2} \parallel + \parallel x_{k+2} - x_{k+1} \parallel \\ &\leq t^\star - t_{k+2} + t_{k+2} - t_{k+1} = t^\star - t_{k+1}, \end{aligned}$$

showing (2.2.18) for all $k \geq 0$.

Lemma 2.2.1 implies that $\{t_n\}$ is a convergent sequence. It then follows from (2.2.17) and (2.2.18) that $\{x_n\}$ is a Cauchy sequence, too. Since B_0 is a closed set, it follows that $\{x_n\}$ converges to some $x^\star \in B_0$. The point $x^\star$ is a zero of F, since

$$0 \leq |F(x_k)| \leq \frac{1}{2} M \, (t_k - t_{k-1})^2 \longrightarrow 0 \quad \text{as} \quad k \to \infty.$$

Moreover, we prove $\nabla F(x^\star) \neq 0$, except if $\parallel x^\star - x_0 \parallel = t^\star$.

Using (2.2.6) and the definition of constant L_0, we obtain

$$\begin{aligned} \parallel \nabla F(x) - \nabla F(x_0) \parallel &\leq M_0 \parallel x - x_0 \parallel \\ &\leq M \, t^\star \leq \mid \nabla F(x_0) \cdot d_0 \mid \leq \parallel \nabla F(x_0) \parallel . \end{aligned}$$

If $\parallel x - x_0 \parallel < t^\star$, then by (2.2.6), we obtain

$$\parallel \nabla F(x) - \nabla F(x_0) \parallel \leq M_0 \parallel x - x_0 \parallel < M_0 t^\star \leq \parallel \nabla F(x_0) \parallel,$$

or

$$\parallel \nabla F(x_0) \parallel > \parallel \nabla F(x) - \nabla F(x_0) \parallel,$$

which shows $\nabla F(x) \neq 0$. □

Remark 2.2.3. *(a) $t^\star$ in (2.2.9) can be replaced by $t^{\star\star}$ given in closed form by (2.2.4).*

(b) The proof followed the corresponding proof in [3], but the new sufficient convergence condition (2.2.8) is weaker than that in [3] (which in turn was shown to be weaker than that in [5]).

To show part (b) of the previous remark, we need the following definition.

Definition 2.2.4. *Let $F : \mathcal{D} \subseteq \mathbb{R}^n \longrightarrow \mathbb{R}$ be a differentiable function, where $\mathcal{D}$ is a convex region. Function ∇F is called Lipschitz continuous, if there exists a constant $M_1 \geq 0$ such that*

$$\parallel \nabla F(x) - \nabla F(y) \parallel \leq M_1 \parallel x - y \parallel \quad \text{for all} \quad x, y \in \mathcal{D}. \qquad (2.2.23)$$

If F is twice differentiable, then we can set

$$\sup_{x \in \mathcal{D}_0} \| F''(x) \| := M_1. \tag{2.2.24}$$

Notice that

$$M_0 \leq M_1 \tag{2.2.25}$$

and

$$M \leq M_1 \quad \text{(since } \mathcal{D}_0 \subseteq \mathcal{D}) \tag{2.2.26}$$

hold in general, and $\dfrac{M_1}{M_0}$ can be arbitrarily large [1–3].

Remark 2.2.5. *If F is twice differentiable, then, in view of the proof of Theorem 2.2.2, constant L_0 can be defined more precisely as*

$$L_0 = \| \nabla F(x_0) \|^{-1} M_0. \tag{2.2.27}$$

Remark 2.2.6. *Our Theorem 2.2.2 improves Theorem 2.2 in [1].*
Theorem 2.2 in [5] uses the condition

$$p = |F(x_0)| \, N \, |\nabla F(x_0) \cdot d_0|^{-2} \leq \frac{1}{2}, \tag{2.2.28}$$

corresponding to our condition (2.2.8), where

$$N = 4M_0 + M_1 + \sqrt{M_1^2 + 8M_0 M_1}.$$

But we have

$$\overline{M} < N. \tag{2.2.29}$$

That is,

$$p \leq \frac{1}{2} \implies p_0 \leq \frac{1}{2}, \tag{2.2.30}$$

but not necessarily vice versa (unless $M_0 = M = M_1$).

Moreover, the corresponding error bounds, as well as the information on the location of the root, are improved if $M_0 < M$ or $M < M_1$. Examples where $M_0 < M < M < M_1$ can be found in [1–4]. The rest of the results in [1] (and consequently in [5]) can be improved along the same lines. It is worth noting that in the proof of Theorem 2.2 in [3], we used M_1 instead of M (i.e., $\mathcal{D}$ instead of $\mathcal{D}_0$). However, the iterates x_n remain in $\mathcal{D}_0$ which is more a precise domain than $\mathcal{D}$, since $\mathcal{D}_0 \subseteq \mathcal{D}$.

REFERENCES

[1] I.K. Argyros, A semilocal convergence analysis for directional Newton methods, Math. Comp. 80 (2011) 327–343.

[2] I.K. Argyros, Convergence and Applications of Newton-Type Iterations, Springer-Verlag, New York, 2008.

[3] I.K. Argyros, S. Hilout, Weaker conditions for the convergence of Newton's method, J. Complexity 28 (2012) 364–387.

[4] A. Ben-Israel, Y. Levin, Maple programs for directional Newton methods, available at ftp://rutcor.rutgers.edu/pub/bisrael/Newton-Dir.mws.

[5] Y. Levin, A. Ben-Israel, Directional Newton methods in n variables, Math. Comp. 71 (237) (2002) 251–262.

[6] G. Lukács, The generalized inverse matrix and the surface–surface intersection problem, in: Theory and Practice of Geometric Modeling, Blaubeuren, 1988, Springer, Berlin, 1989, pp. 167–185.

[7] J.M. Ortega, W.C. Rheinboldt, Iterative Solution of Nonlinear Equations in Several Variables, Academic Press, New York, 1970.

[8] B.T. Polyak, Introduction to Optimization, Transl. Ser. Math. Eng., Optimization Software, Inc., Publications Division, New York, 1987. Translated from Russian. With a foreword by D.P. Bertsekas.

[9] A. Ralston, P. Rabinowitz, A First Course in Numerical Analysis, 2nd edition, McGraw-Hill, 1978.

[10] J. Stoer, K. Bulirsch, Introduction to Numerical Analysis, Springer-Verlag, 1976.

[11] H.F. Walker, L.T. Watson, Least-change Secant update methods, SIAM J. Numer. Anal. 27 (1990) 1227–1262.

Chapter 3

Newton's method

3.1 INTRODUCTION

In this chapter we study the problem of approximating a locally unique solution x^* of equation

$$F(x) = 0, \tag{3.1.1}$$

where F is a Fréchet-differentiable operator defined on an open convex subset D of a Banach space $\mathcal{X}$ with values in a Banach space $\mathcal{Y}$.

Many problems from applied mathematics can be brought finding solutions of equations in a form like (3.1.1) using mathematical modeling [1,2,8,14,21]. Most solutions of these equations are given using iterative methods because the solutions of this kind of equation are rarely found in closed form. A very important problem in the study of iterative procedures is the convergence domain. In general, the convergence domain is small. Therefore, it is important to enlarge the convergence domain without additional hypothesis.

Newton's method, defined for each $n = 0, 1, 2, \ldots$ by

$$x_{n+1} = x_n - F'(x_n)^{-1} F(x_n), \tag{3.1.2}$$

where x_0 is an initial point, is the most popular, studied, and used method for generating a sequence $\{x_n\}$ approximating the solution x^*. There are several convergence results for Newton's method [1–3,7,8,11,13–15,19–21]. Moreover, the authors of this chapter have presented different works related to the convergence of Newton-like methods [4–6,9,10,12,16–18,22,23].

Related to the convergence of Newton's method, we define the conditions (C) for the semilocal convergence:

(C_1) $F : D \subset \mathcal{X} \to \mathcal{Y}$ is Fréchet differentiable and there exist $x_0 \in D$, $\eta \geq 0$ such that $F'(x_0)^{-1} \in L(\mathcal{Y}, \mathcal{X})$ and

$$\|F'(x_0)^{-1} F(x_0)\| \leq \eta.$$

(C_2) There exists $L > 0$ such that for each $x, y \in D$

$$\|F'(x_0)^{-1}(F'(x) - F'(y))\| \leq L \|x - y\|.$$

(C_3) There exists $L_0 > 0$ such that for each $x \in D$

$$\|F'(x_0)^{-1}(F'(x) - F'(x_0))\| \leq L_0 \|x - x_0\|.$$

A Contemporary Study of Iterative Methods. DOI: 10.1016/B978-0-12-809214-9.00003-6

(C_4) There exists $H > 0$ such that for each $\theta \in [0, 1]$

$$\|F'(x_0)^{-1}(F'(x_1) - F'(x_0))\| \leq H\|x_1 - x_0\|,$$

where $x_1 = x_0 - F'(x_0)^{-1}F(x_0)$.
(C_5) There exists $K > 0$ such that

$$\|F'(x_0)^{-1}(F'(x_0 + \theta(x_1 - x_0)) - F'(x_0))\| \leq K\theta\|x_1 - x_0\|.$$

It is easy to see that $(C_2) \implies (C_3) \implies (C_5) \implies (C_4)$. Clearly, we get

$$H \leq K \leq L_0 \leq L \tag{3.1.3}$$

and $\frac{L}{L_0}$ can be arbitrarily large [7].

Throughout the chapter we will respectively denote by $U(z, \varrho)$ and $\bar{U}(z, \varrho)$ the open and closed ball in $\mathcal{X}$ with center $z \in \mathcal{X}$ and radius $\varrho > 0$.

The sufficient convergence criteria for Newton's method using the conditions (C), constants $L, L_0,$ and η given in an affine invariant form are:

- Kantorovich [21]

$$h_K = 2L\eta \leq 1; \tag{3.1.4}$$

- Argyros [7]

$$h_1 = (L_0 + L)\eta \leq 1; \tag{3.1.5}$$

- Argyros [8]

$$h_2 = \frac{1}{4}\left(L + 4L_0 + \sqrt{L^2 + 8L_0L}\right)\eta \leq 1; \tag{3.1.6}$$

- Argyros [13]

$$h_3 = \frac{1}{4}\left(4L_0 + \sqrt{L_0L + 8L_0^2} + \sqrt{L_0L}\right)\eta \leq 1; \tag{3.1.7}$$

- Argyros [15]

$$h_4 = L_4\eta \leq 1, \tag{3.1.8}$$

where

$$L_4^{-1}(t) = L_4^{-1}$$

$$= \begin{cases} \dfrac{1}{L_0 + H}, \text{ if } b = LK + \alpha L_0(K - 2H) = 0, \\[3mm] 2\dfrac{-\alpha(L_0 + H) + \sqrt{\alpha^2(L_0 + H)^2 + \alpha(LK + 2\alpha L_0(K - 2H))}}{LK + 2\alpha(K - 2H)}, \\ \text{ if } b > 0, \\[3mm] -2\dfrac{\alpha(L_0 + H) + \sqrt{\alpha^2(L_0 + H)^2 + \alpha(LK + 2\alpha L_0(K - 2H))}}{LK + 2\alpha(K - 2H)}, \\ \text{ if } b < 0, \end{cases}$$

and

$$\alpha = \frac{2L}{L + \sqrt{L^2 + 8L_0 L}}.$$

If $H = K = L_0 = L$, then (3.1.5)–(3.1.8) coincide with (3.1.4). If $L_0 < L$, then

$$h_K \le 1 \Rightarrow h_1 \le 1 \Rightarrow h_2 \le 1 \Rightarrow h_3 \le 1 \Rightarrow h_4 \le 1,$$

but not vice versa. We also obtain that

$$\frac{h_1}{h_K} \to \frac{1}{2}, \quad \frac{h_2}{h_K} \to \frac{1}{4}, \quad \frac{h_2}{h_1} \to \frac{1}{2},$$
$$\frac{h_3}{h_K} \to 0, \quad \frac{h_3}{h_1} \to 0, \quad \frac{h_3}{h_2} \to 0, \tag{3.1.9}$$
$$\text{as } \frac{L_0}{L} \to 0.$$

Conditions (3.1.9) show by how many times, at most, the better condition improves the worse condition. Therefore the condition to improve is (3.1.8). This is done as follows: Replace condition (C_2) by

$(C_2)'$ There exists $L_1 > 0$ such that

$$\|F'(x_0)^{-1}(F'(x) - F'(y))\| \le L_1 \|x - y\|,$$

for each $x, y \in D_0 := U(x_1, \dfrac{1}{L_0} - \|F'(x_0)^{-1} F(x_0)\|) \cap D$.

Denote the conditions (C_1), (C_3), and $(C_2)'$ by $(C)'$. Clearly, we have that

$$L_1 \le L \tag{3.1.10}$$

holds in general, since $D_0 \subseteq D$. Notice that the iterates $\{x_n\}$ remain in D_0, which is a more precise location than D in all proofs leading to the preceding conditions (3.1.4)–(3.1.8). However, this fact is not used, since D is used instead of the more precise D_0.

Then, by simply replacing (C_2) by $(C_2)'$ and using the conditions $(C)'$ instead of conditions (C), we can replace L by L_1 in all convergence criteria (3.1.4)–(3.1.8). In particular, we get

$$h_5 = L_5 \eta \leq 1, \tag{3.1.11}$$

where $L_5 = L_4(L_1)$. In view of (3.1.8), (3.1.10), and (3.1.11), we obtain

$$h_4 \leq 1 \Rightarrow h_5 \leq 1, \tag{3.1.12}$$

but not necessarily vice versa, unless $L_1 = L$.

The main idea is the introduction of the center Lipschitz condition (C_3). This modification leads not only to weaker sufficient convergence conditions but also to tighter error bounds on the distances $\|x_{n+1} - x_n\|$, $\|x_n - x^*\|$, and at least as precise information on the location of the solution x^*. These advantages are also obtained under the same computational cost, since in practice the computation of constant L requires the computation of the rest of the Lipschtiz constants as special cases.

The rest of the chapter is organized as follows. In Section 3.3 we present the semilocal convergence analysis of Newton's method (3.1.2). Some numerical examples are presented in Section 3.3.

3.2 CONVERGENCE ANALYSIS

In this section, we will present the semilocal convergence analysis of Newton's method using $(C)'$ conditions.

We first must introduce the scalar majorizing sequence $\{s_n\}$ defined by

$$s_0 = 0, \quad s_1 = \eta, \quad s_2 = s_1 + \frac{K(s_1 - s_0)^2}{2(1 - Hs_1)},$$
$$s_{n+2} = s_{n+1} + \frac{L_1(s_{n+1} - s_n)^2}{2(1 - L_0 s_{n+1})}, \quad n = 1, 2, 3, \ldots \tag{3.2.1}$$

Let $s^* = \lim_{n \to \infty} s_n$.

Now we can establish our main semilocal convergence result.

Theorem 3.2.1. *Suppose that $(C)'$ conditions are satisfied and $\bar{U}(x_0, s^*) \subseteq D$. Then:*

(a) *The sequence $\{s_n\}$ generated by (3.2.1) is nondecreasing, bounded above by s^{**} given by*

$$s^{**} = \eta + \left(1 + \frac{\delta_0}{1 - \delta}\right)\frac{H\eta^2}{2(1 - K\eta)},$$

and converges to its unique upper bound s^*, which satisfies

$$\eta \le s^* \le s^{**},$$

where

$$\delta_0 = \frac{L_1(s_2 - s_1)}{2(1 - L_0 s_2)}$$

and

$$\delta = \frac{2L_1}{L_1 + \sqrt{L_1^2 + 8L_0 L_1}}.$$

(b) *The sequence $\{x_n\}$ ($n \ge 1$) generated by Newton's method (3.1.2) is well defined, remains in $\bar{U}(x_1, s^* - \eta)$, and converges to a solution $x^* \in \bar{U}(x_1, s^* - \eta)$ of equation $F(x) = 0$. Moreover, the following estimates hold:*

$$\|x_{n+1} - x_n\| \le s_{n+1} - s_n$$

and

$$\|x_n - x^*\| \le s^* - s_n$$

Furthermore, for $\bar{s}^* \ge s^* - \eta$ such that

$$L_0(\bar{s}^* + s^* - \eta) < 2,$$

the limit point x^* is the only solution of equation $F(x) = 0$ in $\bar{U}(x_1, \bar{s}^*) \cap D$.

Proof. Simply follow the proof of Lemma 2.1 for part (a) and the proof of Theorem 3.2 for part (b) and replace α_0, α, L, and $\{t_n\}$ in [15] by δ_0, δ, L_1, and $\{s_n\}$, respectively (see also (3.2.2)), where

$$\delta_0 \le \alpha_0 = \frac{L(t_2 - t_1)}{2(1 - L_0 s_2)}.$$

Notice also that the iterates $\{x_n\}(n \ge 1) \in D_0$, which is a more precise location than D used in [15], since $D_0 \subset D$. $\qquad\square$

Remark 3.2.2. *(a) The majorizing sequence $\{t_n\}$, and t^*, t^{**} given in [15] under conditions (C) and (3.1.7) are defined by*

$$t_0 = 0, \quad t_1 = \eta, \quad t_2 = t_1 + \frac{K(t_1 - t_0)^2}{2(1 - Ht_1)},$$

$$t_{n+2} = t_{n+1} + \frac{L(t_{n+1} - t_n)^2}{2(1 - L_0 t_{n+1})}, \quad n = 1, 2, \ldots, \tag{3.2.2}$$

and

$$t^* = \lim_{n \to \infty} t_n \le t^{**} = \eta + \left(1 \frac{\alpha_0}{1 - \alpha}\right) H\eta^2)(2(1 - K\eta)).$$

Using a simple inductive argument, from (3.2.1) and (3.2.2) we get for $L_1 < L$ that

$$s_n < t_n, n = 3, 4, \ldots, \tag{3.2.3}$$

$$s_{n+1} - s_n < t_{n+1} - t_n, n = 2, 3, \ldots, \tag{3.2.4}$$

and

$$s^* \le t^*. \tag{3.2.5}$$

Estimates (3.2.3)–(3.2.5) show that the new error bounds are more precise than the old ones and the information on the location of the solution x^* is at least as precise.

Clearly, the new majorizing sequence $\{s_n\}$ is more precise than the corresponding ones associated with conditions (3.1.4), (3.1.5), (3.1.6), and (3.1.7), obtained respectively from $\{s_n\}$ for $L_0 = K = H = L_1 = L$ ($H = L_0$, $K = L$, $L = L_1$), ($H = L_0$, $K = L$, $L = L_1$), and ($H = L_0$, $K = L_0$, $L = L_1$).

(b) Condition $\bar{U}(x_0, s^*) \subseteq D$ can be replaced by $U(x_0, \frac{1}{L_0})$ (or D_0). In this case condition $(C_2)'$ holds for all $x, y \in U(x_0, \frac{1}{L_0})$ (or D_0).

(c) If $L_0 \le L_1$, then by (3.1.11), we have that $x_0 \in \bar{U}(x_1, \frac{1}{L_0} - \|F'(x_0)^{-1}F(x_0)\|)$, since $\bar{U}(x_1, \frac{1}{L_0} - \|F'(x_0)^{-1}F(x_0)\|) \subseteq U(x_0, \frac{1}{L_0})$.

3.3   NUMERICAL EXAMPLES

Example 3.1. Let $\mathcal{X} = \mathcal{Y} = \mathcal{C}[0, 1]$, the space of continuous functions defined on $[0, 1]$, be equipped with the max-norm. Let $D^* = \{x \in \mathcal{C}[0, 1]; \|x\| \le R\}$ for $R > 0$ and consider F defined on D^* by

$$F(x)(s) = x(s) - f(s) - \xi \int_0^1 G(s, t)x(t)^3 \, dt,$$

where

$$x \in C[0, 1]$$

and

$$s \in [0, 1],$$

where $f \in C[0, 1]$ is a given function, ξ is a real constant, and the kernel G is the Green's function

$$G(s, t) = \begin{cases} (1 - s)t, & t \le s, \\ s(1 - t), & s \le t. \end{cases}$$

In this case, for each $x \in D^*$, $F'(x)$ is a linear operator defined on D^* by the following expression:

$$[F'(x)(v)](s) = v(s) - 3\xi \int_0^1 G(s, t)x(t)^2 v(t)\, dt,$$

where

$$v \in C[0, 1]$$

and

$$s \in [0, 1].$$

If we choose

$$x_0(s) = 1$$

and

$$f(s) = 1,$$

it follows that

$$\|I - F'(x_0)\| \le \frac{3|\xi|}{8}.$$

Hence, if

$$|\xi| < \frac{8}{3},$$

$F'(x_0)^{-1}$ is defined,

$$\|F'(x_0)^{-1}\| \le \frac{8}{8 - 3|\xi|},$$

$$\|F(x_0)\| \le \frac{|\xi|}{8},$$

and

$$\eta = \|F'(x_0)^{-1}F(x_0)\| \leq \frac{|\xi|}{8 - 3|\xi|}.$$

Choosing $\xi = 1.00$ and $\lambda = 3$, we have:

$$\eta = 0.2,$$

$$L = 3.8,$$

$$L_0 = 2.6,$$

$$K = 2.28,$$

$$H = 1.28$$

and

$$L_1 = 1.38154\ldots$$

Using this values, we obtain that conditions (3.1.4)–(3.1.7) are not satisfied, since

$$h_K = 1.52 > 1,$$

$$h_1 = 1.28 > 1,$$

$$h_2 = 1.19343\ldots > 1,$$

$$h_3 > 1.07704\ldots > 1,$$

but conditions (3.1.8) and condition (3.1.11) are satisfied, since

$$h_4 = 0.985779\ldots < 1 \quad \text{and} \quad h_5 = 0.97017\ldots < 1.$$

Hence, we can ensure the convergence of the Newton's method by Theorem 3.2.1.

Example 3.2. Let

$$\mathcal{X} = \mathcal{Y} = \mathbb{R},$$

$$x_0 = 1,$$

$$p \in [0, 0.5),$$

$$D = \bar{U}(x_0, 1 - p),$$

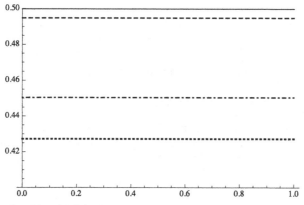

FIGURE 3.1   Condition (3.1.5) in black and dashed, condition (3.1.6) in black and dot-dashed and condition (3.1.7) in black and dotted.

and define function F on D by

$$F(x) = x^3 - p. \tag{3.3.1}$$

Then we have

$$L_0 = 3 - p$$

and

$$L = 2(2 - p).$$

Condition (3.1.4) is not satisfied, since $h_K > 1$ for each $p \in (0, 0.5)$. Conditions (3.1.5), (3.1.6), and (3.1.7) (see Fig. 3.1) are satisfied respectively for

$$p \in [0.494816242, 0.5),$$

$$p \in [0.450339002, 0.5),$$

and

$$p \in [0.4271907643, 0.5).$$

We are now going to consider such an initial point that previous conditions cannot be satisfied but our new criteria are satisfied. Hence, we obtain an improvement with our new weaker criteria.

Moreover, we get that

$$H = \frac{5 + p}{3},$$

$$K = 2,$$

and

$$L_1 = \frac{2}{3(3-p)}(-2p^2 + 5p + 6).$$

Using these values, we obtain that condition (1.8) is satisfied for

$$p \in [0.0984119, 0.5)$$

and condition of Theorem 3.2.1 is satisfied for $p \in [0, 0.5)$, so there exist several values of p for which the previous conditions cannot guarantee the convergence but our new ones can.

REFERENCES

[1] S. Amat, S. Busquier, J.M. Gutiérrez, Geometric constructions of iterative functions to solve nonlinear equations, J. Comput. Appl. Math. 157 (2003) 197–205.

[2] S. Amat, S. Busquier, Third-order iterative methods under Kantorovich conditions, J. Math. Anal. Appl. 336 (2007) 243–261.

[3] S. Amat, S. Busquier, M. Negra, Adaptive approximation of nonlinear operators, Numer. Funct. Anal. Optim. 25 (2004) 397–405.

[4] S. Amat, S. Busquier, C. Bermúdez, Á.A. Magreñán, Expanding the applicability of a third order Newton-type method free of bilinear operators, Algorithms 8 (3) (2015) 669–679.

[5] S. Amat, S. Busquier, C. Bermúdez, Á.A. Magreñán, On the election of the damped parameter of a two-step relaxed Newton-type method, Nonlinear Dynam. 84 (1) (2016) 9–18.

[6] S. Amat, Á.A. Magreñán, N. Romero, On a two-step relaxed Newton-type method, Appl. Math. Comput. 219 (24) (2013) 11341–11347.

[7] I.K. Argyros, On the Newton–Kantorovich hypothesis for solving equations, J. Comput. Math. 169 (2004) 315–332.

[8] I.K. Argyros, A semilocal convergence analysis for directional Newton methods, Math. Comput. 80 (2011) 327–343.

[9] I.K. Argyros, J.A. Ezquerro, M.A. Hernández-Verón, Á.A. Magreñán, Convergence of Newton's method under Vertgeim conditions: new extensions using restricted convergence domains, J. Math. Chem. 55 (7) (2017) 1392–1406.

[10] I.K. Argyros, S. González, Á.A. Magreñán, Majorizing sequences for Newton's method under centred conditions for the derivative, Int. J. Comput. Math. 91 (12) (2014) 2568–2583.

[11] I.K. Argyros, D. González, Extending the applicability of Newton's method for k-Fréchet differentiable operators in Banach spaces, Appl. Math. Comput. 234 (2014) 167–178.

[12] I.K. Argyros, J.M. Gutiérrez, Á.A. Magreñán, N. Romero, Convergence of the relaxed Newton's method, J. Korean Math. Soc. 51 (1) (2014) 137–162.

[13] I.K. Argyros, S. Hilout, Weaker conditions for the convergence of Newton's method, J. Complexity 28 (2012) 364–387.

[14] I.K. Argyros, S. Hilout, Numerical Methods in Nonlinear Analysis, World Scientific Publ. Comp., New Jersey, 2013.

[15] I.K. Argyros, S. Hilout, On an improved convergence analysis of Newton's method, Appl. Math. Comput. 225 (2013) 372–386.

[16] I.K. Argyros, Á.A. Magreñán, Extended convergence results for the Newton–Kantorovich iteration, J. Comput. Appl. Math. 286 (2015) 54–67.

[17] I.K. Argyros, Á.A. Magreñán, L. Orcos, J.A. Sicilia, Local convergence of a relaxed two-step Newton-like method with applications, J. Math. Chem. 55 (7) (2017) 1427–1442, https://www.scopus.com/inward/record.uri?eid=2-.

[18] I.K. Argyros, Á.A. Magreñán, J.A. Sicilia, Improving the domain of parameters for Newton's method with applications, J. Comput. Appl. Math. 318 (2017) 124–135.

[19] J.A. Ezquerro, M.A. Hernández, How to improve the domain of parameters for Newton's method, Appl. Math. Let. 48 (2015) 91–101.
[20] J.M. Gutiérrez, Á.A. Magreñán, N. Romero, On the semilocal convergence of Newton–Kantorovich method under center-Lipschitz conditions, Appl. Math. Comput. 221 (2013) 79–88.
[21] L.V. Kantorovich, G.P. Akilov, Functional Analysis, Pergamon Press, Oxford, 1982.
[22] Á.A. Magreñán, I.K. Argyros, Optimizing the applicability of a theorem by F. Potra for Newton-like methods, Appl. Math. Comput. 242 (2014) 612–623.
[23] Á.A. Magreñán, I.K. Argyros, J.A. Sicilia, New improved convergence analysis for Newton-like methods with applications, J. Math. Chem. 55 (7) (2017) 1505–1520.

Chapter 4

Generalized equations

4.1 INTRODUCTION

In this chapter we consider the problem of solving, in an approximate form, the generalized equation

$$F(x) + Q(x) \ni 0, \tag{4.1.1}$$

where $F : D \longrightarrow H$ is a nonlinear Fréchet-differentiable operator defined on the open subset D of the Hilbert space H, and $Q : H \rightrightarrows H$ is set-valued and maximal monotone. Several problems from applied sciences can be brought into an equation of the form (4.1.1) [17–22,27,28,30,36]. Moreover, if function $\psi : H \longrightarrow (-\infty, +\infty]$ is a strict lower semicontinuous convex function and

$$Q(x) = \partial \psi(x) = \{u \in H : \psi(y) \geq \psi(x) + \langle u, y - x \rangle\}, \quad \text{for all } y \in H,$$

then (4.1.1) becomes the variational inequality problem

$$F(x) + \partial \psi(x) \ni 0,$$

including linear and nonlinear complementary problems (see [1–37]).

In the present chapter we consider Newton's method defined for each $n = 0, 1, 2, \ldots$ by

$$F(x_k) + F'(x_k)(x_{k+1} - x_k) + F(x_{k+1}) \ni 0 \tag{4.1.2}$$

for approximately solving (4.1.1). We will use the idea of restricted convergence domains to present a convergence analysis of (4.1.2). In our analysis we relax the Lipschitz type continuity of the derivative of the operator involved. The main idea beyond this study is to find a larger convergence domain for the Newton's method (4.1.2). Using the restricted convergence domains, we obtained a finer convergence analysis, under the same computational cost, with the following advantages:

- The error estimates on the distances involved are tighter.
- The information on the location of the solution is at least as precise.

4.2 PRELIMINARIES

In order to make the chapter as self-contained as possible, we reintroduce some standard notations and auxiliary results for the monotonicity of set-valued op-

A Contemporary Study of Iterative Methods. DOI: 10.1016/B978-0-12-809214-9.00004-8

erators [18,22,27]. Denote by $U(w, \xi)$ and $\overline{U}(w, \xi)$ the open and closed balls in $\mathbb{B}_1$, respectively, with center $w \in H$ and radius $\xi > 0$.

Next we recall notions of monotonicity for set-valued operators.

Definition 4.2.1. *Let* $Q : H \rightrightarrows H$ *be a set-valued operator. Q is said to be monotone if for any* $x, y \in dom\, Q$ *and* $u \in Q(y), v \in Q(x)$ *implies that the following inequality holds:*

$$\langle u - v, y - x \rangle \geq 0.$$

A subset of $H \times H$ *is monotone if it is the graph of a monotone operator. If* $\varphi : H \longrightarrow (-\infty, +\infty]$ *is a proper function then the subgradient of* φ *is monotone.*

Definition 4.2.2. *Let* $Q : H \rightrightarrows H$ *be monotone. Then Q is maximal monotone if the following holds for all* $x, u \in H$:

$$\langle u - v, y - x \rangle \geq 0 \text{ for each } y \in dom\, Q \text{ and}$$
$$v \in Q(y) \Longrightarrow x \in dom\, Q \text{ and } v \in Q(x).$$

We will be using the following results for proving our results.

Lemma 4.2.3. *Let G be a positive operator (i.e.,* $\langle G(x), x \rangle \geq 0$*). The following statements about G hold:*

- $\|G^2\| = \|G\|^2$;
- *If* G^{-1} *exists, then* G^{-1} *is a positive operator.*

Lemma 4.2.4. *Let G be a positive operator. Suppose that* G^{-1} *exists. Then for each* $x \in H$ *we have*

$$\langle G(x), x \rangle \geq \frac{\|x\|^2}{\|G^{-1}\|}.$$

Lemma 4.2.5. *Let* $B : H \longrightarrow H$ *be a bounded linear operator and* $I : H \longrightarrow H$ *the identity operator. If* $\|B - I\| < 1$ *then B is invertible and* $\|B^{-1}\| \leq \dfrac{1}{(1 - \|B - I\|)}.$

Let $G : H \longrightarrow H$ be a bounded linear operator. Then $\widehat{G} := \frac{1}{2}(G + G^*)$ where G^* is the adjoint of G. Hereafter, we assume that $Q : H \rightrightarrows H$ is a set-valued maximal monotone operator and $F : H \longrightarrow H$ is a Fréchet-differentiable operator.

4.3 LOCAL CONVERGENCE

We present in this section the local convergence study of the Newton's method for solving the generalized equation (4.1.1) based on the partial linearization shown in [30].

Theorem 4.3.1. *Let* $F : D \subset H \longrightarrow H$ *be nonlinear operator with a continuous Fréchet derivative* F', *where* D *is an open subset of* H. *Let* $Q : H \rightrightarrows H$ *be a set-valued operator and* $x^* \in D$. *Suppose that* $0 \in F(x^*) + Q(x^*)$, $F'(x^*)$ *is a positive operator, and* $\widehat{F'(x^*)}^{-1}$ *exists. Let* $R > 0$ *and suppose that there exist* $f_0, f : [0, R) \longrightarrow \mathbb{R}$ *twice continuously differentiable such that*

(h$_0$)

$$\|\widehat{F'(x^*)}^{-1}\| \|F'(x) - F'(x^*)\| \leq f_0'(\|x - x^*\|) - f_0'(0), \qquad (4.3.1)$$

$x \in D$, *and*

$$\|\widehat{F'(x^*)}^{-1}\| \|F'(x) - F'(x^* + \theta(x - x^*))\| \qquad (4.3.2)$$
$$\leq f'(\|x - x^*\|) - f'(\theta\|x - x^*\|)$$

for each $x \in D_0 = D \cap U(x^*, R)$, $\theta \in [0, 1]$.

(h$_1$) $f(0) = f_0(0)$ *and* $f'(0) = f_0'(0) = -1$, $f_0(t) \leq f(t)$, $f_0'(t) \leq f'(t)$ *for each* $t \in [0, R)$.

(h$_2$) f_0', f' *are convex and strictly increasing.*

Let $v := \sup\{t \in [0, R) : f'(t) < 0\}$ *and* $r := \sup\{t \in (0, v) : \dfrac{f(t)}{tf'(t)} - 1 < 1\}$. *Then the sequences with the starting points* $x_0 \in B(x^*, r)/\{x^*\}$ *and* $t_0 = \|x^* - x_0\|$, *respectively, given by*

$$0 \in F(x_k) + F'(x_k)(x_{k+1} - x_k) + Q(x_{k+1}), \quad t_{k+1} = |t_k - \frac{f(t_k)}{f'(t_k)}|, \ k = 0, 1, \ldots,$$
$$(4.3.3)$$

are well defined, $\{t_k\}$ *is strictly decreasing, contained in* $(0, r)$, *and converges to* 0, *while* $\{x_k\}$ *is contained in* $U(x^*, r)$ *and converges to the point* x^*, *which is the unique solution of the generalized equation* $F(x) + Q(x) \ni 0$ *in* $U(x^*, \bar{\sigma})$, *where* $\bar{\sigma} = \min\{r, \sigma\}$ *and* $\sigma := \sup\{0 < t < R : f(t) < 0\}$. *Moreover, the sequence* $\{\dfrac{t_{k+1}}{t_k^2}\}$ *is strictly decreasing,*

$$\|x^* - x_{k+1}\| \leq [\frac{t_{k+1}}{t_k^2}]\|x_k - x^*\|^2, \quad \frac{t_{k+1}}{t_k^2} \leq \frac{f''(t_0)}{2|f'(t_0)|}, \ k = 0, 1, \ldots \quad (4.3.4)$$

If, additionally, $\dfrac{\rho f'(\rho) - f(\rho)}{\rho f'(\rho)} = 1$ *and* $\rho < R$ *then* $r = \rho$ *is the optimal convergence radius. Moreover, for each* $t \in (0, r)$ *and* $x \in \bar{U}(x^*, t)$,

$$\begin{aligned}
\|x_{k+1} - x^*\| \ &\leq\ \frac{e_f(\|x_k - x^*\|, 0)}{|f_0'(\|x_k^* - x^*\|)|} \\
&:=\ \alpha_k \leq \frac{e_f(\|x_k - x^*\|, 0)}{|f'(\|x_k^* - x^*\|)|} \qquad (4.3.5) \\
&\leq\ \frac{|a_f(t)|}{t^2}\|x_k - x^*\|^2,
\end{aligned}$$

where

$$e_f(s, t) := f(t) - (f(s) + f'(s)(t - s))\ \text{for}\ s, t \in [0, R)$$

and

$$a_f(t) := t - \frac{f(t)}{f'(t)}\ \text{for}\ s, t \in [0, v).$$

Lastly, by the second inequality in (4.3.5), there exists $r^* \geq r$ such that

$$\lim_{k \to \infty} x_k = x^*,$$

if

$$x_0 \in U(x^*, r^*) - \{x^*\}.$$

Throughout the rest of the chapter, we assume that the hypotheses of Theorem 4.3.1 hold.

Remark 4.3.2. *The introduction of the center-Lipschitz-type condition (4.3.1) (i.e., function f_0) allows us to introduce the restricted Lipschitz-type condition (4.3.2). The condition used in previous studies such as [33] is given by*

$$\|\widehat{F'(x^*)}^{-1}\| \|F'(x) - F'(x^* + \theta(x - x^*))\| \leq f_1'(\|x - x^*\|) - f_1'(\theta\|x - x^*\|) \tag{4.3.6}$$

for each $x \in D, \theta \in [0, 1]$, where $f_1 : [0, +\infty)\mathbb{R}$ is also twice continuously differentiable. It follows from (4.3.1), (4.3.2), and (4.3.6) that

$$f_0'(t) \leq f_1'(t), \tag{4.3.7}$$

$$f(t) \leq f'(t) \tag{4.3.8}$$

for each $t \in [0, v)$ since $D_0 \subseteq D$.

If $f_0'(t) = f_1'(t) = f'(t)$ for each $t \in [0, v)$, then our results reduce to the corresponding ones in [33]. Otherwise (i.e., if strict inequality holds in (4.3.7) or (4.3.8)), the new results improve the old ones. Indeed, let

$$r_1 := \sup\{t \in (0, \bar{v}) : -\frac{tf_1'(t) - f_1(t)}{tf_1'(t)} < 1\},$$

where $v_1 := \sup\{t \in [0, +\infty) : f_1'(t) < 0\}$. *Then the error bounds are*

$$\|x_{k+1} - x^*\| \quad \leq \quad \frac{e_{f_1}(\|x_k - x^*\|, 0)}{|f_1'(\|x_k^* - x^*\|)|}$$

$$:= \quad \beta_k \leq \frac{|a_{f_1}(t)|}{t^2}\|x_k - x^*\|^2. \tag{4.3.9}$$

In view of the definition of r, r_1 and estimates (4.3.5), (4.3.7), (4.3.8), and (4.3.9), we deduce that

$$r_1 \leq r \tag{4.3.10}$$

and

$$\alpha_k \leq \beta_k, \ k = 1, 2, \ldots \tag{4.3.11}$$

As a consequence, we obtain a larger radius of convergence and tighter error estimates on the distances involved, leading to a wider choice of initial guesses x^* and fewer computations of iterates x_k in order to achieve a desired error tolerance.

Furthermore, it is easy to see that the advantages are obtained under the same computational cost, since in practice the computation of function f_1 requires the computation of function f_0 and f as special cases.

Next, we present an auxiliary Banach lemma relating the operator F with the majorant function f_0.

Lemma 4.3.3. *Let $x^* \in H$ be such that $\widehat{F'(x^*)}$ is a positive operator and $\widehat{F'(x^*)}^{-1}$ exists. If $\|x - x^*\| \leq \min\{R, v\}$, then $\widehat{F'(x^*)}$ is a positive operator and $\widehat{F'(x^*)}^{-1}$ exists. Moreover,*

$$\|\widehat{F'(x)}^{-1}\| \leq \frac{\|\widehat{F'(x^*)}^{-1}\|}{|f_0'(\|x - x^*\|)|}. \tag{4.3.12}$$

Proof. First note that

$$\|\widehat{F'(x)} - \widehat{F'(x^*)}\| \leq \frac{1}{2}\|F'(x) - F'(x^*)\| + \frac{1}{2}\|(F'(x) - F'(x^*))^*\|$$

$$= \|F'(x) - F'(x^*)\|. \tag{4.3.13}$$

Take $x \in U(x^*, r)$. Since $r < v$ we have $\|x - x^*\| < v$. Thus $f'(\|x - x^*\|) < 0$ which, together with (4.3.13), imply that for all $x \in U(x^*, r)$,

$$\|\widehat{F'(x^*)}^{-1}\|\|\widehat{F'(x)} - \widehat{F'(x^*)}\|$$

$$\leq \|\widehat{F'(x^*)}^{-1}\|\|F'(x) - F'(x^*)\| \leq f_0'(\|x - x^*\|) - f_0'(0) < 1. \tag{4.3.14}$$

Using Banach lemma, we conclude that $\widehat{F'(x^*)}^{-1}$ exists. Moreover, by the above inequality,

$$
\begin{aligned}
\|\widehat{F'(x)}^{-1}\| &\leq \frac{\|\widehat{F'(x^*)}^{-1}\|}{1 - \|\widehat{F'(x^*)}^{-1}\|\,\|F'(x) - F'(x^*)\|} \\
&\leq \frac{\|\widehat{F'(x^*)}^{-1}\|}{1 - (f_0'(\|x - x^*\|) - f_0(0))} \\
&= \frac{\|\widehat{F'(x^*)}^{-1}\|}{|f_0'(\|x - x^*\|)|}.
\end{aligned}
$$

The last step follows from the fact that $r = \min\{R, v\}$. On the other hand, using (4.3.14) we have

$$
\|\widehat{F'(x)} - \widehat{F'(x^*)}\| \leq \frac{1}{\|\widehat{F'(x^*)}^{-1}\|}. \tag{4.3.15}
$$

Take $y \in H$. Then it follows from the above inequality that

$$
\begin{aligned}
\langle (\widehat{F'(x^*)} - \widehat{F'(x)})y, y \rangle &\leq \|\widehat{F'(x^*)} - \widehat{F'(x)}\|\,\|y\|^2 \\
&\leq \frac{\|y\|^2}{\|\widehat{F'(x^*)}^{-1}\|},
\end{aligned}
$$

which implies, after a simple manipulation, that

$$
\widehat{F'(x^*)}y, y \rangle - \frac{\|y\|^2}{\|\widehat{F'(x^*)}^{-1}\|} \leq \langle \widehat{F'(x)}y, y \rangle.
$$

Since $\widehat{F'(x^*)}$ is a positive operator and $\widehat{F'(x^*)}^{-1}$ exists by assumption, we obtain by Lemma 4.2.5 that

$$
\langle \widehat{F'(x^*)}y, y \rangle \geq \frac{\|y\|^2}{\|\widehat{F'(x^*)}^{-1}\|}.
$$

Therefore, combining the two last inequalities, we conclude that $\langle \widehat{F'(x)}y, y \rangle \geq 0$, i.e., $\widehat{F'(x)}$ is a positive operator.

Lemma 4.2.4 shows that $\widehat{F'(x)}$ is a positive operator and $\widehat{F'(x)}^{-1}$ exists, thus by Lemma 4.2.3 we have that for any $y \in H$

$$
\langle \widehat{F'(x)}y, y \rangle \geq \frac{\|y\|^2}{\|\widehat{F'(x)}^{-1}\|}. \qquad \square
$$

Note that $\langle \widehat{F'(x)}y, y \rangle = \langle F'(x)y, y \rangle$, thus by the second part of Lemma 4.2.4 we conclude that Newton's method is well-defined. Let us call N_{F+Q}, the Newton's method for $f + F$ in that region, namely, $N_{F+Q} : U(x^*, r) \longrightarrow H$ being defined by

$$0 \in F(x) + F'(x)(N_{F+Q}(x) - x) + Q(N_{F+Q}(x)), \text{ for each } x \in U(x^*, r).$$
$$(4.3.16)$$

Remark 4.3.4. *Under condition (4.3.6), it was shown in [33] that*

$$\|\widehat{F'(x)}^{-1}\| \leq \frac{\|\widehat{F'(x^*)}^{-1}\|}{|f_1'(\|x - x^*\|)|} \tag{4.3.17}$$

instead of (4.3.12). However, we have that (4.3.12) gives a tighter error estimate than (4.3.13), since $|f_1'(t)| \leq |f_0'(t)|$. This is a crucial difference in the proof of Theorem 4.3.1.

Proof of Theorem 4.3.1. Simply follow the proof of Theorem 4 in [33] but notice that the iterates x_k lie in D_0 which is a more precise location than D (used in [33]), allowing the usage of tighter function f than f_1, and also the usage of tighter function f_0 than f_1 for the computation of the upper bounds of the inverses $\|\widehat{F'(x)}^{-1}\|$ (i.e., we use (4.3.12) instead of (4.3.17)). □

4.4 SPECIAL CASES

Although in Remark 4.3.4 we have shown the advantages of our new approach over earlier ones, we also compare our results in the special case of the Kantorovich theory [8,11,27,29] with the corresponding ones in [16,29], Traub and Wozniakowski [35], when $F \equiv \{0\}$. Similar favorable comparisons can be given in the special case of Smale's theory [34] or Wang's theory [37].

Let the functions f_0, f, and f_1 be defined by

$$f_0'(t) = \frac{L_0}{2}t^2 - t,$$

$$f'(t) = \frac{L}{2}t^2 - t,$$

and

$$f_1'(t) = \frac{L_1}{2}t^2 - t$$

for some positive constants L_0, L, and L_1 to be determined using a specialized operator F.

4.5 NUMERICAL EXAMPLE

Example 4.5.1. *Let*

$$X = Y = \mathbb{R}^3,$$

$$D = \bar{U}(0, 1),$$

and

$$x^* = \begin{pmatrix} 0 \\ 0 \\ 0 \end{pmatrix}.$$

Define function F on D for

$$w = \begin{pmatrix} x \\ y \\ z \end{pmatrix}$$

by

$$F(w) = \begin{pmatrix} e^x - 1 \\ \dfrac{e-1}{2}y^2 + y \\ z \end{pmatrix}.$$

Then the Fréchet derivative is given by

$$F'(v) = \begin{bmatrix} e^x & 0 & 0 \\ 0 & (e-1)y+1 & 0 \\ 0 & 0 & 1 \end{bmatrix}.$$

Notice that

$$L_0 = e - 1,$$

$$L = e^{\frac{1}{L_0}},$$

$$L_1 = e,$$

and consequently,

$$f_0(t) < f(t) < f_1(t).$$

Therefore we have that the conditions of Theorem 4.3.1 hold. Moreover, we obtain

$$r_1 := \frac{2}{3L_1},$$

$$r := \frac{2}{3L},$$

and

$$r^* := \frac{2}{2L_0 + L}.$$

Consequently, we have that

$$r_1 < r < r^*.$$

Furthermore, the corresponding error bounds are:

$$\|x_{k+1} - x^*\| \leq \frac{L\|x_k - x^*\|^2}{2(1 - L_0\|x_k - x^*\|)},$$

$$\|x_{k+1} - x^*\| \leq \frac{L\|x_k - x^*\|^2}{2(1 - L\|x_k - x^*\|)},$$

and

$$\|x_{k+1} - x^*\| \leq \frac{L_1\|x_k - x^*\|^2}{2(1 - L_1\|x_k - x^*\|)}.$$

REFERENCES

[1] I.K. Argyros, A semilocal convergence analysis for directional Newton methods, Math. Comp. 80 (2011) 327–343.

[2] I.K. Argyros, Computational theory of iterative methods, in: K. Chui, L. Wuytack (Eds.), Stud. Comput. Math., vol. 15, Elsevier, New York, USA, 2007.

[3] I.K. Argyros, Concerning the convergence of Newton's method and quadratic majorants, J. Appl. Math. Comput. 29 (2009) 391–400.

[4] I.K. Argyros, D. González, Local convergence analysis of inexact Gauss–Newton method for singular systems of equations under majorant and center-majorant condition, SeMA J. 69 (1) (2015) 37–51.

[5] I.K. Argyros, S. Hilout, On the Gauss–Newton method, J. Appl. Math. (2010) 1–14.

[6] I.K. Argyros, S. Hilout, Extending the applicability of the Gauss–Newton method under average Lipschitz-conditions, Numer. Alg. 58 (2011) 23–52.

[7] I.K. Argyros, S. Hilout, On the solution of systems of equations with constant rank derivatives, Numer. Alg. 57 (2011) 235–253.

[8] I.K. Argyros, Y.J. Cho, S. Hilout, Numerical Methods for Equations and Its Applications, CRC Press/Taylor and Francis Group, New York, 2012.

[9] I.K. Argyros, S. Hilout, Improved local convergence of Newton's method under weaker majorant condition, J. Comput. Appl. Math. 236 (7) (2012) 1892–1902.

[10] I.K. Argyros, S. Hilout, Weaker conditions for the convergence of Newton's method, J. Complexity 28 (2012) 364–387.

[11] I.K. Argyros, S. Hilout, Computational Methods in Nonlinear Analysis, World Scientific Publ. Comp., New Jersey, 2013.

[12] L. Blum, F. Cucker, M. Shub, S. Smale, Complexity and Real Computation, Springer-Verlag, New York, 1998. With a foreword by Richard M. Karp.

[13] L. Blum, F. Cucker, M. Shub, S. Smale, Complexity and Real Computation, Springer-Verlag, New York, 1997.

[14] D.C. Chang, J. Wang, J.C. Yao, Newtons method for variational inequality problems: Smale's point estimate theory under the γ-condition, Appl. Anal. 94 (1) (2015) 44–55.

[15] J.E. Dennis Jr., R.B. Schnabel, Numerical Methods for Unconstrained Optimization and Nonlinear Equations, Classics Appl. Math., SIAM, Philadelphia, 1996.

[16] P. Deuflhard, G. Heindl, Affine invariant convergence for Newton's method and extensions to related methods, SIAM J. Numer. Anal. 16 (1) (1979) 1–10.

[17] A.L. Dontchev, R.T. Rockafellar, Implicit Functions and Solution Mappings. A View From Variational Analysis, Springer Monogr. Math., Springer, Dordrecht, 2009.

[18] S.P. Dokov, A.L. Dontchev, Robinson's strong regularity implies robust local convergence of Newton's method optimal control, in: Gainesville, FL, 1997, in: Appl. Optim., vol. 15, Kluwer Acad. Publ., Dordrecht, 1998, pp. 116–129.

[19] A.L. Dontchev, Local analysis of a Newton-type method based on partial linearization, in: The Mathematics of Numerical Analysis, Park City, UT, 1995, Amer. Math. Soc., Providence, RI, 1996.

[20] A.L. Dontchev, Local convergence of the Newton method for generalized equations, C. R. Acad. Sci. Paris Ser. I Math. 322 (4) (1996) 327–331.

[21] A.L. Dontchev, R.T. Rockafellar, Implicit Functions and Solution Mappings. A View From Variational Analysis, Springer Monogr. Math., Springer, Dordrecht, 2009.

[22] A.L. Dontchev, R.T. Rockafellar, Newton's method for generalized equations: a sequential implicit function theorem, Math. Program. 123 (1, Ser. B) (2010) 139–159.

[23] O. Ferreira, Local convergence of Newton's method in Banach space from the viewpoint of the majorant principle, IMA J. Numer. Anal. 29 (3) (2009) 746–759.

[24] O.P. Ferreira, M.L.N. Gonçalves, P.R. Oliveira, Local convergence analysis of the Gauss–Newton method under a majorant condition, J. Complexity 27 (1) (2011) 111–125.

[25] O.P. Ferreira, B.F. Svaiter, Kantorovich's majorants principle for Newton's method, Comput. Optim. Appl. 42 (2) (2009) 213–229.

[26] O.P. Ferreira, A robust semi-local convergence analysis of Newton's method for cone inclusion problems in Banach spaces under affine invariant majorant condition, J. Comput. Appl. Math. 279 (2015) 318–335.

[27] M. Josephy, Newton's Method for Generalized Equations and the PIES Energy Model, University of Wisconsin–Madison, 1979.

[28] A. Pietrus, C. Jean-Alexis, Newton-secant method for functions with values in a cone, Serdica Math. J. 39 (3–4) (2013) 271–286.

[29] L.B. Rall, A note on the convergence of Newton's method, SIAM J. Numer. Anal. 11 (1) (1974) 34–36.

[30] S.M. Robinson, Extension of Newton's method to nonlinear functions with values in a cone, Numer. Math. 19 (1972) 341–347.

[31] S.M. Robinson, Strongly regular generalized equations, Math. Oper. Res. 5 (1) (1980) 43–62.

[32] S.M. Robinson, Normed convex processes, Trans. Amer. Math. Soc. 174 (1972) 127–140.

[33] G.N. Silva, Convergence of the Newton's method for generalized equations under the majorant condition, arXiv:1603.05280v1 [math.NA], March 2016.

[34] S. Smale, Newton method estimates from data at one point, in: R. Ewing, K. Gross, C. Martin (Eds.), The Merging of Disciplines: New Directions in Pure, Applied and Computational Mathematics, Springer-Verlag, New York, 1986, pp. 185–196.

[35] J.F. Traub, H. Wózniakowski, Convergence and complexity of Newton iteration for operator equations, J. Assoc. Comput. Mach. 26 (2) (1979) 250–258.

[36] L.U. Uko, Generalized equations and the generalized Newton method, Math. Program. 73 (3, Ser. A) (1996) 251–268.

[37] X. Wang, Convergence of Newton's method and uniqueness of the solution of equation in Banach space, IMA J. Numer. Anal. 20 (2000) 123–134.

Chapter 5

Gauss–Newton method

5.1 INTRODUCTION

In this chapter we are concerned with approximately finding a solution $x^\star$ of the nonlinear least squares problem

$$\min_{x \in \mathcal{D}} \frac{1}{2} F(x)^T F(x) \qquad (5.1.1)$$

where $F : \mathcal{D} \longrightarrow \mathbb{R}^m$ is a continuously differentiable mapping from $\mathcal{D}$ to $\mathbb{R}^m$; $\mathcal{D} \subseteq \mathbb{R}^n$ is an open set, and $m \geq n$. A solution $x^\star \in \mathcal{D}$ of (5.1.1) is also called a least squares solution of the equation $F(x) = 0$ [1,2,8,11,12,15–17]. In this chapter we use the Gauss–Newton method defined by

$$x_{k+1} = x_k - \left[F'(x_k)^T F'(x_k) \right]^{-1} F'(x_k)^T F(x_k) \quad \text{for each} \quad k = 0, 1, \ldots, \qquad (5.1.2)$$

where $x_0 \in \mathcal{D}$ is an initial guess to generate a sequence $\{x_k\}$ approximating the solution $x^\star$.

If $x^\star$ is a solution of (5.1.1), $F(x^\star) = 0$, and $F'(x^\star)$ is invertible, then the theories of Gauss–Newton methods merge into those of Newton's method. Convergence results for Gauss–Newton-type methods under various Lipschitz-type conditions can be found in several works by different authors [1–3,5–7,9,10,13, 14].

In the present chapter we are motivated by the work in [4,9] and optimization considerations. Using the idea of restricted convergence domains a more precise majorant condition and functions, we provide a local convergence analysis for the Gauss–Newton method and the following advantages are obtained:

- Larger radius of convergence,
- Tighter error estimates on the distances $\| x_k - x^\star \|$ for each $k \geq 0$,
- A clearer relationship between the majorant function and the associated least squares problems (5.1.1).

These advantages are obtained under the same computational cost because the restricted convergence domains lead to more precise location of the iterates $\{x_n\}$ than in previous works. Even more, these advantages have wider importance in computational mathematics, since we have a wider choice of initial guesses x_0 and fewer computations to obtain a desired error tolerance on the distances $\| x_k - x^\star \|$ for each $k = 0, 1, \ldots$

A Contemporary Study of Iterative Methods. DOI: 10.1016/B978-0-12-809214-9.00005-X

61

The chapter is organized as follows: Section 5.2 contains the local convergence analysis of Gauss–Newton method. Some proofs are abbreviated to avoid repetitions with the corresponding ones in [4,9]. Special cases and applications are given in the concluding Section 5.3.

5.2 LOCAL CONVERGENCE ANALYSIS

We study the local convergence of Gauss–Newton method. Let $\mathcal{D} \subseteq \mathbb{R}^n$ be open, $F : \mathcal{D} \to \mathbb{R}^m$ be a continuously differentiable mapping, and $m \geq n$. Let $U(z, \rho)$ and $\bar{U}(z, \rho)$ stand respectively for the open and closed balls in $\mathbb{R}^n$ with center $z \in \mathbb{R}^n$ and radius $\rho > 0$.

Let $x^\star \in \mathcal{D}$, $R > 0$,

$$c = \| F(x^\star) \|,$$

and

$$\beta = \| \left[F'(x^\star)^T F'(x^\star) \right]^{-1} F'(x^\star)^T \| .$$

Suppose that $x^\star$ is a solution of (5.1.1), $F'(x^\star)$ has full rank, and there exist continuously differentiable functions $f_0 : [0, R) \longrightarrow [0, +\infty)$, $f : [0, R) \longrightarrow [0, +\infty)$ such that the following assumptions hold:

$(\mathcal{H}_1)$

$$\| F'(x) - F'(x^\star) \| \leq f_0'(\| x - x^\star \|) - f_0'(0) \quad \text{for each } x \in U(x^\star, \kappa),$$
$$(5.2.1)$$

where $\kappa = \sup\{t \in [0, R) : U(x^\star, t) \subseteq \mathcal{D}\}$.

$(\mathcal{H}_2)$ For $R_0 = \sup\{t \in [0, R) : f_0'(t) < 0\}$,

$$\| F'(x) - F'(x^\star + \tau (x - x^\star)) \| \leq f'(\| x - x^\star \|) - f'(\tau \| x - x^\star \|),$$
$$(5.2.2)$$

for all $x \in D_0 := U(x^\star, R_0) \cap U(x^\star, \kappa)$ and $\tau \in [0, 1]$;

$(\mathcal{H}_3)$ $f_0(0) = f(0) = 0$ and $f_0'(0) = f'(0) = -1$;

$(\mathcal{H}_4)$ f_0', f' are strictly increasing,

$$f_0(t) \leq f(t) \quad \text{and} \quad f_0'(t) \leq f'(t)$$
$$\text{for each} \quad t \in [0, R_1), \quad R_1 = \min\{R_0, \kappa\};$$

$(\mathcal{H}_5)$

$$\alpha_0 = \sqrt{2} c \beta^2 D^+ f'(0) < 1. \tag{5.2.3}$$

Moreover, define the parameters v_0, ρ_0, and r^0 by

$$v_0 := \sup\{t \in [0, R) : \beta (f_0'(t) + 1) < 1\}, \tag{5.2.4}$$

$$\rho_0 := \sup\{t \in [0, v_0) : \beta \, \frac{t \, f'(t) - f(t) + \sqrt{2} \, c \, \beta \, (f_0'(t) + 1)}{t \, (1 - \beta \, (f_0'(t) + 1))} < 1\}, \quad (5.2.5)$$

and

$$r^0 := \min\{R_0, \rho_0\}. \quad (5.2.6)$$

It is also clear that under (5.2.1) function f_0' is defined and therefore R_0 is at least as small as R. Therefore the majorant function satisfying (5.2.2) (i.e., f') is at least as small as the majorant function denoted by g satisfying (5.2.2) (i.e., g') but on $U(x^*, \kappa)$, leading to the advantages of the new approach over the approach in [4,9], since $D_0 \subseteq U(x^*, \kappa)$.

Moreover, if

$(\mathcal{H}_6)$ $2c\beta_0 D^+ f'(0) < 1$,

then the point x^* is the unique solution of (5.1.1) in $U(x^*, \sigma_0)$, where

$$0 < \sigma_0 := \sup\{t \in (0, \kappa) : \beta \left(\frac{f(t)}{t} + 1 \right) + c\beta_0 \left(\frac{f'(t) + 1}{t} \right) < 1\},$$

$$\beta_0 := \| \left[F'(x^*)^T F'(x^*) \right]^{-1} \|.$$

Our main result is presented as the following

Theorem 5.2.1. *Let $F : \mathcal{D} \subseteq \mathbb{R}^n \longrightarrow \mathbb{R}^m$ be a continuously differentiable operator. Let v_0, ρ_0, and r^0 be as defined in (5.2.4), (5.2.5), and (5.2.6), respectively. Suppose that $(\mathcal{H}_0)$–$(\mathcal{H}_3)$ hold. Then sequence $\{x_k\}$ generated by (5.1.2) and starting at $x_0 \in U(x^*, r^0) \setminus \{x^*\}$ is well defined, remains in $U(x^*, r^0)$ for all $n \geq 0$, and converges to x^*. Moreover, the following estimate holds for each $k = 0, 1, \dots$:*

$$\| x_{k+1} - x^* \| \leq \Xi_k \, \| x_k - x^* \| < \| x_n - x^* \| < r^0 \quad (5.2.7)$$

where

$$\Xi_k = \beta \, \frac{f'(\| x^* - x_k \|) \, \| x^* - x_k \| - f(\| x^* - x_k \|)}{\| x^* - x_k \|^2 \, (1 - \beta \, (f_0'(\| x^* - x_k \|) + 1))} \, \| x^* - x_k \| + \frac{\sqrt{2} \, c \, \beta^2 \, (f_0'(\| x_k - x^* \|) + 1)}{\| x^* - x_k \| \, (1 - \beta \, (f_0'(\| x_k - x^* \|) + 1))}.$$

$$(5.2.8)$$

Proof. Simply replace f by g in the proof of Theorem 3.1 in [4] (see also the proof in [9]), where $g : [0, \kappa) \longrightarrow (-\infty, +\infty)$ is continuously differentiable, satisfies

$$\| F'(x) - F'(x^* + \tau(x - x^*)) \| \leq g'(\| x - x^* \|) - g'(\tau \| x - x^* \|) \quad (5.2.9)$$

for all $x \in U(x^*, \kappa)$ and $\tau \in [0, 1]$, $g(0) = 0$, $g'(0) = -1$, and is strictly increasing on the interval $[0, \kappa)$. Iterates x_n lie in D_0, which is a more precise location than $U(x^*, \kappa)$ used in the previous works [4,9], if $R_0 < \kappa$. $\square$

Moreover, we have the following lemma.

Lemma 5.2.2. *If, in addition to $(\mathcal{H}_0)$–$(\mathcal{H}_3)$, condition $(\mathcal{H}_4)$ holds, then the point x^* is the unique solution of (5.1.1) in $U(x^*, \sigma_0)$.*

Remark 5.2.3. *If $f(t) = g(t) = f_0(t)$ for each $t \in [0, R_0)$ and $R_0 = R$, then Theorem 5.2.1 reduces to the corresponding theorem in [9]. Moreover, if $f_0(t) \leq f(t) = g(t)$, we obtain the results in [4]. Notice that, in view of (5.2.1), (5.2.2) and (5.2.9), we have*

$$f_0'(t) \leq g'(t) \text{ for each } t \in [0, \kappa) \qquad (5.2.10)$$

and

$$f'(t) \leq g'(t) \text{ for each } t \in [0, \kappa). \qquad (5.2.11)$$

Therefore, if

$$f_0'(t) \leq f'(t) < g'(t) \text{ for each } t \in [0, R_1) \qquad (5.2.12)$$

then the following advantages, under less computational cost (since in practice the computation of function g requires the computation of functions f_0 and f as special cases), are obtained:

- *weaker sufficient convergence criteria,*
- *tighter error bounds on the distances $\|x_n - x^*\|$, $\|x_{n+1} - x_n\|$,*
- *at least as precise information on the location of the solution x^*.*

5.3   NUMERICAL EXAMPLES

Specialization of Theorem 5.2.1 to some interesting cases such as Smale's α-theory (see also Wang's α-theory) and Kantorovich theory have been reported in [9], if $f_0'(t) = f'(t) = g'(t)$ for each $t \in [0, R)$ with $R_0 = R$, and in [4], if $f_0'(t) < f'(t) = g'(t)$ for each $t \in [0, R)$. Next, we present examples where $f_0'(t) < f'(t) < g'(t)$ for each $t \in [0, R_1)$ to show the advantages of the new approach over those in [4,9].

Example 5.3.1. *Let*

$$X = Y = \mathbb{R}^3,$$

$$D = \bar{U}(0, 1),$$

and

$$x^* = \begin{pmatrix} 0 \\ 0 \\ 0 \end{pmatrix}.$$

Define function F on D for

$$w = \begin{pmatrix} x \\ y \\ z \end{pmatrix}$$

by

$$F(w) = \begin{pmatrix} e^x - 1 \\ \dfrac{e-1}{2}y^2 + y \\ z \end{pmatrix}.$$

Then the Fréchet derivative is given by

$$F'(v) = \begin{bmatrix} e^x & 0 & 0 \\ 0 & (e-1)y+1 & 0 \\ 0 & 0 & 1 \end{bmatrix}.$$

Let

$$f_0(t) = \frac{L_0}{2}t^2 - t,$$

$$f(t) = \frac{L}{2}t^2 - t,$$

and

$$g(t) = \frac{L_1}{2}t^2 - t,$$

where

$$m = 3,$$

$$n = 3,$$

$$c = 0,$$

$$L_0 = e - 1,$$

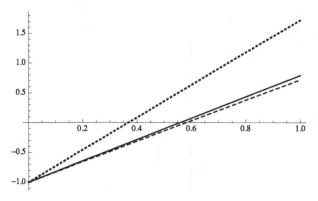

FIGURE 5.1 $f_0'(t)$ in dashed, $f'(t)$ in black and $g'(t)$ in dotted.

$$L = e^{\frac{1}{L_0}},$$

and

$$L_1 = e.$$

Then, as it can be seen in Fig. 5.1,

$$f_0'(t) = (e - 1)t - 1 \le f'(t) = e^{\frac{1}{e-1}}t - 1 < g'(t) = et - 1$$

holds as a strict inequality and as a consequence (5.2.10) also holds. Therefore, the new results have the advantages ($\mathcal{A}$) over the corresponding ones in [4,9].

REFERENCES

[1] I.K. Argyros, Convergence and Applications of Newton-Type Iterations, Springer-Verlag Publ., New York, 2008.
[2] I.K. Argyros, Y.J. Cho, S. Hilout, Numerical Methods for Equations and Its Applications, Science Publishers, New Hampshire, USA, 2012.
[3] I.K. Argyros, S. Hilout, Extending the applicability of the Gauss–Newton method under average Lipschitz-type conditions, Numer. Algorithms 58 (2011) 23–52.
[4] I.K. Argyros, A.A. Magreñán, Improved local convergence analysis of the Gauss–Newton method under a majorant condition, Comput. Optim. Appl. 60 (2) (2015) 423–439.
[5] J. Chen, The convergence analysis of inexact Gauss–Newton methods for nonlinear problems, Comput. Optim. Appl. 40 (2008) 97–118.
[6] J. Chen, W. Li, Convergence of Gauss–Newton's method and uniqueness of the solution, Appl. Math. Comput. 170 (2005) 686–705.
[7] J. Chen, W. Li, Local convergence results of Gauss–Newton's like method in weak conditions, J. Math. Anal. Appl. 324 (2006) 1381–1394.
[8] J.E. Dennis Jr., R.B. Schnabel, Numerical Methods for Unconstrained Optimization and Nonlinear Equations, Classics in Appl. Math., vol. 16, SIAM, Philadelphia, PA, 1996 (Corrected reprint of the 1983 original).
[9] O.P. Ferreira, M.L.N. Gonçalves, P.R. Oliveira, Local convergence analysis of Gauss–Newton's method under majorant condition, J. Complexity 27 (1) (2011) 111–125.

[10] W.M. Häubler, A Kantorovich-type convergence analysis for the Gauss–Newton method, Numer. Math. 48 (1986) 119–125.
[11] J.B. Hiriart-Urruty, C. Lemaréchal, Convex Analysis and Minimization Algorithms. I. Fundamentals. II. Advanced Theory and Bundle Methods, vols. 305, 306, Springer-Verlag, Berlin, 1993.
[12] L.V. Kantorovich, G.P. Akilov, Functional Analysis, Pergamon Press, Oxford, 1982.
[13] C. Li, N. Hu, J. Wang, Convergence behavior of Gauss–Newton's method and extensions to the Smale point estimate theory, J. Complexity 26 (2010) 268–295.
[14] C. Li, W-H. Zhang, X–Q. Jin, Convergence and uniqueness properties of Gauss–Newton's method, Comput. Math. Appl. 47 (2004) 1057–1067.
[15] P.D. Proinov, General local convergence theory for a class of iterative processes and its applications to Newton's method, J. Complexity 25 (2009) 38–62.
[16] S. Smale, Newton's method estimates from data at one point, in: R. Ewing, K. Gross, C. Martin (Eds.), The Merging of Disciplines: New Directions in Pure, Applied and Computational Mathematics, Springer, New York, 1986, pp. 185–196.
[17] J.F. Traub, Iterative Methods for the Solution of Equations, Prentice Hall, Englewood Cliffs, New Jersey, 1964.

Chapter 6

Gauss–Newton method for convex optimization

6.1 INTRODUCTION

In this chapter we are concerned with solving convex composite optimizations problem. In this chapter we are motivated by previous results that can be found in [23,17]. We present a convergence analysis of Gauss–Newton method in Section 6.2. The convergence of Gauss–Newton method is based on the majorant function in [17]. The same formulation using the majorant function provided in [23] (see [23,21,28,29]) is used. In [3,5,8] a convergence analysis in a Banach space setting was given for (GNM) defined by

$$x_{k+1} = x_k - \left[F'(x_k)^+ F'(x_k) \right]^{-1} F'(x_k)^+ F(x_k)$$

for each $k = 0, 1, 2, \ldots,$

where x_0 is an initial guess and $F'(x)^+$ in the Moore–Penrose inverse [11–13, 19,26] of operator $F'(x)$ with $F : \mathbb{R}^n \to \mathbb{R}^m$ being continuously differentiable. In [23] a semilocal convergence analysis using a combination of a majorant and a center-majorant function was given with the following advantages, under the same computational cost:

- the error estimates on the distances involved are tighter;
- the information on the location of the solution is at least as precise.

The chapter is organized as follows: Section 6.2 contains the definition of Gauss–Newton algorithm (from now on referred as GNA). In order to make the chapter as self-contained as possible, the notion of quasi-regularity is also reintroduced (see, e.g., [12,17,21]). The semilocal convergence analysis of GNA is presented in Section 6.3. Numerical examples and applications where theoretical results are proved and favorable comparisons to earlier studies (see, e.g., [12,17,18,21,22]) are presented in the concluding Section 6.4.

A Contemporary Study of Iterative Methods. DOI: 10.1016/B978-0-12-809214-9.00006-1

6.2   GAUSS–NEWTON ALGORITHM AND QUASI-REGULARITY CONDITION

6.2.1   Gauss–Newton algorithm

Using the idea of restricted convergence domains, we study the convex composite optimization problem

$$\min_{x \in \mathbb{R}^n} p(x) := h(F(x)), \tag{6.2.1}$$

where $h : \mathbb{R}^m \longrightarrow \mathbb{R}$ is convex, $F : \mathbb{R}^n \longrightarrow \mathbb{R}^m$ is a Fréchet-differentiable operator, and $m, l \in \mathbb{N}^\star$. The importance of (6.2.1) can be found in [2,10,23,12,19, 21,22,25,27]. We assume that the minimum h_{min} of the function h is attained. Problem (6.2.1) is related to

$$F(x) \in \mathcal{C}, \tag{6.2.2}$$

where

$$\mathcal{C} = \operatorname{argmin} h \tag{6.2.3}$$

is the set of all minimum points of h.

Let $\xi \in [1, \infty[, \Delta \in]0, \infty]$ and, for each $x \in \mathbb{R}^n$, define $\mathcal{D}_\Delta(x)$ by

$$
\begin{aligned}
\mathcal{D}_\Delta(x) \quad = \quad & \{d \in \mathbb{R}^n : \| d \| \le \Delta, \; h(F(x) + F'(x)d) \le h(F(x) + F'(x)d') \\
& \text{for all } d' \in \mathbb{R}^n \text{ with } \| d' \| \le \Delta \}.
\end{aligned}
\tag{6.2.4}
$$

Let $x_0 \in \mathbb{R}^n$ be an initial point. The GNA associated with (ξ, Δ, x_0) as defined in [12] (see also [17]) is as follows:

Algorithm GNA: (ξ, Δ, x_0)

INITIALIZATION. Take $\xi \in [1, \infty)$, $\Delta \in (0, \infty]$ and $x_0 \in \mathbb{R}^n$, set $k = 0$.
STOP CRITERION. Compute $\mathcal{D}_\Delta(x_k)$. If $0 \in \mathcal{D}_\Delta(x_k)$, STOP. Otherwise.
ITERATIVE STEP. Compute d_k satisfying $d_k \in \mathcal{D}_\Delta(x_k)$, $\|d_k\| \le \xi d(0, D\Delta(x_k))$,

Then set $x_{k+1} = x_k + d_k$, $k = k + 1$ and GO TO STOP CRITERION.

Here $d(x, W)$ denotes the distance from x to W in a finite-dimensional Banach space containing W. Note that the set $\mathcal{D}_\Delta(x)$ $(x \in \mathbb{R}^n)$ is nonempty and is the solution of the following convex optimization problem:

$$\min_{d \in \mathbb{R}^n, \|d\| \le \Delta} h(F(x) + F'(x)d), \tag{6.2.5}$$

which can be solved by well known methods such as the subgradient, cutting plane, or bundle methods (see, e.g., [12,19,25–27]).

Throughout this chapter, we will denote by $U(x, r)$ the open ball in $\mathbb{R}^n$ (or $\mathbb{R}^m$) centered at x and of radius $r > 0$, and by $\overline{U}(x, r)$ its closure. Let W be a closed convex subset of $\mathbb{R}^n$ (or $\mathbb{R}^m$). The negative polar of W denoted by $W^{\ominus}$ is defined as

$$W^{\ominus} = \{z : < z, w > \le 0 \quad \text{for each} \quad w \in W\}. \tag{6.2.6}$$

6.2.2 Quasi-regularity

In order to make the chapter as self-contained as possible, in this subsection, we mention some concepts and results on regularities which can be found in [12] (see also, e.g., [23,17,21,22,25]). For a set-valued mapping $T : \mathbb{R}^n \rightrightarrows \mathbb{R}^m$ and for a set A in $\mathbb{R}^n$ or $\mathbb{R}^m$, we denote by

$$D(T) = \{x \in \mathbb{R}^n : Tx \ne \emptyset\}, \quad R(T) = \bigcup_{x \in D(T)} Tx, \tag{6.2.7}$$

$$T^{-1}y = \{x \in \mathbb{R}^n : y \in Tx\}, \quad \text{and} \quad \| A \| = \inf_{a \in A} \| a \|.$$

Consider the inclusion

$$F(x) \in C, \tag{6.2.8}$$

where C is a closed convex set in $\mathbb{R}^m$. Let $x \in \mathbb{R}^n$ and

$$\mathcal{D}(x) = \{d \in \mathbb{R}^n : F(x) + F'(x)d \in C\}. \tag{6.2.9}$$

Definition 6.2.1. *Let $x_0 \in \mathbb{R}^n$.*

(a) *x_0 is a quasi-regular point of (6.2.8) if there exist $R_0 \in]0, +\infty[$ and an increasing positive function β on $[0, R_0)$ such that*

$$\mathcal{D}(x) \ne \emptyset \text{ and } d(0, \mathcal{D}(x)) \le \beta(\| x - x_0 \|) d(F(x), C) \tag{6.2.10}$$
$$\text{for all } x \in U(x_0, R_0).$$

$\beta(\| x - x_0 \|)$ is an "error bound" in determining how far away the origin is from the solution set of (6.2.8).

(b) *x_0 is a regular point of (6.2.8) if*

$$\ker(F'(x_0)^T) \cap (C - F(x_0))^{\ominus} = \{0\}. \tag{6.2.11}$$

Proposition 6.2.2. *(see, e.g., [12,17,21,25]) Let x_0 be a regular point of (6.2.8). Then there are constants $R_0 > 0$ and $\beta > 0$ such that (6.2.10) holds for R_0 and $\beta(\cdot) = \beta$. Therefore, x_0 is a quasi-regular point with the quasi-regular radius $R_{x_0} \ge R_0$ and the quasi-regular bound function $\beta_{x_0} \le \beta$ on $[0, R_0]$.*

Remark 6.2.3. **(a)** $\mathcal{D}(x)$ *can be considered as the solution set of the linearized problem associated to (6.2.8), namely*

$$F(x) + F'(x)d \in C. \tag{6.2.12}$$

(b) *If C defined in (6.2.8) is the set of all minimum points of h and if there exists $d_0 \in \mathcal{D}(x)$ with $\| d_0 \| \leq \Delta$, then $d_0 \in \mathcal{D}_\Delta(x)$ and for each $d \in \mathbb{R}^n$, we have the following equivalence:*

$$d \in \mathcal{D}_\Delta(x) \Longleftrightarrow d \in \mathcal{D}(x) \Longleftrightarrow d \in \mathcal{D}_\infty(x). \tag{6.2.13}$$

(c) *Let R_{x_0} denote the supremum of R_0 such that (6.2.10) holds for some function β defined in Definition 6.2.1. Let $R_0 \in [0, R_{x_0}]$ and $\mathcal{B}_R(x_0)$ denote the set of function β defined on $[0, R_0)$ such that (6.2.10) holds. Define*

$$\beta_{x_0}(t) = \inf\{\beta(t) : \beta \in \mathcal{B}_{R_{x_0}}(x_0)\} \quad \text{for each} \quad t \in [0, R_{x_0}). \tag{6.2.14}$$

All function $\beta \in \mathcal{B}_R(x_0)$ with $\lim\limits_{t \to R^-} \beta(t) < +\infty$ can be extended to an element of $\mathcal{B}_{R_{x_0}}(x_0)$ and we have that

$$\beta_{x_0}(t) = \inf\{\beta(t) : \beta \in \mathcal{B}_R(x_0)\} \quad \text{for each} \quad t \in [0, R_0). \tag{6.2.15}$$

R_{x_0} and β_{x_0} are called the quasi-regular radius and the quasi-regular function of the quasi-regular point x_0, respectively.

Definition 6.2.4. (a) *A set-valued mapping $T : \mathbb{R}^n \rightrightarrows \mathbb{R}^m$ is convex if the following properties hold:*
  (i) $Tx + Ty \subseteq T(x + y)$ *for all $x, y \in \mathbb{R}^n$.*
  (ii) $T\lambda x = \lambda Tx$ *for all $\lambda > 0$ and $x \in \mathbb{R}^n$.*
  (iii) $0 \in T0.$

6.3   SEMILOCAL CONVERGENCE

Now we present the semilocal convergence of GNA. We begin by studying the convergence of majorizing sequences for GNA. Then we study the convergence of GNA. We need the definition of the center-majorant function and the definition of the majorant function for F.

Definition 6.3.1. *Let $R > 0$, $x_0 \in \mathbb{R}^n$, and $F : \mathbb{R}^n \to \mathbb{R}^m$ be continuously Fréchet-differentiable. A twice-differentiable function $f_0 : [0, R) \to \mathbb{R}$ is called a center-majorant function for F on $U(x_0, R)$, if for each $x \in U(x_0, R)$,*

(h_0^0) $\| F'(x) - F'(x_0)\| \leq f_0'(\|x - x_0\|) - f_0'(0);$
(h_1^0) $f_0(0) = 0, f_0'(0) = -1;$
    and
(h_2^0) f_0' *is convex and strictly increasing.*

Definition 6.3.2. *[5,8,17] Let $x_0 \in \mathbb{R}^n$ and $F : \mathbb{R}^n \to \mathbb{R}^m$ be continuously differentiable. Define $R_0 = \sup\{t \in [0, R) : f_0'(t) < 0\}$. A twice-differentiable function $f : [0, R_0) \to \mathbb{R}$ is called a majorant function for F on $U(x_0, R_0)$, if for each $x, y \in U(x_0, R_0)$, $\|x - x_0\| + \|y - x\| < R_0$,*

(h_0) $\|F'(y) - F'(x)\| \le f'(\|y - x\| + \|x - x_0\|) - f'(\|x - x_0\|)$;
(h_1) $f(0) = 0$, $f'(0) = -1$;
   and
(h_2) f' *is convex and strictly increasing.*

Furthermore, let us suppose that the following condition is satisfied:

(h_3) $f_0(t) \le f(t)$ *and* $f_0'(t) \le f'(t)$ *for each* $t \in [0, R_0)$.

Now we can present the following remark.

Remark 6.3.3. *Suppose that $R_0 < R$. If $R_0 \ge R$, then we do not need to introduce Definition 6.3.2.*

Let $\xi > 0$ and $\alpha > 0$ be fixed and define auxiliary function $\varphi : [0, R_0) \to \mathbb{R}$ by

$$\varphi(t) = \xi + (\alpha - 1)t + \alpha f(t). \tag{6.3.1}$$

We shall use the following hypotheses:

(h_4) *there exists $s^* \in (0, R)$ such that for each $t \in (0, s^*)$, $\varphi(t) > 0$ and $\varphi(s^*) = 0$;*
(h_5) $\varphi(s^*) < 0$.

From now on we assume the hypotheses (h_0)–(h_4) and (h_0^0)–(h_2^0) which will be called hypotheses (H).

Next we present the main semilocal convergence result of the Gauss–Newton method generated by GNA.

Theorem 6.3.4. *Suppose that hypotheses (H) are satisfied. Then*

(i) Sequence $\{s_k\}$ generated by the Gauss–Newton method for $s_0 = 0$, $s_{k+1} = s_k - \frac{\varphi(s_k)}{\varphi'(s_k)}$ for solving equation $\psi(t) = 0$ is well defined, strictly increasing, remains in $[0, s^*)$, and converges Q-linearly to s^*.
   Let $\eta \in [1, \infty]$, $\Delta \in (0, \infty]$, and let $h : \mathbb{R}^m \to \mathbb{R}$ be real-valued convex with minimizer set C such that $C \neq \emptyset$.
(ii) Suppose that $x_0 \in \mathbb{R}^n$ is a quasi-regular point of the inclusion

$$F(x) \in C,$$

with the quasi-regular radius r_{x_0} and the quasi-regular bound function β_{x_0} defined by (2.14) and (2.15), respectively. If $d(F(x_0), C) > 0$, $s^* \le r_{x_0}$, $\Delta \ge \xi \ge \eta\beta_{x_0}(0)d(F(x_0), C)$, and

$$\alpha \ge \sup\left\{ \frac{\eta\beta_{x_0}(t)}{\eta\beta_{x_0}(t)(1 + f'(t)) + 1} : \xi \le t < s^* \right\},$$

then sequence $\{x_k\}$ generated by GNA is well defined, remains in $U(x_0, s^*)$ for each $k = 0, 1, 2, \ldots$, such that

$$F(x_k) + F'(x_k)(x_{k+1} - x_k) \in C \text{ for each } k = 0, 1, 2 \ldots \quad (6.3.2)$$

Moreover, the following estimates hold:

$$\|x_{k+1} - x_k\| \leq s_{k+1} - s_k, \quad (6.3.3)$$

$$\|x_{k+1} - x_k\| \leq \frac{s_{k+1} - s_k}{(s_k - s_{k-1})^2} \|x_k - x_{k-1}\|^2, \quad (6.3.4)$$

for each $k = 0, 1, 2 \ldots$, and $k = 1, 2, \ldots$, respectively, and sequence $\{x_k\}$ converges to a point $x^* \in U(x_0, s^*)$ satisfying $F(x^*) \in C$ and

$$\|x^* - x_k\| \leq t^* - s_k \text{ for each } k = 0, 1, 2, \ldots \quad (6.3.5)$$

The convergence is R-linear. If hypothesis (h_5) holds, then the sequences $\{s_k\}$ and $\{x_k\}$ converge Q-quadratically and R-quadratically to s^* and x^*, respectively. Furthermore, if

$$\alpha > \bar{\alpha} := \sup \left\{ \frac{\eta_{\beta_{x_0}}(t)}{\eta_{\beta_{x_0}}(t)(1 + f'(t)) + 1} : \xi \leq t < s^* \right\},$$

then the sequence $\{x_k\}$ converges R-quadratically to x^*.

Proof. Simply replace function g in [23] (see also [17]) by function f in the proof, where g is a majorant function for F on $U(x_0, R)$. That is, we have instead of (h_0):
(h'_0)

$$\|F'(y) - F'(x)\| \leq g'(\|y - x\| + \|x - x_0\|) - g'(\|x - x_0\|) \quad (6.3.6)$$

for each $x, y \in U(x_0, R)$ with $\|x - x_0\| + \|y - x\| < R$. The iterates x_n lie in $U(x_0, R_0)$ which is a more precise location than $U(x_0, R)$. □

Remark 6.3.5. (a) *If $f(t) = g(t) = f_0(t)$ for each $t \in [0, R_0)$ and $R_0 = R$, then Theorem 6.3.4 reduces to the corresponding theorem in [17]. Moreover, if $f_0(t) \leq f(t) = g(t)$, we obtain the results in [23]. Notice that we have*

$$f'_0(t) \leq g'(t) \text{ for each } t \in [0, R) \quad (6.3.7)$$

and

$$f'(t) \leq g'(t) \text{ for each } t \in [0, R_0). \quad (6.3.8)$$

Therefore, if

$$f_0'(t) \leq f'(t) < g'(t) \text{ for each } t \in [0, R_0),$$ (6.3.9)

then the following advantages, using less computational cost (since in practice the computation of function g requires the computation of functions f_0 and f as special cases), are obtained:

- *weaker sufficient convergence criteria, tighter error bounds on the distances $\|x_n - x^*\|$, $\|x_{n+1} - x_n\|$,*
- *at least as precise information on the location of the solution x^*.*

Moreover, it is clear that under (h_0^0) function f_0' is defined. Therefore R_0 is at least as small as R. Therefore the majorant function satisfying (h_0^0) (i.e., f') is at least as small as the majorant function satisfying (h_0') (i.e., g'), leading to the advantages of the new approach over the approach in [17] or [23]. Indeed, we have that if function ψ has a solution t^*, since $\varphi(t^*) \leq \psi(t^*) = 0$ and $\varphi(0) = \psi(0) = \xi > 0$. Then, we get that function φ has a solution r^* such that

$$r^* \leq t^*,$$ (6.3.10)

but not necessarily vice versa. If also follows from (6.3.10) that the new information about the location of the solution x^* is at least as precise as the one given in [18].

(b) *Let us specialize conditions (6.2.8)–(6.2.10) even further in the case when L_0, K, and L are constant functions, and $\alpha = 1$. Then the functions corresponding to (6.3.1) reduce to*

$$\psi(t) = \frac{L}{2}t^2 - t + \xi$$ (6.3.11)

[1,17,23] and

$$\varphi(t) = \frac{K}{2}t^2 - t + \xi,$$ (6.3.12)

respectively. In this case the convergence criteria become respectively

$$h = L\xi \leq \frac{1}{2}$$ (6.3.13)

and

$$h_1 = K\xi \leq \frac{1}{2}.$$ (6.3.14)

Notice that

$$h \leq \frac{1}{2} \Longrightarrow h_1 \leq \frac{1}{2}$$ (6.3.15)

but not necessarily vice versa. Unless, of course, $K = L$. Condition (6.3.13) is the Kantorovich hypothesis for the semilocal convergence of Newton's method to a solution x^* of equation $F(x) = 0$ [20]. In the case of Wang's condition [29], we have

$$\varphi(t) = \frac{\gamma t^2}{1 - \gamma t} - t + \xi,$$

$$\psi(t) = \frac{\beta t^2}{1 - \beta t} - t + \xi,$$

$$L(u) = \frac{2\gamma}{(1 - \gamma u)^3}, \quad \gamma > 0, \ 0 \le t \le \frac{1}{\gamma},$$

and

$$K(u) = \frac{2\beta}{(1 - \beta u)^3}, \quad \beta > 0, \ 0 \le t \le \frac{1}{\beta},$$

with convergence criteria given respectively by

$$H = \gamma \xi \le 3 - 2\sqrt{2}, \tag{6.3.16}$$

$$H_1 = \beta \xi \le 3 - 2\sqrt{2}. \tag{6.3.17}$$

Then again we have that

$$H \le 3 - 2\sqrt{2} \Longrightarrow H_1 \le 3 - 2\sqrt{2}$$

but not necessarily vice versa, unless $\beta = \gamma$.

(c) *Regarding the error bounds and the limit of majorizing sequence, let us define majorizing sequence $\{r_{\alpha,k}\}$ by*

$$r_{\alpha,0} = 0; \ r_{\alpha,k+1} = r_{\alpha,k} - \frac{\varphi(r_{\alpha,k})}{\varphi'_{\alpha,0}(r_{\alpha,k})}$$

for each $k = 0, 1, 2, \ldots$, where

$$\varphi_{\alpha,0}(t) = \xi - t + \alpha \int_0^t L_0(t)(t - u)\,du.$$

Suppose that

$$-\frac{\varphi(r)}{\varphi'_{\alpha,0}(r)} \le -\frac{\varphi(s)}{\varphi'(s)}$$

for each $r, s \in [0, R_0]$ with $r \le s$. According to the proof of Theorem 6.3.4, sequence $\{r_{\alpha,k}\}$ is also a majorizing sequence for GNA.

Moreover, using mathematical induction, we obtain that

$$r_k \leq s_k,$$

$$r_{k+1} - r_k \leq s_{k+1} - s_k,$$

and

$$r^* = \lim_{k \to \infty} r_k \leq s^*.$$

Furthermore, the first two preceding inequalities are strict for $n \geq 2$, if

$$L_0(u) < K(u) \text{ for each } u \in [0, R_0].$$

Similarly, suppose that

$$-\frac{\varphi(s)}{\varphi'(s)} \leq -\frac{\psi(t)}{\psi'(t)} \tag{6.3.18}$$

for each $s, t \in [0, R_0]$ with $s \leq t$. Then we have

$$s_{\alpha,k} \leq t_{\alpha,k},$$

$$s_{\alpha,k+1} - s_{\alpha,k} \leq t_{\alpha,k+1} - t_{\alpha,k},$$

and

$$s^* \leq t^*.$$

The first two preceding inequalities are also strict for $k \geq 2$, if a strict inequality holds in (6.3.18).

6.4 NUMERICAL EXAMPLE

Specializations of results to some interesting cases such as Smale's α-theory (see also Wang's α-theory) and Kantorovich theory have been reported in [17, 23,25], if

$$f_0'(t) = f'(t) = g'(t)$$

for each $t \in [0, R)$ with $R_0 = R$, and in [23], if

$$f_0'(t) < f'(t) = g'(t)$$

for each $t \in [0, R_0)$. Next we present examples where

$$f_0'(t) < f'(t) < g'(t)$$

for each $t \in [0, R_0)$ to show the advantages of the new approach over those in [17,24,26]. We choose for simplicity $m = n = \alpha = 1$.

Example 6.4.1. *Let*

$$x_0 = 1,$$
$$D = U(1, 1 - q),$$
$$q \in [0, \frac{1}{2}),$$

and define function F on D by

$$F(x) = x^3 - q. \tag{6.4.1}$$

Then we have

$$\xi = \frac{1}{3}(1 - q),$$
$$L_0 = 3 - q,$$
$$L = 2(2 - q),$$

and

$$K = 2(1 + \frac{1}{L_0}).$$

The Newton–Kantorovich condition (6.3.13) is not satisfied, since

$$h > \frac{1}{2}, \tag{6.4.2}$$

for each

$$q \in [0, \frac{1}{2}).$$

Hence, there is no guarantee by the Newton–Kantorovich theorem [4,6,7,9, 14–17] that Newton's method (6.2.1) converges to a zero of operator F. On the other hand, by (6.3.14) we have

$$h_1 \leq \frac{1}{2}, \tag{6.4.3}$$

if

$$0.461983163 < q < \frac{1}{2}.$$

Hence, we have proved the improvements using this example.

REFERENCES

[1] S. Amat, S. Busquier, J.M. Gutiérrez, Geometric constructions of iterative functions to solve nonlinear equations, J. Comput. Appl. Math. 157 (2003) 197–205.

[2] I.K. Argyros, Computational theory of iterative methods, in: K. Chui, L. Wuytack (Eds.), Stud. Comput. Math., vol. 15, Elsevier, New York, USA, 2007.

[3] I.K. Argyros, Concerning the convergence of Newton's method and quadratic majorants, J. Appl. Math. Comput. 29 (2009) 391–400.

[4] I.K. Argyros, S. Hilout, On the Gauss–Newton method, J. Appl. Math. (2010) 1–14.

[5] I.K. Argyros, S. Hilout, Extending the applicability of the Gauss–Newton method under average Lipschitz-conditions, Numer. Alg. 58 (2011) 23–52.

[6] I.K. Argyros, A semilocal convergence analysis for directional Newton methods, Math. Comp. 80 (2011) 327–343.

[7] I.K. Argyros, Y.J. Cho, S. Hilout, Numerical Methods for Equations and Its Applications, CRC Press/Taylor and Francis Group, New York, 2012.

[8] I.K. Argyros, S. Hilout, Improved local convergence of Newton's method under weaker majorant condition, J. Comput. Appl. Math. 236 (7) (2012) 1892–1902.

[9] I.K. Argyros, S. Hilout, Weaker conditions for the convergence of Newton's method, J. Complexity 28 (2012) 364–387.

[10] I.K. Argyros, S. Hilout, Computational Methods in Nonlinear Analysis, World Scientific Publ. Comp, New Jersey, 2013.

[11] A. Ben-Israel, T.N.E. Greville, Generalized inverses, in: Theory and Applications, second edition, in: CMS Books Math/Ouvrages Math SMC, vol. 15, Springer-Verlag, New York, 2003.

[12] J.V. Burke, M.C. Ferris, A Gauss–Newton method for convex composite optimization, Math. Program. Ser A. 71 (1995) 179–194.

[13] J.E. Dennis Jr., R.B. Schnabel, Numerical Methods for Unconstrained Optimization and Nonlinear Equations, Classics in Appl. Math., vol. 16, SIAM, Philadelphia, PA, 1996.

[14] O.P. Ferreira, M.L.N. Gonçalves, P.R. Oliveira, Local convergence analysis of the Gauss–Newton method under a majorant condition, J. Complexity 27 (1) (2011) 111–125.

[15] O.P. Ferreira, B.F. Svaiter, Kantorovich's theorem on Newton's method in Riemannian manifolds, J. Complexity 18 (1) (2002) 304–329.

[16] O.P. Ferreira, B.F. Svaiter, Kantorovich's majorants principle for Newton's method, Comput. Optim. Appl. 42 (2009) 213–229.

[17] O.P. Ferreira, M.L.N. Gonçalves, P.R. Oliveira, Convergence of the Gauss–Newton method for convex composite optimization under a majorant condition, SIAM J. Optim. 23 (3) (2013) 1757–1783.

[18] W.M. Häussler, A Kantorovich-type convergence analysis for the Gauss–Newton method, Numer. Math. 48 (1986) 119–125.

[19] J.B. Hiriart-Urruty, C. Lemaréchal, Convex Analysis and Minimization Algorithms. I. Fundamentals. II. Advanced Theory and Bundle Methods, vols. 305, 306, Springer-Verlag, Berlin, 1993.

[20] L.V. Kantorovich, G.P. Akilov, Functional Analysis, Pergamon Press, Oxford, 1982.

[21] C. Li, K.F. Ng, Majorizing functions and convergence of the Gauss–Newton method for convex composite optimization, SIAM J. Optim. 18 (2) (2007) 613–692.

[22] C. Li, X.H. Wang, On convergence of the Gauss–Newton method for convex composite optimization, Math. Program. Ser A. 91 (2002) 349–356.

[23] A. Magrenán, I.K. Argyros, Expanding the applicability of the Gauss–Newton method for convex optimization under a majorant condition, SeMA J. 65 (2014) 37–56.

[24] F.A. Potra, Sharp error bounds for a class of Newton-like methods, Libertas Math. 5 (1985) 71–84.

[25] S.M. Robinson, Extension of Newton's method to nonlinear functions with values in a cone, Numer. Math. 19 (1972) 341–347.

[26] R.T. Rockafellar, Monotone Processes of Convex and Concave Type, Mem. Amer. Math. Soc., vol. 77, American Mathematical Society, Providence, RI, 1967.

[27] R.T. Rockafellar, Convex Analysis, Princeton Math. Ser., vol. 28, Princeton University Press, Princeton, NJ, 1970.

[28] S. Smale, Newton's method estimates from data at one point, in: The Merging of Disciplines: New Directions in Pure, Applied, and Computational Mathematics, Laramie, WY, 1985, Springer, New York, 1986, pp. 185–196.

[29] X.H. Wang, Convergence of Newton's method and uniqueness of the solution of equations in Banach space, IMA J. Numer. Anal. 20 (2000) 123–134.

Chapter 7

Proximal Gauss–Newton method

7.1 INTRODUCTION

Let H_1 and H_2 be Hilbert spaces. Let $\Omega \subseteq H_1$ be open and $F : \Omega \longrightarrow H_2$ continuously Fréchet-differentiable. In this chapter we are concerned with the problem of approximating a solution $x^\star$ of the penalized nonlinear least squares problem

$$\min_{x\in\Omega} \| F(x) \|^2 + J(x), \tag{7.1.1}$$

where $J : \Omega \to \mathbb{R} \cup \{\infty\}$ is proper, convex and lower semicontinuous. A solution $x^\star \in \Omega$ of (7.1.1) is also called a least squares solution of the equation $F(x) = 0$. If $J = 0$, the well known Gauss–Newton method defined by

$$x_{n+1} = x_n - F'(x_n)^+ F(x_n) \qquad \text{for each} \quad n = 0, 1, 2, \ldots, \tag{7.1.2}$$

where $x_0 \in \Omega$ is an initial point [11] and $F'(x_n)^+$ is the Moore–Penrose inverse of the linear operator $F'(x_n)$ is considered for finding an approximation for x^*. In the present chapter we use the proximal Gauss–Newton method for solving the penalized nonlinear least squares problem (7.1.1). Several convergence results under various Lipschitz-type conditions for Gauss–Newton-type methods can be found in [2,5,9,10,12–14]. The convergence of these methods requires as one of the hypotheses that F' satisfies a Lipschitz condition or F'' is bounded in Ω. In the last decades, some authors have relaxed these hypotheses [4,8–10, 13].

In [6], motivated by the work presented in [9] and optimization considerations, the authors presented a new local convergence analysis for inexact Gauss–Newton-like methods by using a majorant and center-majorant function instead of just a majorant function with the following advantages (under the same computational cost, since the computation of the majorant function requires the computation of the center-majorant function as a special case):

- Larger radius of convergence and tighter error estimates on the distances $\| x_n - x^\star \|$ for each $n = 0, 1, \ldots$;
- A clearer relationship between the majorant function and the associated least squares problems (7.1.1).

In the present chapter, we obtain the same advantages over the earlier works by using the idea of restricted convergence domains. This way we find a more

A Contemporary Study of Iterative Methods. DOI: 10.1016/B978-0-12-809214-9.00007-3

precise location of where the iterates lie than in earlier studies [1,6–9], leading to tighter majorizing functions.

The chapter is organized as follows. In order to make the chapter as self-contained as possible, we provide the necessary background in Section 7.2. Section 7.3 contains the local convergence analysis of inexact Gauss–Newton-like methods. A numerical example is given in the concluding Section 7.4.

7.2 BACKGROUND

Some commonly used concepts are reintroduced in order to make the chapter as self-contained as possible. Let $A : H_1 \longrightarrow H_2$ be continuous, linear, and injective with closed image, the Moore–Penrose inverse [3] $A^+ : H_2 \longrightarrow H_1$ is defined by $A^+ = (A^\star A)^{-1} A^\star$. $\mathcal{I}$ denotes the identity operator on H_1 (or H_2). Let $\mathcal{L}(H_1, H_2)$ be the space of bounded linear operators from H_1 into H_2. Let $M \in \mathcal{L}(H_1, H_2)$, by $Ker(M)$ and $Im(M)$ we denote the kernel and image of M, respectively, and by M^* its adjoint operator. Let $M \in \mathcal{L}(H_1, H_2)$ with a closed image. Recall that the Moore–Penrose inverse of M is the linear operator $M^+ \in \mathcal{L}(H_2, H_1)$ which satisfies

$$M M^+ M = M, \ M^+ M M^+ = M, \ (M M^+)^* = M M^+, \ (M^+ M)^* = M^+ M. \tag{7.2.1}$$

It follows from (7.2.1) that if $\prod_S$ denotes the projection of X onto subspace S, then

$$M^+ M = I_X - \prod_{Ker(M)}, \quad M M^+ = \prod_{Im(M)}. \tag{7.2.2}$$

Moreover, if M is injective, then

$$M^+ = (M^* M)^{-1} M^*, \ M^+ M = I_X, \ \|M^+\|^2 = \|(M^* M)^{-1}\|. \tag{7.2.3}$$

The semilocal convergence of proximal Gauss–Newton method using Wang's condition was introduced in [14]. Next, in order to make the chapter as self-contained as possible, we briefly illustrate how this method is defined. Let $Q : H_1 \to H_1$ be continuous, positive, self-adjoint, and bounded from below. It follows that $Q^{-1} \in \mathcal{L}(H_1, H_1)$. Define a scalar product on X by $< u, v > = < u, Q_v >$. Then the corresponding induced norm $\| \cdot \|_Q$ is equivalent to the given norm on X, since $\dfrac{1}{\| Q^{-1} \|} \| x \| \leq \| x \|_Q^2 \leq Q \| \| x \|^2$. The Moreau approximation of J [14] with respect to the scalar product induced by Q is the functional $\Gamma : H_1 \to \mathbb{R}$ defined by

$$\Gamma(y) = \inf_{x \in X} \left\{ J(x) + \frac{1}{2} \| x - y \|_Q^2 \right\}. \tag{7.2.4}$$

It follows from the properties of J that the infimum in (7.2.4) is obtained at a unique point. Let us denote by $\text{prox}\,{}_J^Q(y)$ the proximity operator:

$$
\begin{aligned}
\text{prox}\,{}_J^Q: \quad & H_1 \quad \rightarrow \quad H_1 \\[4pt]
& y \quad \rightarrow \quad \Gamma(y) = \text{argmin}_{x \in H_1}\left\{ J(x) + \frac{1}{2}\,\| x - y \|_Q^2 \right\}.
\end{aligned}
\tag{7.2.5}
$$

The first optimality condition for (7.2.4) leads to

$$
\begin{aligned}
z = \text{prox}\,{}_J^Q(y) \quad &\Leftrightarrow \quad 0 \in \partial J(z) + Q(z - y) \\
&\Leftrightarrow \quad Q(z) \in (\partial I + Q)(z),
\end{aligned}
$$

which leads to

$$
\text{prox}\,{}_J^Q(y) = (\partial I + Q)^{-1}(Q(y)),
$$

by using the minimum in (7.2.4). In view of the above, we can define the proximal Gauss–Newton method by

$$
x_{n+1} = \text{prox}\,{}_J^{H(x_n)}(x_n - F'(x_n)^+ F(x_n)) \text{ for each } n = 0, 1, 2, \ldots
\tag{7.2.6}
$$

where x_0 is an initial point, $H(x_n) = F'(x_n)^* F'(x_n)$, and $\text{prox}\,{}_J^{H(x_n)}$ is defined in (7.2.5).

7.3 LOCAL CONVERGENCE ANALYSIS OF THE PROXIMAL GAUSS–NEWTON METHOD

Throughout this chapter we will respectively denote by $U(x, r)$ and $\overline{U}(x, r)$ the open and closed ball in H_1 with center $x \in \Omega$ and radius $r > 0$. We shall prove the main local convergence results under the (H) conditions:

(H_0) Let $D \subseteq X$ be open; $J : D \rightarrow \mathbb{R} \cup \{+\infty\}$ be proper, convex and lower semicontinuously Fréchet-differentiable such that F' has a closed image in D;

(H_1) Let $x^* \in D$, $R > 0$, $\alpha := \| F(x^*) \|$, $\beta := \| F'(x^*)^+ \|$, and $\gamma := \beta \| F'(x^*) \|$. Operator $-F'(x^*)^* F(x^*) \in \partial J(x^*)$, $F'(x)$ is injective, and there exist continuously differentiable f_0, $f : [0, R) \rightarrow \mathbb{R}$ such that

$$
\begin{aligned}
&\beta \| F'(x) - F'(x^*) \| \le f_0'(\lambda(x)) - f_0'(0) \\
&\text{for each } x \in D, \, \theta \in [0, 1] \text{ and } \lambda(x) = \| x - x^* \|,
\end{aligned}
\tag{7.3.1}
$$

and

$$
\begin{aligned}
&\beta \| F'(x) - F'(x^* + \theta(x - x^*)) \| \le f_0'(\lambda(x)) - f_0'(\theta \lambda(x)) \\
&\text{for each } x \in D_0 = D \cap U(x^*, R);
\end{aligned}
\tag{7.3.2}
$$

(H_2) $f_0(0) = f(0) = 0$ and $f_0'(0) = f'(0) = -1$;

(H_3) f_0', f' are strictly increasing and for each $t \in [0, R_0)$,

$$f_0(t) \leq f(t) \text{ and } f_0'(t) \leq f'(t);$$

(H_4) $\left[\left(1 + \sqrt{2}\right)\gamma + 1\right]\alpha\beta D^+ f_0'(0) < 1;$

Let positive constants v, ρ, r and function Ψ be defined by

$$v := \sup\{t \in [0, R) : f'(t) < 0\},$$
$$\rho := \sup\{t \in [0, v) : \psi(t) < 1\},$$
$$r := \min\{v, \rho\},$$

and

$$\psi(t)$$
$$:= \frac{(f_0'(t) + 1 + \gamma)\left[tf'(t) - f(t) + \alpha\beta\left(1 + \sqrt{2}\right)(f_0'(t) + 1)\right] + \alpha\beta(f_0'(t) + 1)}{t\left[f_0'(t)\right]^2}.$$

Now we can state our main result.

Theorem 7.3.1. *Under hypotheses* (H), *let* $x_0 \in U(x^*, r) \setminus \{x^*\}$. *Then sequence* $\{x_n\}$ *generated by proximal Gauss–Newton method (7.2.6) for solving penalized nonlinear least squares problem (7.1.1) is well defined, remains in* $U(x^*, r)$, *and converges to* x^*. *Moreover, the following estimates hold for each* $n = 0, 1, 2, \dots$:

$$\lambda_{n+1} = \lambda(x_{n+1}) \leq \varphi_{n+1} := \varphi(\lambda_0, \lambda_n, f, f_0'), \qquad (7.3.3)$$

where

$$\varphi(\lambda_0, \lambda_n, f, f', f_0') = \frac{(f_0'(\lambda_0) + 1 + \gamma)[f_0'(\lambda_0)\lambda_0 - f(\lambda_0)]}{(\lambda_0 f_0'(\lambda_0))^2}\lambda_n^2$$
$$+ \frac{\left(1 + \sqrt{2}\right)\alpha\beta(f_0'(\lambda_0) + 1)^2}{(\lambda_0 f_0'(\lambda_0))^2}\lambda_n^2$$
$$+ \frac{\alpha\beta\left[\left(1 + \sqrt{2}\right)\gamma + 1\right][f_0'(\lambda_0) + 1]}{\lambda_0(f_0'(\lambda_0))^2}\lambda_n.$$

Proof. Simply replace function g in [6,9] by function f in the proof, where g is a majorant function for F on D. That is, we have instead of (h_0):

$$\beta\|F'(y) - F'(x)\| \leq f_1'(\|y - x\| + \|x - x_0\|) - f_1'(\|x - x_0\|) \qquad (7.3.4)$$

for each $x, y \in D$ with $\|x - x_0\| + \|y - x\| < R$. The iterates x_n lie in D_0, which is a more precise location than D, since $D_0 \subseteq D$. $\qquad\square$

We can now present the following remark.

Remark 7.3.2. *If $f(t) = f_1(t) = f_0(t)$ for each $t \in [0, R)$, then the previous Theorem 7.3.1 reduces to the corresponding theorem in [9]. Moreover, if $f_0(t) \leq f(t) \leq f_1(t)$, we obtain the results in [6]. Notice that, in view of (7.3.1), (7.3.2) and (H_3), we have*

$$f_0'(t) \leq f_1'(t) \text{ for each } t \in [0, R) \tag{7.3.5}$$

and

$$f'(t) \leq f_1'(t) \text{ for each } t \in [0, R). \tag{7.3.6}$$

Therefore, if

$$f_0'(t) < f'(t) < f_1'(t) \text{ for each } t \in [0, R) \tag{7.3.7}$$

then the following advantages are obtained:

- *weaker sufficient convergence criteria, tighter error bounds on the distances,*
- *at least as precise information on the location of the solution x^*.*

It is easy to see that under (7.3.1) function f_0' is defined. Therefore the majorant function satisfying (7.3.2) (i.e., f') is at least as small as the majorant function satisfying (7.3.4) (i.e., f_1'), leading to the advantages of the new approach over the approach in [6] or [9].

7.4 NUMERICAL EXAMPLES

Specialization of results to some interesting cases such as Smale's α-theory [15] (see also Wang's α-theory [16]) and Kantorovich theory have been reported in [9], if $f_0'(t) = f'(t) = g'(t)$ for each $t \in [0, R)$, and in [6], if $f_0'(t) < f(t) = f_1'(t)$ for each $t \in [0, R)$. Next we present an example where $f_0'(t) < f'(t) < f_1'(t)$ for each $t \in [0, R)$.

Example 7.4.1. *Let*

$$X = Y = \mathbb{R}^3,$$

$$D = \bar{U}(0, 1),$$

and

$$x^* = \begin{pmatrix} 0 \\ 0 \\ 0 \end{pmatrix}.$$

Define function F on D for

$$w = \begin{pmatrix} x \\ y \\ z \end{pmatrix}$$

by

$$F(w) = \begin{pmatrix} e^x - 1 \\ \dfrac{e-1}{2}y^2 + y \\ z \end{pmatrix}.$$

Then the Fréchet derivative is given by

$$F'(v) = \begin{bmatrix} e^x & 0 & 0 \\ 0 & (e-1)y+1 & 0 \\ 0 & 0 & 1 \end{bmatrix}.$$

Let

$$f_0(t) = \frac{L_0}{2}t^2 - t,$$

$$f(t) = \frac{L}{2}t^2 - t,$$

and

$$g(t) = \frac{L_1}{2}t^2 - t,$$

where

$$m = 3,$$
$$n = 3,$$
$$c = 0,$$
$$L_0 = e - 1,$$

$$L = e^{\frac{1}{L_0}}$$

and

$$L_1 = e.$$

Then (7.3.7) holds as a strict inequality. Therefore, the new results have the advantages presented in the introduction over the corresponding results in [6,9, 13].

REFERENCES

[1] G.B. Allende, M.L.N. Gonçalves, Local convergence analysis of a proximal Gauss–Newton method under a majorant condition, preprint, http://arxiv.org/abs/1304.6461.

[2] S. Amat, S. Busquier, J.M. Gutiérrez, Geometric constructions of iterative functions to solve nonlinear equations, J. Comput. Appl. Math. 157 (2003) 197–205.

[3] I.K. Argyros, Convergence and Applications of Newton-Type Iterations, Springer-Verlag, New York, 2008.

[4] I.K. Argyros, S. Hilout, Extending the applicability of the Gauss–Newton method under average Lipschitz-type conditions, Numer. Algorithms 58 (2011) 23–52.

[5] I.K. Argyros, S. Hilout, Improved local convergence of Newton's method under weak majorant condition, J. Comput. Appl. Math. 236 (2012) 1892–1902.

[6] I.K. Argyros, S. Hilout, Improved local convergence analysis of inexact Gauss–Newton like methods under the majorizing condition in Banach space, J. Franklin Inst. 350 (2013) 1531–1544.

[7] I.K. Argyros, A.A. Magreñán, Local convergence analysis of proximal Gauss–Newton method for penalized nonlinear least square problems, Appl. Math. Comput. 241 (15) (2014) 401–408.

[8] J. Chen, W. Li, Local convergence results of Gauss–Newton's like method in weak conditions, J. Math. Anal. Appl. 324 (2006) 1381–1394.

[9] O.P. Ferreira, M.L.N. Gonçalves, P.R. Oliveira, Local convergence analysis of Gauss–Newton like methods under majorant condition, J. Complexity 27 (2011) 111–125.

[10] O.P. Ferreira, M.L.N. Gonçalves, P.R. Oliveira, Local convergence analysis of inexact Gauss–Newton like methods under majorant condition, J. Comput. Appl. Math. 236 (2012) 2487–2498.

[11] J.M. Gutiérrez, M.A. Hernández, Newton's method under weak Kantorovich conditions, IMA J. Numer. Anal. 20 (2000) 521–532.

[12] W.M. Häubler, A Kantorovich-type convergence analysis for the Gauss–Newton method, Numer. Math. 48 (1986) 119–125.

[13] C. Li, N. Hu, J. Wang, Convergence bahavior of Gauss–Newton's method and extensions to the Smale point estimate theory, J. Complexity 26 (2010) 268–295.

[14] S. Salzo, S. Villa, Convergence analysis of a proximal Gauss–Newton method, Comput. Optim. Appl. 53 (2012) 557–589.

[15] S. Smale, Newton's method estimates from data at one point, in: The Merging of Disciplines: New Directions Pure, Applied, and Computational Mathematics, Laramie, WY, 1985, Springer, New York, 1986, pp. 185–196.

[16] X.H. Wang, Convergence of Newton's method and uniqueness of the solution of equations in Banach spaces, IMA J. Numer. Anal. 20 (2000) 123–134.

Chapter 8

Multistep modified Newton–Hermitian and Skew-Hermitian Splitting method

8.1 INTRODUCTION

Let $F : D \subseteq \mathbb{C}^n \to \mathbb{C}^n$ be Gateaux-differentiable and D be an open set. Let also $x_0 \in D$ be a point at which $F'(x)$ is continuous and positive definite. Suppose that $F'(x) = H(x) + S(x)$, where

$$H(x) = \frac{1}{2}(F'(x) + F'(x)^*)$$

and

$$S(x) = \frac{1}{2}(F'(x) - F'(x)^*)$$

are the Hermitian and Skew-Hermitian parts of the Jacobian matrix $F'(x)$, respectively. Several problems from applied sciences can be written as

$$F(x) = 0, \tag{8.1.1}$$

using mathematical modeling [1–22]. Due to the fact, that the solution of equations like (8.1.1) can rarely be found in closed form, most solution methods of this equation are iterative. In particular, Hermitian and Skew-Hermitian Splitting (HSS) methods have been shown to be very efficient in solving large sparse non-Hermitian positive definite systems of linear equations [11,12,17,19,22].

In this chapter, we study the semilocal convergence of the multistep modified Newton-HSS (MMN-HSS) method defined by

$$x_k^{(0)} = x_k,$$
$$x_k^{(i)} = x_k^{(i-1)} - (I - T(\alpha;x)^{l_k^{(i)}})F'(x_k)^{-1}F(x_k^{(i-1)}), \ 1 \le i \le m,$$
$$x_{k+1} = x_k^{(m)}, \ i = 1, 2, \ldots, m, \ k = 0, 1, \ldots, \tag{8.1.2}$$

A Contemporary Study of Iterative Methods. DOI: 10.1016/B978-0-12-809214-9.00008-5

where $x_0 \in D$ is an initial guess, $T(\alpha; x) = (\alpha I + S(x))^{-1}(\alpha I - H(x))(\alpha I + H(x))^{-1}(\alpha I - H(x))$, $l_k^{(i)}$ is a sequence of positive integers, α and *tol* are positive constants,

$$\|F(x_k)\| \leq tol \|F(x_0)\|,$$

and

$$\|F(x_k) + F'(x_k)d_{k,l_k}\| \leq \eta_k \|F(x_k)\|, \quad \eta_k \in [0, 1), \quad \eta_k \leq \eta \leq 1.$$

The local and semilocal convergence analysis of method (8.1.2) was given in [19] using Lipschitz continuity conditions on F. Some years later, authors extended the local convergence of method (8.1.2) using generalized Lipschitz continuity conditions [8].

In the present chapter, we show that the results in [19] can be extended as those for MN-HSS in [8]. Using generalized Lipschitz-type conditions, we present a new semilocal convergence analysis with advantages, under the same computational cost as before:

- Larger radius of convergence,
- More precise error estimates on $\|x_k - x_*\|$,
- More solvable situations.

The rest of the chapter is structured as follows: Section 8.2 contains the semilocal convergence analysis of the MMN-HSS method. Section 8.3 contains the numerical examples.

8.2 SEMILOCAL CONVERGENCE

Throughout this chapter we will denote the following as (H) hypotheses:

(H1) Let $x_0 \in \mathbb{C}^n$. There exist $\beta_1 > 0$, $\beta_2 > 0$, $\gamma > 0$, and $\mu > 0$ such that

$$\|H(x_0)\| \leq \beta_1, \quad \|S(x_0)\| \leq \beta_2, \quad \|F'(x_0)^{-1}\| \leq \gamma, \quad \|F(x_0)\| \leq \mu.$$

(H2) There exist continuous and nondecreasing functions $v_1 : [0, +\infty) \to \mathbb{R}$, $v_2 : [0, +\infty) \to \mathbb{R}$, with $v_1(0) = v_2(0) = 0$, such that for each $x, y \in D$

$$\|H(x) - H(x_0)\| \leq v_1(\|x - x_0\|),$$

$$\|S(x) - S(x_0)\| \leq v_2(\|x - x_0\|).$$

Define functions w and v by $w(t) = w_1(t) + w_2(t)$ and $v(t) = v_1(t) + v_2(t)$. Let

$$r_0 = \sup\{t \geq 0 : \gamma v(t) < 1\}$$

and set

$$D_0 = D \cap U(x_0, r_0).$$

(H3) There exist continuous and nondecreasing functions $w_1 : [0, +\infty) \to \mathbb{R}$, $w_2 : [0, +\infty) \to \mathbb{R}$, with $w_1(0) = w_2(0) = 0$, such that for each $x, y \in D_0$

$$\|H(x) - H(y)\| \leq w_1(\|x - y\|),$$

$$\|S(x) - S(y)\| \leq w_2(\|x - y\|).$$

Using the previous condition and the following auxiliary results, we will give the main semilocal convergence result.

Lemma 8.2.1. *Under the (H) hypotheses, the following items hold for each* $x, y \in D_0$:

$$\|F'(x) - F'(y)\| \leq w(\|x - y\|), \tag{8.2.1}$$

$$\|F'(x) - F'(x_0)\| \leq v(\|x - y\|), \tag{8.2.2}$$

$$\|F'(x)\| \leq v(\|x - y\|) + \beta_1 + \beta_2, \tag{8.2.3}$$

$$\|F'(x) - F(y) - F'(y)(x - y)\| \leq \int_0^1 w(\|x - y\|\xi)d\xi \|x - y\|, \tag{8.2.4}$$

and

$$\|F'(x)^{-1}\| \leq \frac{\gamma}{1 - \gamma v(\|x - x_0\|)}. \tag{8.2.5}$$

Proof. By hypothesis (H_3) and since $F'(x) = H(x) + S(x)$, we have that

$$\begin{aligned}\|F'(x) - F'(y)\| &= \|(H(x) - H(y)) + (S(x) - S(y))\| \\ &\leq \|H(x) - H(y)\| + \|S(x) - S(y)\| \\ &\leq w_1(\|x - y\|) + w_2(\|x - y\| = w(\|x - y\|),\end{aligned}$$

and by (H_2)

$$\begin{aligned}\|F'(x) - F'(x_0)\| &\leq \|H(x) - H(x_0)\| + \|S(x) - S(x_0)\| \\ &\leq v_1(\|x - x_0\|) + v_2(\|x - x_0\|) \\ &= v(\|x - x_0\|),\end{aligned}$$

which show (8.2.1) and (8.2.2), respectively.

Then by (H_1) and (H_3) we get

$$\begin{aligned}\|F'(x)\| &= \|(F'(x) - F(x_0)) + F'(x_0)\| \\ &\leq \|F'(x) - F'(x_0)\| + \|H(x_0)\| + \|S(x_0)\| \\ &\leq v(\|x - x_0\|) + \beta_1 + \beta_2,\end{aligned}$$

which shows (8.2.3). Using (H_3), we obtain

$$\|F(x) - F(y) - F'(y)(x - y)\| = \left\| \int_0^1 F'(y + \xi(x - y) - F'(y))d\xi(x - y) \right\|$$

$$\leq \int_0^1 w(\|x - y\|\xi)d\xi \|x - y\|,$$

which shows (8.2.4). By (H_1), (H_2), and (8.2.2), we get in turn that for $x \in D_0$,

$$\|F'(x_0)^{-1}\|\|F'(x) - F'(x_0)\| \leq \gamma(\|x - x_0\|) \leq \gamma v(r_0) < 1. \tag{8.2.6}$$

It follows from (8.2.6) and the Banach lemma on invertible operators [4] that $F'(x)^{-1}$ exists so that (8.2.5) is satisfied. $\qquad\square$

We must define some scalar functions and parameters to be used in the semilocal convergence analysis. Let $t_0 = 0$ and $s_0^{(1)} = (1 + \eta)\gamma\mu$. Define scalar sequences $\{t_k\}, \{s_k^{(1)}\}, \ldots, \{s_k^{(m-1)}\}$ by the following schemes:

$$t_0 = 0, \ s_k^{(0)} = t_k, \ t_{k+1} = s_k^m,$$

$$s_k^{(i)} = s_k^{(i-1)}$$
$$+ \frac{[(\int_0^1 w((s_k^{(i-1)} - s_k^{(i-2)})\xi)d\xi + w(s_k^{(i-2)} - t_k))(1 + \eta)\gamma + \eta(1 - \gamma v(t_k))](s_k^{(i-1)} - s_k^{(i-2)})}{1 - \gamma v(t_k)}$$

$$t_{k+1} = s_k^{(i)} + (1 - \gamma v(t_k))(s_k^{(i)} - s_k^{(i-1)}),$$
$$i = 0, 1, 2, \ldots, m - 1, \ k = 0, 1, 2, \ldots \tag{8.2.7}$$

Moreover, define functions q and h_q on the interval $[0, r_0)$ by

$$q(t) = \frac{(1 + \eta)\gamma \int_0^1 w((1 + \eta)\gamma\mu\xi)d\xi + (1 + \eta)\gamma w(t) + \eta(1 - \gamma v(t))}{1 - \gamma v(t)}$$

and

$$h_q(t) = q(t) - 1.$$

We have that $h_q(0) = \eta - 1 < 0$ and $h_q \to \infty$ as $t \to r_0^-$. It follows from the intermediate value theorem that function h_q has zeros in the interval $(0, r_0)$. Denote by r_q the smallest such zero. Then we have that for each $t \in [0, r_0)$

$$0 \leq q(t) \leq 1. \tag{8.2.8}$$

$\qquad\square$

Lemma 8.2.2. *Suppose that equation*

$$t(1 - q(t)) - \left((1 + \eta)\gamma\mu + (1 + \eta)\gamma \int_0^1 w((1 + \eta)\gamma\mu\xi)d\xi + \eta \right) = 0$$

$$\tag{8.2.9}$$

has zeros in the interval $(0, r_q)$. Denote by r the smallest such zero. Then sequence $\{t_k\}$, generated by (8.2.7), is nondecreasing, bounded from above by r_q, and converges to its unique least upper bound r^* which satisfies

$$0 < r^* \leq r < r_q. \tag{8.2.10}$$

Proof. Equation (8.2.9) can be written as

$$\frac{t_1 - t_0}{1 - q(r)} = r, \tag{8.2.11}$$

since by (8.2.7)

$$t_1 = (1 + \eta)\gamma\mu + (1 + \eta)\gamma \int_0^1 w((1 + \eta)\gamma\mu\tau)d\tau + \eta + w((1 + \eta)\gamma\mu)$$

and r solves (8.2.9). It follows from the definition of sequence $\{t_k\}$, functions w_1, w_2, v_1, v_2 and (8.2.8) that

$$0 \leq t_0 \leq s_0 \leq t_1 \leq s_1 \leq \cdots \leq t_k \leq s_k \leq t_{k+1} < r,$$

$$t_{k+2} - t_{k+1} = q(r)(t_{k+1} - t_k) \leq q(r)^{k+1}(t_1 - t_0),$$

and

$$t_{k+2} \leq t_{k+1} + q(r)^{k+1}(t_1 - t_0) \leq t_k + q(r)^k(t_1 - t_0) + q(r)^{k+1}(t_1 - t_0)$$

$$\leq \cdots \leq t_1 + q(r)(t_1 - t_0) + \cdots + q(r)^{k+1}(t_1 - t_0)$$

$$\leq \frac{t_1 - t_0}{1 - q(r)}(1 - q(r)^{k+2}) < \frac{t_1 - t_0}{1 - q(r)} = r.$$

Therefore, sequence $\{t_k\}$ converges to r^* which satisfies (8.2.10). $\qquad\square$

Next our main semilocal result is presented.

Theorem 8.2.3. *Suppose that the hypotheses (H) and hypotheses of Lemma 8.2.2 hold. Define $\bar{r} = \min\{r_1^+, r^*\}$, where r_1^+ is defined in ([7], Theorem 2.1) and r^* is given in Lemma 8.2.2. Let $u = \min\{m_*, l_*\}$, $m_* = \liminf_{k\to\infty} m_k$, $l_* = \liminf_{k\to\infty} l_k$. Furthermore, suppose that*

$$u > \left\lfloor \frac{\ln \eta}{\ln((\tau + 1)\theta)} \right\rfloor, \tag{8.2.12}$$

where the symbol $\lfloor \cdot \rfloor$ denotes the smallest integer no less than the corresponding real number, $\tau \in (0, \dfrac{1 - \theta}{\theta})$, and

$$\theta := \theta(\alpha; x_0) = \|T(\alpha; x_0\| < 1. \tag{8.2.13}$$

Then the sequence $\{x_k\}$ generated by the MNN-HSS method is well defined, remains in $U(x_0, \bar{r})$ for each $k = 0, 1, 2, \ldots$, and converges to a solution x_* of equation $F(x) = 0$.

Proof. Notice that we showed in [8] that for each $x \in U(x_0, \bar{r})$,

$$\|T(\alpha; x)\| \leq (\tau + 1)\theta < 1. \tag{8.2.14}$$

The following statements will be shown using mathematical induction:

$$
\begin{cases}
\|x_k - x_0\| \leq t_k - t_0, \\[2mm]
\|F(x_k)\| \leq \dfrac{1}{(1+\eta)\gamma}\phi(t_k), \\[2mm]
\|x_k^{(1)} - x_k\| \leq s_k^{(1)} - t_k \\[2mm]
\|F(x_k^{(i)})\| \leq \dfrac{1}{(1+\eta)\gamma}\phi(s_k^{(i)}), \\[2mm]
\|x_k^{(i+1)} - x_k^{(i)}\| \leq s_k^{(i+1)} - s_k^{(i)}, \ i = 1, 2, \ldots, m-2, \\[2mm]
\|F(x_k^{(m-1)})\| \leq \dfrac{1}{(1+\eta)\gamma}\phi(s_k^{(m-1)}), \\[2mm]
\|x_{k+1} - x_k^{(m-1)}\| \leq t_{k+1} - s_k^{(m-1)}.
\end{cases}
\tag{8.2.15}
$$

We have for $k = 0$:

$$\|x_0 - x_0\| = 0 \leq t_0 - t_0,$$

$$\|F(x_0)\| \leq \delta \leq \frac{\gamma(1 - \gamma v(t_0))(s_0^1 - t_0)}{\gamma(1+\eta)},$$

$$\|x_0^{(1)} - x_0\| \leq \|I - T(\alpha; x_0)^{l_0^{(1)}}\| \, \|F'(x_0)^{-1}\| \, \|F(x_0)\|$$

$$\leq (1 + \theta^{l_0^{(1)}}) < (1+\eta)\gamma\delta = s_0^{(1)}.$$

Suppose the following inequalities hold for each $i < m - 1$:

$$
\begin{cases}
\|F(x_0^{(i)})\| \leq \dfrac{1}{(1+\eta)\gamma}(1 - \gamma v(t_0))(s_0^{(i+1)} - s_0^{(i)}), \\[2mm]
\|x_0^{(i+1)} - x_0^{(i)}\| \leq s_0^{(i+1)} - s_0^{(i)}, \ i = 1, 2, \ldots, m-2.
\end{cases}
\tag{8.2.16}
$$

We shall prove inequalities (8.2.16) for $m - 1$. Using the (H) conditions, we obtain

$$\|F(x_0^{(m-1)})\| \leq \|F(x_0^{(m-1)}) - F(x_0^{(m-2)}) - F'(x_0)(x_0^{(m-1)} - x_0^{(m-2)})\|$$

$$+ \|F(x_0^{(m-2)}) + F'(x_0)(x_0^{(m-1)} - x_0^{(m-2)})\|$$

$$\leq \| F(x_0^{(m-1)}) - F(x_0^{(m-2)}) - F'(x_0^{(m-2)})(x_0^{(m-1)} - x_0^{(m-2)}) \|$$
$$+ \| F'(x_0^{(m-2)}) - F'(x_0) \| \| (x_0^{(m-1)} - x_0^{(m-2)}) \| + \eta \| F(x_0^{(m-2)}) \|$$
$$\leq \int_0^1 w(\| (x_0^{(m-1)} - x_0^{(m-2)}) \| \xi) d\xi \| \| (x_0^{(m-1)} - x_0^{(m-2)}) \|$$
$$+ w(\| x_0^{(m-2)} - x_0 \|) \| (x_0^{(m-1)} - x_0^{(m-2)}) \| + \eta \| F(x_0^{(m-2)}) \|. \tag{8.2.17}$$

Then we also get

$$\| (x_0^{(m-1)} - x_0^{(m-2)}) \| \leq s_0^{(m-1)} - s_0^{(m-2)},$$

$$\| x_0^{(m-2)} - x_0 \| \leq \| x_0^{(m-2)} - x_0^{(m-3)} \| + \cdots + \| x_0^{(1)} - x_0 \|$$
$$\leq (s_0^{(m-2)} - s_0^{(m-3)}) + \cdots + (s_0^{(1)} - t_0)$$
$$\leq s_0^{(m-2)} - t_0 = s_0^{(m-2)},$$

and

$$\| F(x_0^{(m-2)}) \| \leq \frac{1}{(1+\eta)\gamma} (1 - \gamma v(t_0))(s_0^{(m-1)} - s_0^{(m-2)}).$$

As a consequence, from (8.2.17) we obtain that

$$\| F(x_0^{(m-1)}) \| \leq \int_0^1 w((s_0^{(m-1)} - s_0^{(m-2)})\xi) d\xi (s_0^{(m-1)} - s_0^{(m-2)})$$
$$+ w(s_0^{(m-2)} - t_0)(s_0^{(m-1)} - s_0^{(m-2)})$$
$$+ \frac{\eta(1 - \gamma v(t_0))}{(1+\eta)\gamma} (s_0^{(m)} - s_0^{(m-1)})$$
$$\leq \frac{1 - \gamma v(t_0)}{(1+\eta)\gamma} (s_0^{(m)} - s_0^{(m-1)}). \tag{8.2.18}$$

Then we have by (8.2.7) that

$$\| x_1 - x_0^{(m-1)} \| \leq \| I - T(\alpha; x_0)^{l_0^{(m)}} \| \| F'(x_0)^{-1} \| \| F(x_0^{(m-1)}) \|$$
$$\leq (1 + ((\tau + 1)\theta)^{l_0^{(m)}})\gamma \frac{1}{(1+\eta)\gamma} (1 - \gamma v(t_0))(s_0^{(m)} - s_0^{(m-1)}))$$
$$= t_1 - s_0^{(m-1)}$$

holds, and the inequalities (8.2.15) hold for $k = 0$. Suppose that the inequalities (8.2.15) hold for all nonnegative integers less than k. Next, we prove (8.2.15) for k.

By the induction hypotheses, we obtain

$$\| x_k - x_0 \| \leq \| x_k - x_{k-1}^{(m-1)} \| + \| x_{k-1}^{(m-1)} - x_{k-1}^{(m-2)} \| + \cdots + \| x_{k-1}^{(1)} - x_{k-1}^{(0)} \|$$

$$+ \|x_{k-1} - x_0\|$$
$$\leq (t_k - s_{k-1}^{(m-1)}) + (s_{k-1}^{(m-1)} - s_{k-1}^{(m-2)}) + \cdots + (s_{k-1}^{(1)} - t_{k-1})$$
$$+ (t_{k-1} - t_0)$$
$$= t_k - t_0 < r_* < r.$$

In view of $x_{k-1}, x_{k-1}^{(1)}, \ldots, x_{k-1}^{(m-1)} \in \mathbb{U}(x_0, r)$, we have

$$\|F(x_k)\| \leq \|F(x_k) - F(x_{k-1}^{(m-1)}) - F'(x_{k-1})(x_k - x_{k-1}^{(m-1)})\|$$
$$+ \|F(x_{k-1}^{(m-1)}) + F'(x_{k-1})(x_k - x_{k-1}^{(m-1)})\|$$
$$\leq \|F(x_k) - F(x_{k-1}^{(m-1)}) - F'(x_{k-1}^{(m-1)})(x_k - x_{k-1}^{(m-1)})\|$$
$$+ \|F'(x_{k-1}^{(m-1)}) - F'(x_{k-1})\|\|(x_k - x_{k-1}^{(m-1)})\| + \eta\|F(x_{k-1}^{(m-1)})\|$$
$$\leq \int_0^1 w(\|(x_k - x_{k-1}^{(m-1)})\|\xi)d\xi \|\|(x_k - x_{k-1}^{(m-1)})\|$$
$$+ w(\|x_{k-1}^{(m-1)} - x_{k-1}\|)\|x_k - x_{k-1}^{(m-1)}\|$$
$$+ \frac{\eta(1 - \gamma v(t_{k-1}))}{(1 + \eta)\gamma}(s_{k-1}^{(m)} - s_{k-1}^{(m-1)})$$
$$\leq \frac{(1 - \gamma v(t_k))}{(1 + \eta)\gamma}(s_k^{(m)} - s_k^{(m-1)}). \qquad (8.2.19)$$

We also obtain

$$\|x_k - x_{k-1}^{(m-1)}\| \leq t_k - s_{k-1}^{(m-1)}, \qquad (8.2.20)$$

$$\|x_{k-1}^{(m-1)} - x_{k-1}\| \leq \|x_{k-1}^{m-1} - x_{k-1}^{(m-2)}\| + \cdots + \|x_{k-1}^{(1)} - x_{k-1}\|$$
$$\leq (s_{k-1}^{(m-1)} - s_{k-1}^{(m-2)}) + \cdots + (s_{k-1}^{(1)} - t_{k-1})$$
$$\leq s_{k-1}^{(m-1)} - t_{k-1}, \qquad (8.2.21)$$

and

$$\|F(x_{k-1}^{(m-1)})\| \leq \frac{1}{1 + \eta)\gamma}(1 - \gamma v(t_{k-1}))(s_{k-1}^{(m)} - s_{k-1}^{(m-1)}). \qquad (8.2.22)$$

It follows that

$$\|x_k^{(1)} - x_k\| \leq \|I - T(\alpha; x_k)^{l_k^{(1)}}\|\|F'(x_k)^{-1}\|\|F(x_k)\|$$
$$\leq (1 + \theta^{l_k^{(1)}})\frac{\gamma}{1 - \gamma v(t_k)}\frac{1 - \gamma v(t_k)}{(1 + \eta)\gamma}(s_k^{(1)} - t_k)$$
$$\leq s_k^{(1)} - t_k.$$

Suppose the following inequalities hold for any positive integers less than $m - 1$:

$$\begin{cases} \|F(x_0^{(i)})\| \leq \dfrac{1}{(1+\eta)\gamma}(1 - \gamma v(t_k))(s_k^{(i+1)} - s_k^{(i)}), \\ \|x_k^{(i+1)} - x_k^{(i)}\| \leq s_k^{(i+1)} - s_k^{(i)}, \; i = 1, 2, \ldots, m-2. \end{cases} \tag{8.2.23}$$

We will prove (8.2.23) for $m - 1$. As in (8.2.19) we have that

$$\begin{aligned} \|F(x_k^{(m-1)})\| \leq & \|F(x_k^{(m-1)}) - F(x_{k-1}^{(m-2)}) - F'(x_k)(x_k^{(m-1)} - x_k^{(m-2)})\| \\ & + \|F(x_k^{(m-2)}) + F'(x_k)(x_k^{(m-1)} - x_k^{(m-2)})\| \\ \leq & \|F(x_k^{(m-1)}) - F(x_k^{(m-2)}) - F'(x_k^{(m-2)})(x_k^{(m-1)} - x_k^{(m-2)})\| \\ & + \|F'(x_k^{(m-2)}) - F'(x_k)\|\|x_k^{(m-1)} - x_{k-1}^{(m-2)}\| + \eta\|F(x_k^{(m-2)})\| \\ \leq & \dfrac{1}{(1+\eta)\gamma}(1 - \gamma v(t_k))(s_k^{(m)} - s_k^{(m-1)}). \end{aligned} \tag{8.2.24}$$

Then we obtain

$$\|x_k^{(m-1)} - x_k^{(m-2)}\| \leq \|s_k^{(m-1)} - s_k^{(m-2)}\|, \tag{8.2.25}$$

$$\begin{aligned} \|x_k^{(m-2)} - x_k\| &\leq \|x_k^{m-2} - x_{k-1}^{(m-3)}\| + \cdots + \|x_k^{(1)} - x_k\| \\ &\leq (s_k^{(m-2)} - s_k^{(m-3)}) + \cdots + (s_k^{(1)} - t_k) \\ &\leq s_k^{(m-2)} - t_k, \end{aligned} \tag{8.2.26}$$

and

$$\|F(x_k^{(m-2)})\| \leq \dfrac{1}{(1+\eta)\gamma}(1 - \gamma v(t_k))(s_k^{(m-1)} - s_k^{(m-2)}). \tag{8.2.27}$$

Therefore,

$$\begin{aligned} \|x_{k+1} - x_k^{(m-1)}\| &\leq \|I - T(\alpha; x_k)^{l_k^{(m)}}\|\|F'(x_k)^{-1}\|\|F(x_k^{(m-1)})\| \\ &\leq (1 + \theta^{l_k^{(m)}})\dfrac{\gamma(1 - \gamma v(t_k))(s_k^{(m)} - s_k^{(m-1)})}{(1 - \gamma v(t_k))(1+\eta)\gamma} \\ &\leq t_{k+1} - s_k^{(m-1)} \end{aligned} \tag{8.2.28}$$

holds. The induction for (8.2.15) Δ is completed. The sequences $\{t_k\}, \{s_k\}, \ldots,$ $\{s_k^{(m-1)}\}$ converge to r^*, and

$$\begin{aligned} \|x_{k+1} - x_0\| &\leq \|x_{k+1} - x_k^{(m-1)}\| + \|x_k^{(m-1)} - x_k^{(m-2)}\| + \cdots + \|x_k^{(1)} - x_k^{(0)}\| \\ &\quad + \|x_k - x_0\|, \\ &\leq (t_{k+1} - s_k^{(m-1)}) + (s_k^{(m-1)} - s_k^{(m-2)}) + \cdots + (s_k^{(1)} - t_k) \end{aligned}$$

$$+ (t_k - t_0),$$
$$= t_{k+1} - t_0 < r^* < r. \tag{8.2.29}$$

Then, the sequence $\{x_k\}$ also converges to some $x \in \overline{U}(x^*, r)$. By letting $k \to \infty$ in (8.2.19), we get that

$$F(x_*) = 0. \tag{8.2.30}$$

$\square$

Remark 8.2.4. **(a)** *Let us choose functions* w_1, w_2, v_1, v_2 *as* $w_1(t) = L_1 t$, $w_2(t) = L_2 t$, $v_1(t) = K_1 t$, $v_2(t) = K_2 t$ *for some positive constants* K_1, K_2, L_1, L_2 *and set* $L = L_1 + L_2$, $K = K_1 + K_2$. *Suppose that* $D_0 = D$. *Then notice that*

$$K \leq L, \tag{8.2.31}$$

since

$$K_1 \leq L_1, \tag{8.2.32}$$

$$K_2 \leq L_2, \tag{8.2.33}$$

$$\beta_1 \leq \beta, \tag{8.2.34}$$

and

$$\beta_2 \leq \beta, \tag{8.2.35}$$

where $\beta := \max\{\|H(x_0)\|, \|S(x_0)\|\}$.
Notice that in [19], $K_1 = L_1$, $K_2 = L_2$, and $\beta = \beta_1 = \beta_2$. Therefore, if strict inequality holds in any of (8.2.32), (8.2.33), (8.2.34), or (8.2.35), the present results improve those in [19].
(b) *The set D_0 in (H_3) can be replaced by $D_1 = D \cap U(x_1, r_0 - \|x_1 - x_0\|)$, leading to even smaller "w" and "v" functions, since $D_1 \subset D_0$.*

8.3   NUMERICAL EXAMPLES

Example 8.3.1. *Consider the system of nonlinear equation $F(X) = 0$, wherein*

$$F = \begin{pmatrix} F_1 \\ \vdots \\ F_n \end{pmatrix}$$

and

$$X = \begin{pmatrix} x_1 \\ \vdots \\ x_n \end{pmatrix},$$

with

$$F_i(X) = (3 - 2x_i)x_i^{3/2} - x_{i-1} - 2x_{i+1} + 1, \quad i = 1, 2, \dots, n,$$

where $x_0 = x_{n+1} = 0$ by convention. This system has a complex solution. Therefore, we consider the complex initial guess

$$X_0 = \begin{pmatrix} -i \\ \vdots \\ -i \end{pmatrix}.$$

The derivative $F'(X)$ is given by

$F'(X) =$

$$\begin{bmatrix} \frac{3}{2}(3 - 2x_1)\sqrt{x_1} - 2x_1^{\frac{3}{2}} & -2 & \dots & 0 & 0 \\ -1 & \frac{3}{2}(3 - 2x_2)\sqrt{x_2} - 2x_2^{\frac{3}{2}} & \dots & 0 & 0 \\ \vdots & \vdots & \ddots & \vdots & \vdots \\ 0 & 0 & \dots & -1 & \frac{3}{2}(3 - 2x_n)\sqrt{x_n} - 2x_n^{\frac{3}{2}} \end{bmatrix}.$$

In Table 8.2, 3MN-HSS and 4MN-HSS correspond to three- and four-step MN-HSS methods.

It is clear that $F'(X)$ is sparse and positive definite. Now we solve this nonlinear problem by the Newton-HSS method (N-HSS) (see [10]), modified Newton-HSS method (MN-HSS) (see [22]), three-step modified Newton-HSS (3MN-HSS) and four-step modified Newton-HSS (4MN-HSS) method. The methods are compared with respect to error estimates, CPU time (CPU-time), and the number of iterations (IT). We use experimentally optimal parameter values of α for the methods corresponding to the problem dimension $n = 100, 200, 500, 1000$; see Table 8.1. The numerical results are displayed in Table 8.2. From numerical results we observe that MN-HSS outperforms N-HSS with respect to CPU time and the number of iterations. Note that in this example, the results in [19] cannot be applied since the operators involved are

TABLE 8.1 Optimal values of α for N-HSS and MN-HSS methods.

n	100	200	500	1000
N-HSS	4.1	4.1	4.2	4.1
MN-HSS	4.4	4.4	4.3	4.3
MMN-HSS	4.4	4.4	4.3	4.3

TABLE 8.2 Numerical results.

n	Method	Error estimates	CPU-time	IT
100	N-HSS	3.98×10^{-6}	1.744	5
	MN-HSS	4.16×10^{-8}	1.485	4
	3MN-HSS	8.28×10^{-5}	1.281	3
	4MN-HSS	1.12×10^{-6}	1.327	3
200	N-HSS	3.83×10^{-6}	6.162	5
	MN-HSS	5.46×10^{-8}	4.450	4
	3MN-HSS	7.53×10^{-5}	4.287	3
	4MN-HSS	9.05×10^{-7}	4.108	3
500	N-HSS	4.65×10^{-6}	32.594	5
	MN-HSS	4.94×10^{-8}	24.968	4
	3MN-HSS	7.69×10^{-5}	21.250	3
	4MN-HSS	9.62×10^{-7}	20.406	3
1000	N-HSS	4.29×10^{-6}	119.937	5
	MN-HSS	5.32×10^{-8}	98.203	4
	3MN-HSS	9.16×10^{-5}	89.018	3
	4MN-HSS	8.94×10^{-7}	91.000	3

not Lipschitz. However, our results can be applied by choosing "w" and "v" functions appropriately. We leave these details to the interested readers.

Example 8.3.2. *Suppose that the motion of an object in three dimensions is governed by the system of differential equations:*

$$f_1'(x) - f_1(x) - 1 = 0,$$
$$f_2'(y) - (e - 1)y - 1 = 0,$$
$$f_3'(z) - 1 = 0. \tag{8.3.1}$$

with x, y, $z \in \Omega$ *for* $f_1(0) = f_2(0) = f_3(0) = 0$. *Then the solution of the system is given for*

$$w = \begin{pmatrix} x \\ y \\ z \end{pmatrix}$$

by the function $F := (f_1, f_2, f_3) : \Omega \to \mathbb{R}^3$ *defined by*

$$F(w) = \begin{pmatrix} e^x - 1 \\ \dfrac{e-1}{2}y^2 + y \\ z \end{pmatrix}. \tag{8.3.2}$$

Then the Fréchet-derivative is given by

$$F'(v) = \begin{bmatrix} e^x & 0 & 0 \\ 0 & (e-1)y + 1 & 0 \\ 0 & 0 & 1 \end{bmatrix}, \tag{8.3.3}$$

and we have

$$x^* = \begin{pmatrix} 0 \\ 0 \\ 0 \end{pmatrix},$$

$$w(t) = w_1(t) + w_2(t),$$
$$v(t) = v_1(t) + v_2(t),$$
$$w_1(t) = L_1 t,$$
$$w_2(t) = L_2 t,$$
$$v_1(t) = K_1 t,$$
$$v_2(t) = K_2 t,$$

where

$$L_1 = e - 1,$$
$$L_2 = e,$$
$$K_1 = e - 2,$$

$$K_2 = e,$$

$$\eta = 0.001,$$

$$\gamma = 1,$$

and

$$\mu = 0.01.$$

After solving the equation $h_q(t) = 0$, we obtain the root $r_q = 0.124067$. Similarly, the roots of equation (8.2.9) are 0.0452196 and 0.0933513. So,

$$r = \min\{0.0452196, 0.0933513\} = 0.0452196.$$

Therefore,

$$r = 0.0452196 < r_q = 0.124067.$$

Also we have

$$r^* = 0.0452196$$

and (see [7])

$$r_1^+ = 0.020274.$$

So,

$$\bar{r} = \min\{r_1^+, r^*\} = \min\{0.020274, 0.0452196\} = 0.020274.$$

It follows that sequence $\{x_k\}$ converges, $\{t_k\} \to r^*$ in D, and as such it converges to $x_* \in U(x_0, \bar{r}) = U(0, 0.020274)$.

REFERENCES

[1] S. Amat, S. Busquier, S. Plaza, J.M. Guttérrez, Geometric constructions of iterative functions to solve nonlinear equations, J. Comput. App. Math. 157 (2003) 197–205.
[2] S. Amat, S. Busquier, S. Plaza, Dynamics of the King and Jarratt iterations, Aequationes Math. 69 (2005) 212–223.
[3] S. Amat, M.A. Hernández, N. Remero, A modified Chebyshev's iterative method with at least sixth order of convergence, Appl. Math. Comput. 206 (2008) 164–174.
[4] I.K. Argyros, Convergence and Applications of Newton-Type Iterations, Springer-Verlag, New York, 2008.
[5] I.K. Argyros, S. Hilout, Computational Methods in Nonlinear Analysis, World Scientific Publ. Comp., New Jersey, 2013.
[6] I.K. Argyros, Á.A. Magreñán, Ball convergence theorems and the convergence plans of an iterative methods for nonlinear equations, SeMA J. 71 (2015) 39–55.
[7] I.K. Argyros, J.R. Sharma, D. Kumar, Local convergence of Newton-HSS methods with positive definite Jacobian matrices under generalized conditions, SeMA J. (2017), https://doi.org/10.1007/s40324-017-0116-2.
[8] I.K. Argyros, J.R. Sharma, D. Kumar, Extending the applicability of the MMN-HSS method for solving systems of nonlinear equations under generalized conditions, Algorithms 10 (2017) 54.

[9] O. Axelsson, Iterative Solution Methods, Cambridge University Press, Cambridge, 1994.

[10] Z.-Z. Bai, X.P. Guo, The Newton-HSS methods for systems of nonlinear equations with positive definite Jacobian matrices, J. Comput. Math. 28 (2010) 235–260.

[11] Z.-Z. Bai, G.H. Golub, J.Y. Pan, Preconditioned Hermitian and skew-Hermitian splitting methods for non-Hermitian positive semidefinite linear systems, Numer. Math. 98 (2004) 1–32.

[12] Z.-Z. Bai, G.H. Golub, M.K. Ng, Hermitian and skew-Hermitian splitting methods for non-Hermitian positive definite Linear systems, SIAM J. Matrix Anal. Appl. 24 (2003) 603–626.

[13] A. Cordero, J.A. Ezquerro, M.A. Hernández-Veron, J.R. Torregrosa, On the local convergence of a fifth-order iterative method in Banach spaces, Appl. Math. Comput. 251 (2015) 396–403.

[14] R.S. Dembo, S.C. Eisenstat, T. Steihaug, Inexact Newton methods, SIAM J. Numer. Anal. 19 (1982) 400–408.

[15] J.A. Ezquerro, M.A. Hernández, New iterations of R-order four with reduced computational cost, BIT Numer. Math. 49 (2009) 325–342.

[16] J.M. Gutiérrez, M.A. Hernández, Recurrence relations for the super Halley method, Comput. Math. Appl. 36 (1998) 1–8.

[17] X.-P. Guo, I.S. Duff, Semilocal and global convergence of Newton-HSS method for systems of nonlinear equations, Numer. Linear Algebra Appl. 18 (2011) 299–315.

[18] M.A. Hernández, E. Martínez, On the semilocal convergence of a three steps Newton-type process under mild convergence conditions, Numer. Algor. 70 (2015) 377–392.

[19] Y. Li, X.-P. Guo, Semilocal convergence analysis of MMN-HSS methods under the Hölder conditions, Int. J. Comp. Math. (2016).

[20] J.M. Ortega, W.C. Rheinboldt, Iterative Solution of Nonlinear Equations in Several Variables, Academic Press, New York, 1970.

[21] W.-P. Shen, C. Li, Convergence criterion of inexact methods for operators with Hölder continuous derivatives, Taiwanese J. Math. 12 (2008) 1865–1882.

[22] Q.-B. Wu, M.-H. Chen, Convergence analysis of modified Newton-HSS method for solving systems of nonlinear equations, Numer. Algor. 64 (2013) 659–683.

Chapter 9

Secant-like methods in chemistry

9.1 INTRODUCTION

In this chapter we are concerned with the problem of approximating a solution x^* of equation

$$F(x) = 0, \tag{9.1.1}$$

where F is a Fréchet-differentiable operator defined on a convex subset $\mathcal{D}$ of a Banach space $\mathcal{X}$ with values in a Banach space $\mathcal{Y}$.

Several problems from applied mathematics can be modeled in a form like (9.1.1) using mathematical modeling [6,7,21,32]. Except in special cases, the solutions of these equations cannot be found in closed form. This is the main reason why the most commonly used solution methods are iterative. The convergence analysis of iterative methods is usually divided into two categories: semilocal and local convergence analysis. In the semilocal convergence analysis one derives convergence criteria from the information around an initial point whereas in the local analysis one finds estimates of the radii of convergence balls from the information around a solution.

We present a unifying local, as well as a semilocal, convergence analysis for the secant-like method defined for each $n = 0, 1, 2, \ldots$ by

$$x_{n+1} = x_n - A_n^{-1} F(x_n), \quad A_n = \delta F(y_n, z_n), \tag{9.1.2}$$

where $x_{-1}, x_0 \in \mathcal{D}$ are initial points, $y_n = g_1(x_{n-1}, x_n)$, $z_n = g_2(x_{n-1}, x_n)$, $\delta F : \mathcal{D} \times \mathcal{D} \to \mathcal{L}(\mathcal{X}, \mathcal{Y})$ and $g_1, g_2 : \mathcal{D} \times \mathcal{D} \to \mathcal{D}$. Mapping δF may be a divided difference of order one at the points $(x, y) \in \mathcal{D} \to \mathcal{D}$ with $x \neq y$ satisfying

$$\delta F(x, y)(x - y) = F(x) - F(y), \tag{9.1.3}$$

and if operator F is Fréchet-differentiable, then $F'(x) = \delta F(x, x)$ for each $x \in \mathcal{D}$. Many methods involving divided differences are special cases of method (9.1.2):

- **Secant method** [5–7,16,21,30–32]

$$x_{n+1} = x_n - \delta F(x_{n-1}, x_n)^{-1} F(x_n). \tag{9.1.4}$$

Choose $g_1(x, y) = x$ and $g_2(x, y) = y$.

A Contemporary Study of Iterative Methods. DOI: 10.1016/B978-0-12-809214-9.00009-7

- **Amat et al. method** ($\mathcal{X} = \mathcal{Y} = \mathbb{R}^m$) [4]

$$x_{n+1} = x_n - \delta F(y_n, z_n)^{-1} F(x_n). \tag{9.1.5}$$

Choose $g_1(x, y) = x - aF(x)$ and $g_2(x, y) = x + bF(x)$ for $a, b \in [0, +\infty)$.
- **Ezquerro et al. method** [17]

$$x_{n+1} = x_n - \delta F(y_n, x_n)^{-1} F(x_n). \tag{9.1.6}$$

Choose $g_1(x, y) = \lambda y + (1 - \lambda)x$ and $g_2(x, y) = y$ for all $\lambda \in [0, 1]$.
- **Kurchatov method** [6,23,36]

$$x_{n+1} = x_n - \delta F(2x_n - x_{n-1}, x_{n-1})^{-1} F(x_n). \tag{9.1.7}$$

Choose $g_1(x, y) = 2y - x$ and $g_2(x, y) = x$.

If $x_{-1} = x_0$ and $z_n = y_n = x_n$, for each $n = 0, 1, 2 \ldots$, then we obtain Newton-like methods [4–8,10,16,32]. Several other choices are also possible [5–7,16,20,32].

One of the most important problems in the study of iterative procedures is the convergence domain. In general, the convergence domain is small. As a consequence, it is important to enlarge the convergence domain without additional hypotheses. Another important problem related to the convergence of iterative methods is to find more precise error estimates on the distances involved, namely $\|x_{n+1} - x_n\|$ and $\|x_n - x^\star\|$.

The simplified secant method defined for each $n = 0, 1, 2, \ldots$ by

$$x_{n+1} = x_n - A_0^{-1} F(x_n) \quad (x_{-1}, x_0 \in D)$$

was first studied by S. Ulm [37]. The first semilocal convergence analysis was given by P. Laasonen [23]. His results were improved by F. A. Potra and V. Pták [30–32]. A semilocal convergence analysis for general secant-type methods was given by J. E. Dennis [16], Potra [30–32], Argyros [5–8,10], Hernández et al. [20,21], and others [4], [17], [19] have provided sufficient convergence conditions for the secant method based on Lipschitz-type conditions on δF.

Moreover, the use of Lipschitz and center-Lipschitz conditions is one way to enlarge the convergence domain of different methods. This technique consists of using both conditions together, instead of just the Lipschitz one, which allow us to find a finer majorizing sequence, that is, a larger convergence domain. It has been used in order to find weaker convergence criteria for Newton's method by different authors [1–8,10,19,24]. Moreover, we refer the reader to the results presented by the authors of this chapter [9,11–15,25–29].

In this chapter, using both Lipschitz and center-Lipschitz conditions, we provide a new semilocal convergence analysis for (9.1.2). It turns out that our error bounds and the information on the location of the solution are more precise (under the same convergence condition) than the old ones given in earlier studies such as [1–38].

The rest of the chapter is organized as follows: The semilocal convergence analysis of the secant method is presented in Section 9.2. The local convergence analysis of the secant method is presented in Section 9.3. Numerical examples and applications to chemistry validating the theoretical results are also presented.

9.2 CONVERGENCE ANALYSIS OF THE SECANT METHOD

We present now the conditions that will be used to guarantee the convergence.
The conditions (C) are:

(C_1) F is a Fréchet-differentiable operator defined on a convex subset $\mathcal{D}$ of a Banach space $\mathcal{X}$ with values in a Banach space $\mathcal{Y}$.

(C_2) x_{-1} and x_0 are two points belonging in $\mathcal{D}$ satisfying the inequality

$$\|x_{-1} - x_0\| \le c.$$

(C_3) There exist continuous and nondecreasing functions $h_1, h_2 : [0, +\infty) \times [0, +\infty) \to [0, +\infty)$, functions $g_1, g_2 : \mathcal{D} \times \mathcal{D} \to \mathcal{D}$ such that for each $u, v \in \mathcal{D}$ there exist $y = g_1(u, v) \in \mathcal{D}$ and $z = g_2(u, v) \in \mathcal{D}$ satisfying

$$\|y - x_0\| \le h_1(\|u - x_0\|, \|v - x_0\|)$$

and

$$\|z - x_0\| \le h_2(\|u - x_0\|, \|v - x_0\|).$$

(C_4) There exist a mapping $\delta F :\to \times \mathcal{D} \to \mathcal{L}(\mathcal{X}, \mathcal{Y})$, continuous, nondecreasing functions $f_1, f_2 : [0, +\infty) \times [0, +\infty) \to [0, +\infty)$ with $f_1(0, 0) = f_2(0, 0) = 0$, $n \ge 0$, and $x_{-1}, x_0, y_0 = g_1(x_{-1}, x_0), z_0 = g_2(x_{-1}, x_0) \in \mathcal{D}$ such that $F'(x_0)^{-1}, A_0^{-1} \in \mathcal{L}(\mathcal{Y}, \mathcal{X})$,

$$\|A_0^{-1} F(x_0)\| \le \eta$$

and for each $x, y, z \in \mathcal{D}$

$$\|F'(x_0)^{-1}(\delta F(x, y) - F'(x_0))\| \le f_0(\|x - x_0\|, \|y - x_0\|)$$

and

$$\|F'(x_0)^{-1}(\delta F(x, y) - F'(z))\| \le f_1(\|x - z\|, \|y - z\|).$$

(C_5) Define functions $\varphi, \psi : [0, +\infty) \to [0, +\infty)$ by

$$\varphi(t) = f_0(h_1(t, t), h_2(t, t))$$

and

$$\psi(t) = \int_0^1 f_1(t + \theta\eta + h_1(t + c, t), t + \theta\eta + h_2(t + c, t))d\theta.$$

Moreover, define function q by

$$q(t) = \frac{\psi(t)}{1 - \varphi(t)}.$$

Suppose that equation

$$t(1 - q(t)) - \eta = 0 \tag{9.2.1}$$

has at least one positive zero and the smallest positive zero, denoted by r, satisfies

$$\varphi(r) < 1,$$

$$h_1(c, 0) \leq \bar{r}, \quad h_1(r, r) \leq \bar{r},$$

and

$$h_2(c, 0) \leq \bar{r}, \quad h_2(r, r) \leq \bar{r}, \text{ for some } \bar{r} \geq r.$$

Set $q = q(r)$.
(C_6) $\bar{U}(x_0, \bar{r}) \subseteq \mathcal{D}$.
(C_7) There exists $r_1 \geq r$ such that

$$\int_0^1 f_0(\theta r + (1 - \theta) r_1, \theta r + (1 - \theta) r_1) d\theta < 1.$$

Using these conditions we can state the following result.

Theorem 9.2.1. *Suppose that the conditions (C) hold. Then the sequence $\{x_n\}$ generated by (9.1.2) starting at x_{-1} and x_0 is well defined, remains in $U(x_0, r)$ for all $n \geq 0$, and converges to a solution $x^* \in \bar{U}(x_0, r)$ of equation $F(x) = 0$, which is unique in $\bar{U}(x_0, r_1) \cap \mathcal{D}$.*

Proof. We use mathematical induction to show that sequence $\{x_n\}$ is well defined and belongs in $U(x_0, r)$. By (C_2)–(C_4), $y_0 = g_1(x_{-1}, x_0) \in \mathcal{D}$, $z_0 = g_2(x_{-1}, x_0) \in \mathcal{D}$, and $A_0^{-1} \in \mathcal{L}(\mathcal{Y}, \mathcal{X})$. It follows that x_1 is well defined and $\|x_1 - x_0\| = \|A_0^{-1} F(x_0)\| < \eta < r$, since r is a solution of equation (9.2.1). Then we have that $x_1 \in U(x_0, r)$. Moreover, $y_1 = g_1(x_0, x_1)$ and $z_1 = g_2(x_0, x_1)$ are well defined by the last two hypotheses in (C_5). By the second condition in (C_4), we get that

$$
\begin{aligned}
\|F'(x_0)^{-1}(A_1 - F'(x_0))\| &\leq f_0(\|y_1 - x_0\|, \|z_1 - x_0\|) \\
&\leq f_0(h_1(\|x_0 - x_0\|, \|x_1 - x_0\|), \\
&\qquad h_2(\|x_0 - x_0\|, \|x_1 - x_0\|)) \\
&\leq f_0(h_1(0, \eta), h_2(0, \eta)) \leq \varphi(r) < 1.
\end{aligned}
\tag{9.2.2}
$$

Next, from (9.2.2) and the Banach lemma on invertible operators [7,22,32], we get that $\mathcal{A}_1^{-1} \in \mathcal{L}(\mathcal{Y}, \mathcal{X})$ and

$$\|A_1^{-1} F'(x_0)\| \leq \frac{1}{1 - f_0(h_1(0, \eta), h_2(0, \eta))} \leq \frac{1}{1 - \varphi(r)}. \qquad (9.2.3)$$

As a consequence, x_2 is well defined.

We can write using (9.1.2) for $n = 0$ and (C_1) that

$$
\begin{aligned}
F(x_1) &= F(x_1) - F(x_0) - A_0(x_1 - x_0) \\
&= \int_0^1 \left[F'(x_0 + \theta(x_1 - x_0)) - A_0 \right] d\theta (x_1 - x_0).
\end{aligned}
\qquad (9.2.4)
$$

Then, using the first and second condition in (C_4), (C_3), and (9.2.4), we obtain

$$\|F'(x_0)^{-1} F(x_1)\| = \left\| \int_0^1 F'(x_0)^{-1} \left[F'(x_0 + \theta(x_1 - x_0)) - A_0 \right] d\theta (x_1 - x_0) \right\|$$

$$\leq \int_0^1 f_1(\|x_0 - x_0\| + \theta\|x_1 - x_0\| + \|y_0 - x_0\|,$$

$$\|x_0 - x_0\| + \theta\|x_1 - x_0\| + \|z_0 - x_0\|) d\theta \|x_1 - x_0\|$$

$$\leq \int_0^1 f_1(\theta\eta + h_1(\|x_{-1} - x_0\|, \|x_0 - x_0\|)\|x_1 - x_0\|$$

$$+ \theta\eta + h_2(\|x_{-1} - x_0\|, \|x_0 - x_0\|)) d\theta \|x_1 - x_0\|$$

$$\leq \int_0^1 f_1(\theta\eta + h_1(c, 0), \theta\eta + h_2(c, 0)) d\theta \|x_1 - x_0\|$$

$$\leq \psi(r) \|x_1 - x_0\|.$$

$$(9.2.5)$$

By (9.1.2), (9.2.3), (C_5), and (9.2.5), we get that

$$
\begin{aligned}
\|x_2 - x_1\| &= \|A_1^{-1} F(x_1)\| \leq \|A_1^{-1} F'(x_0)\| \|F'(x_0)^{-1} F(x_1)\| \\
&\leq \frac{\psi(r) \|x_1 - x_0\|}{1 - \varphi(r)} \leq q \|x_1 - x_0\| < \eta
\end{aligned}
\qquad (9.2.6)
$$

and

$$\|x_2 - x_0\| = \|x_2 - x_1\| + \|x_1 - x_0\| \leq \frac{1 - q^2}{1 - q} \|x_1 - x_0\| < \frac{\eta}{1 - q} = r. \qquad (9.2.7)$$

That is, $x_2 \in U(x_0, r)$. Moreover, $y_2 = g_1(x_1, x_2)$ and $z_2 = g_2(x_1, x_2)$ are well defined by the last two hypotheses in (C_5). Then, as in (9.2.2), we have

$$\|F'(x_0)^{-1}(A_2 - F'(x_0))\| \leq f_0(h_1(\eta, r), h_2(\eta, r)) \leq \varphi(r) < 1.$$

Consequently, we obtain

$$\|A_2^{-1}F'(x_0)\| \leq \frac{1}{1 - f_0(h_1(\eta, r), h_2(\eta, r))} \leq \frac{1}{1 - \varphi(r)}. \qquad (9.2.8)$$

Hence, x_3 is well defined. Moreover, $y_2 = g_1(x_1, x_2)$ and $z_2 = g_2(x_1, x_2)$ are well defined. Then, by (9.1.2) for $n = 0$ and (C_1), we have

$$F(x_2) = F(x_2) - F(x_1) - A_1(x_2 - x_1)$$
$$= \int_0^1 (F'(x_1 + \theta(x_2 - x_1)) - A_1)d\theta(x_2 - x_1). \qquad (9.2.9)$$

Then, using the first and second hypotheses in (C_4), (C_3), and (9.2.4), we get

$$
\begin{aligned}
\|F'(x_0)^{-1}F(x_2)\| &\leq \int_0^1 f_1(\|x_1 - x_0\| + \theta\|x_2 - x_1\| + \|y_1 - x_0\|, \\
&\quad \|x_1 - x_0\| + \theta\|x_2 - x_1\| + \|z_1 - x_0\|)d\theta\|x_2 - x_1\| \\
&\leq \int_0^1 f_1(\eta + \theta\eta + h_1(\|x_0 - x_0\|, \|x_1 - x_0\|), \\
&\quad \eta + \theta\eta + h_2(\|x_0 - x_0\|, \|x_1 - x_0\|))d\theta\|x_2 - x_1\| \\
&\leq \int_0^1 f_1((1 + \theta)\eta + h_1(c, \eta), (1 + \theta)\eta + \\
&\quad h_2(c, \eta))d\theta\|x_2 - x_1\| \\
&\leq \psi(r)\|x_2 - x_1\|.
\end{aligned}
$$
$$(9.2.10)$$

Hence we get in turn that

$$
\begin{aligned}
\|x_3 - x_2\| &= \|A_2^{-1}F(x_0)\|\|F'(x_0)^{-1}F(x_2)\| \\
&\leq \frac{\psi(r)}{1 - \varphi(r)}\|x_2 - x_1\| \leq q\|x_2 - x_1\| < q^2\|x_1 - x_0\| < \eta
\end{aligned}
$$

and

$$\|x_3 - x_0\| = \|x_3 - x_2\| + \|x_2 - x_1\| + \|x_1 - x_0\| \leq \frac{1-q^3}{1-q}\eta < \frac{\eta}{1-q} = r.$$

In a similar way, we have that y_3, z_3 are well defined and

$$\|A_3^{-1}F(x_0)\|$$

$$\leq \frac{1}{1 - (f_0(h_1(\|x_2 - x_0\|, \|x_3 - x_0\|), h_2(\|x_2 - x_0\|, \|x_3 - x_0\|)))} \quad (9.2.11)$$

$$\leq \frac{1}{1 - f_0(h_1(r,r), h_2(r,r))} \leq \frac{1}{1 - \varphi(r)}.$$

Moreover, we have

$$\|F'(x_0)^{-1}F(x_3)\| \leq \int_0^1 f_1(\|x_2 - x_0\| + \theta\|x_3 - x_2\| + \|y_2 - x_0\|,$$

$$\|x_2 - x_0\| + \theta\|x_3 - x_2\| + \|z_2 - x_0\|)d\theta\|x_3 - x_2\|$$

$$\leq \int_0^1 f_1(r + \theta\eta + h_1(\eta, r),$$

$$r + \theta\eta + h_2(\eta, r))d\theta\|x_3 - x_2\|$$

$$\leq \psi(r)\|x_3 - x_2\|.$$

$$(9.2.12)$$

In view of (9.2.11) and (9.2.12), we get that

$$\|x_4 - x_3\| = \|A_3^{-1}F'(x_0)\|\|F'(x_0)^{-1}F(x_3)\|$$

$$\leq \frac{\psi(r)}{1 - \varphi(r)}\|x_3 - x_2\| \leq q\|x_3 - x_2\| < q^3\|x_1 - x_0\| < \eta$$

and

$$\|x_4 - x_0\| = \|x_4 - x_3\| + \|x_3 - x_2\| + \|x_2 - x_1\| + \|x_1 - x_0\|$$

$$\leq \frac{1-q^4}{1-q}\eta < \frac{\eta}{1-q} = r.$$

Furthermore, we have that y_4, z_4 are well defined,

$$\|A_4^{-1}F'(x_0)\| \leq \frac{1}{1 - \varphi(r)}, \quad (9.2.13)$$

and

$$\|F'(x_0)^{-1}F(x_4)\| \leq \int_0^1 f_1(r+\theta\eta+h_1(r,r),r$$

$$+\theta\eta+h_2(r,r))d\theta\|x_4-x_3\| \qquad (9.2.14)$$

$$\leq \psi(r)\|x_3-x_2\|.$$

Then from (9.2.14) we get

$$\|x_5-x_4\| = \|A_4^{-1}F'(x_0)\|\,\|F'(x_0)^{-1}F(x_4)\|$$

$$\leq \frac{\psi(r)}{1-\varphi(r)}\|x_4-x_3\| \leq q\|x_4-x_3\| < q^4\|x_1-x_0\| < \eta$$

and

$$\|x_5-x_0\| = \|x_5-x_4\|+\|x_4-x_3\|+\|x_3-x_2\|+\|x_2-x_1\|+\|x_1-x_0\|$$

$$\leq \frac{1-q^5}{1-q}\eta < \frac{\eta}{1-q} = r.$$

Suppose that for $i = 1, 2, \ldots, k-1$,
- The operator $A_k^{-1} \in \mathcal{L}(\mathcal{Y}, \mathcal{X})$ and

$$\|A_k^{-1}F'(x_0)\| \leq \frac{1}{1-\varphi(r)},$$

- $\|x_{k+1}-x_k\| \leq q\|x_k-x_{k-1}\|$,
- $\|x_{k+1}-x_0\| \leq \frac{1-q^{k+1}}{1-q}\eta < r.$

Then, in an analogous way we obtain

$$\|A_{k+1}^{-1}F'(x_0)\| \leq \frac{1}{1-\varphi(r)},$$

$$\|x_{k+2}-x_{k+1}\| \leq q^{k+1}\eta, \qquad (9.2.15)$$

and

$$\|x_{k+2}-x_0\| \leq \frac{1-q^{k+2}}{1-q}\eta < r.$$

Next, it is easy to see that $x_{k+2} \in U(x_0, r)$. That is, sequence $\{x_n\}$ is well defined. Using (9.2.15) we get

$$\|x_{k+j} - x_k\| = \|x_{k+j} - x_{k+j-1}\| + \|x_{k+j-1} - x_{k+j-2}\| + \cdots + \|x_{k+1} - x_k\|$$

$$\leq (q^{j-1} + q^{j-2} + \cdots + q + 1)\|x_{k+1} - x_k\|$$

$$= \frac{1-q^j}{1-q}\|x_{k+1} - x_k\| < \frac{q^k}{1-q}\|x_1 - x_0\|$$

(9.2.16)

for $j = 1, 2, \ldots$ and $q < 1$. Now from (9.2.16) sequence $\{x_k\}$ is Cauchy in a Banach space $\mathcal{X}$ and as such it converges to some $x^* \in \bar{U}(x_0, r)$ (since $\bar{U}(x_0, r)$ is a closed set). Taking into account the estimate

$$\|F'(x_0)^{-1} F(x_k)\| \leq \psi(r)\|x_k - x_{k-1}\|$$

and letting $k \to \infty$, we deduce that $F(x^*) = 0$. Finally, we shall prove the uniqueness assertion. Let $y^* \in \bar{U}(x_0, r)$ be such that $F(y^*) = 0$. Then, using the last condition in (C_4) and (C_7), we get in turn that for $Q = \int_0^1 F'(y^* + \theta(x^* - y^*))d\theta$,

$$\|F'(x_0)^{-1}(F'(x_0) - Q)\|$$

$$\leq \int_0^1 f_0(\|y^* + \theta(x^* - y^*) - x_0\|, \|y^* + \theta(x^* - y^*) - x_0\|)d\theta$$

$$\leq \int_0^1 f_0(\theta\|x^* - x_0\| + (1-\theta)\|y^* - x_0\|, \theta\|x^* - x_0\| + (1-\theta)\|y^* - x_0\|)d\theta$$

$$\leq \int_0^1 f_0(\theta r + (1-\theta)r_1, \theta r + (1-\theta)r_1)d\theta < 1.$$

(9.2.17)

If follows from (9.2.17) and the Banach's lemma on invertible operators that $Q^{-1} \in \mathcal{L}(\mathcal{Y}, \mathcal{X})$. Then, using the identity $0 = F(x^*) - F(y^*) = Q(y^* - x^*)$, we conclude that $y^* = x^*$. $\square$

Next we have the following interesting remarks.

Remark 9.2.2. *(a) The condition $A_0^{-1} \in \mathcal{L}(\mathcal{Y}, \mathcal{X})$ can be dropped as follows. Suppose that*

$$f_0(h_1(c, 0), h_2(c, 0)) < 1.$$

(9.2.18)

Then we have by the third condition in (C_4) that

$$\|F'(x_0)^{-1}(A_0 - F'(x_0))\| \leq f_0(\|y_0 - x_0\|, \|z_0 - x_0\|)$$

$$\leq f_0(h_1(\|x_{-1} - x_0\|, \|x_0 - x_0\|),$$

$$h_2(\|x_{-1} - x_0\|, \|x_0 - x_0\|))$$

$$\leq f_0(h_1(c, 0), h_2(c, 0)) < 1.$$

It follows from (9.2.18) that $A_0^{-1} \in \mathcal{L}(\mathcal{Y}, \mathcal{X})$ and

$$\|A_0 F'(x_0)^{-1}\| \leq \frac{1}{1 - f_0(h_1(c, 0), h_2(c, 0))}. \qquad (9.2.19)$$

Then, due to the estimate

$$\|x_1 - x_0\| \leq \|A_0^{-1} F'(x_0)\| \|F'(x_0)^{-1} F(x_0)\| \leq \frac{\|F'(x_0)^{-1} F(x_0)\|}{1 - f_0(h_1(c, 0), h_2(c, 0))},$$

we can define

$$\eta = \frac{\|F'(x_0)^{-1} F(x_0)\|}{1 - f_0(h_1(c, 0), h_2(c, 0))}. \qquad (9.2.20)$$

Then the conclusions of Theorem 9.2.1 hold with (9.2.18) replacing condition $A_0^{-1} \in \mathcal{L}(\mathcal{Y}, \mathcal{X})$ and with η given by (9.2.20).

(b) *In view of the second and third condition in (C_4), we have that*

$$f_0(s_1, s_2) \leq f_1(s_3, s_4) \quad \text{for each } s_i \geq 0, i = 1, 2, 3, 4$$
$$\text{with } s_1 \leq s_3 \text{ and } s_2 \leq s_4.$$

Now we define conditions (H) as follows:

(H_1) F is a continuous operator defined on a convex subset $\mathcal{D}$ of a Banach space $\mathcal{X}$ with values in a Banach space $\mathcal{Y}$.

(H_2) x_{-1} and x_0 are two points belonging in $\mathcal{D}$ satisfying the inequality

$$\|x_{-1} - x_0\| \leq c.$$

(H_3) There exist continuous and nondecreasing functions $h_1, h_2 : [0, +\infty) \times [0, +\infty) \to [0, +\infty)$, functions $g_1, g_2 : \mathcal{D} \times \mathcal{D} \to \mathcal{D}$ such that for each $u, v \in \mathcal{D}$ there exist $y = g_1(u, v) \in \mathcal{D}$ and $z = g_2(u, v) \in \mathcal{D}$ satisfying

$$\|y - x_0\| \leq h_1(\|u - x_0\|, \|v - x_0\|)$$

and

$$\|z - x_0\| \leq h_2(\|u - x_0\|, \|v - x_0\|).$$

(H_4) There exist a divided difference of order one, δF, continuous non-decreasing functions $f_1, f_2 : [0, +\infty) \times [0, +\infty) \rightarrow [0, +\infty)$, $n \geq 0$, and $x_{-1}, x_0, y_0 = g_1(x_{-1}, x_0), z_0 = g_2(x_{-1}, x_0) \in \mathcal{D}$ such that $A_0^{-1} \in \mathcal{L}(\mathcal{Y}, \mathcal{X})$,

$$\|A_0^{-1} F(x_0)\| \leq \eta$$

and for each $x, y, u, v \in \mathcal{D}$

$$\|A_0^{-1}(\delta F(x, y) - A_0)\| \leq f_0(\|x - y_0\|, \|y - z_0\|)$$

and

$$\|A_0^{-1}(\delta F(x, y) - \delta F(u, v))\| \leq f_1(\|x - u\|, \|y - v\|).$$

(H_5) Define functions $\varphi, \psi : [0, +\infty) \rightarrow [0, +\infty)$ by

$$\varphi(t) = f_0(h_1(t, t) + h_1(c, 0), h_2(t, t) + h_2(c, 0)),$$

$$\psi(t) = f_1(t + h_1(t + c, t), t + h_2(t + c, t)),$$

and

$$q(t) = \frac{\psi(t)}{1 - \varphi(t)}.$$

Suppose that equation

$$t(1 - q(t)) - \eta = 0$$

has at least one positive zero and the smallest positive zero, denoted by R, satisfies

$$\varphi(R) < 1,$$

$$h_1(c, 0) \leq \bar{R}, \quad h_1(R, R) \leq \bar{R},$$

and

$$h_2(c, 0) \leq \bar{R}, \quad h_2(R, R) \leq \bar{R}, \text{ for some } \bar{R} \geq R.$$

Set $q = q(R)$.
(H_6) $\bar{U}(x_0, \bar{R}) \subseteq \mathcal{D}$.
(H_7) There exists $R_1 \geq R$ such that

$$f_0(R_1 + h_1(c, 0), R_1 + h_2(c, 0)) < 1.$$

Using the above conditions we can present the following result.

Theorem 9.2.3. *Suppose that conditions (H) hold. Then the sequence $\{x_n\}$ generated by (9.1.2) starting at x_{-1} and x_0 is well defined, remains in $U(x_0, R)$ for all $n \geq 0$, and converges to a solution $x^* \in \bar{U}(x_0, R)$ of equation $F(x) = 0$, which is unique in $\bar{U}(x_0, R_1) \cap \mathcal{D}$.*

Remark 9.2.4. *Notice that the second and third conditions in (H_4) do not necessarily imply the Fréchet-differentiability of F.*

9.2.1   Semilocal example 1

We study

$$x(s) = 1 + \int_0^1 G(s, t)x(t)^2 dt, \quad s \in [0, 1], \tag{9.2.21}$$

where $x \in C[0, 1]$ and the kernel G is the Green function in $[0, 1] \times [0, 1]$.

Using a discretization process and transform equation (9.2.21) into a finite dimensional problem. For this, we follow the process described in the second section in [17] with $m = 8$ and we obtain the following system of nonlinear equations:

$$F(x) \equiv x - 1 - Av_x = 0, \quad F : \mathbb{R}^8 \to \mathbb{R}^8, \tag{9.2.22}$$

where

$$a = \begin{pmatrix} x_1 \\ x_2 \\ \vdots \\ x_n \end{pmatrix},$$

$$1 = \begin{pmatrix} 1 \\ 1 \\ \vdots \\ 1 \end{pmatrix},$$

$$A = (a_{ij})_{i,j=1}^8,$$

and

$$v_x = \begin{pmatrix} x_1^2 \\ x_2^2 \\ \vdots \\ x_n^2 \end{pmatrix}.$$

We use the divided difference of first order of F as $[u, v; F] = I - B$, where $B = (b_{ij})_{i,j=1}^8$ with $b_{ij} = a_{ij}(u_j + v_j)$.

Choosing the starting points

$$x_{-1} = \begin{pmatrix} \dfrac{7}{10} \\ \dfrac{7}{10} \\ \vdots \\ \dfrac{7}{10} \end{pmatrix}$$

and

$$x_0 = \begin{pmatrix} \dfrac{18}{10} \\ \dfrac{18}{10} \\ \vdots \\ \dfrac{18}{10} \end{pmatrix}$$

method (9.1.6) with

$$\lambda = \frac{1}{2},$$

and the max-norm, we obtain

$$c = \frac{11}{10},$$

$$y_0 = \begin{pmatrix} \dfrac{5}{4} \\[2mm] \dfrac{5}{4} \\[2mm] \vdots \\[2mm] \dfrac{5}{4} \end{pmatrix}$$

$$\beta = 1.555774\ldots,$$

$$\eta = 0.7839875\ldots,$$

$$m = 0.257405\ldots,$$

and that the polynomial

$$p(t) = -0.789051 + t\left(1 - \frac{0.257406}{1 - 0.19223(0.55 + 2t)}\right)$$

has no real roots, so the conditions in [17] are not satisfied and we cannot ensure the convergence of method (9.1.6). Next, we are going to use our (C)
conditions in order to ensure the convergence of the methods. We consider
$g_1(x, y) = \dfrac{y + x}{2}$ and $g_2(x, y) = y$, and we define the functions

$$h_1(s, t) = \frac{s}{2} + \frac{t}{2},$$

$$h_2(s, t) = t,$$

$$f_0(s, t) = 0.0103507\ldots(s + t)$$

and

$$f_1(s, t) = 0.047827\ldots(s + t),$$

which satisfy the (C_3) and (C_4) conditions. Moreover, we define

$$\varphi(t) = 0.0207015t,$$

$$\psi(t) = 0.0640438 + 0.191311t,$$

and

$$q(t) = \frac{\phi(t)}{1 - \varphi(t)} = \frac{0.0640438 + 0.191311t}{1 - 0.0207015t}.$$

The solutions of equation (9.2.1) are

$$r_1 = 1.09603\ldots \quad \text{and} \quad r_2 = 3.39564\ldots$$

Then, by denoting $r = 1.09603\ldots$ and $\bar{r} = r$, it is easy to see that the following conditions are verified:

$$\varphi(r) = 0.0226895\ldots < 1,$$

$$h_1(c, 0) = 0.55 \leq r, \quad h_1(r, r) = r \leq \bar{r},$$

and

$$h_2(c, 0) = 0 \leq \bar{r}, \quad h_2(r, r) = r \leq \bar{r}.$$

Finally, to show that condition (C_7) is verified, we choose $r_1 = 1.25$ and obtain

$$\int_0^1 f_0(\theta r + (1-\theta)r_1, \theta r + (1-\theta)r_1)d\theta = 0.0258769\ldots < 1.$$

So all the conditions of Theorem 9.2.1 are satisfied, and as a consequence we can ensure the convergence of method (9.1.6).

9.2.2   Semilocal Example 2

We consider the same problem as before but now using method (9.1.5). If we choose the starting points

$$x_{-1} = \begin{pmatrix} \dfrac{12}{10} \\ \dfrac{12}{10} \\ \vdots \\ \dfrac{12}{10} \end{pmatrix}$$

and

$$x_0 = \begin{pmatrix} \dfrac{14}{10} \\ \dfrac{14}{10} \\ \ldots \\ \dfrac{14}{10} \end{pmatrix},$$

method (9.1.5) with

$$a = 0.8,$$
$$b = 0.75$$

and the max-norm, we obtain

$$c = 0.2$$

$$y_0 = \begin{pmatrix} 1.11459 \\ 1.16864 \\ 1.23603 \\ 1.28163 \\ 1.28163 \\ 1.23603 \\ 1.16864 \\ 1.11459 \end{pmatrix},$$

$$z_0 = \begin{pmatrix} 1.70443 \\ 1.64679 \\ 1.5749 \\ 1.52626 \\ 1.52626 \\ 1.5749 \\ 1.64679 \\ 1.70443 \end{pmatrix}$$

$$\beta = 1.49123\ldots,$$
$$\delta = 0.380542\ldots,$$
$$\eta = 0.394737\ldots,$$
$$\gamma = 1.67305\ldots,$$

and

$$M = 0.47606\ldots$$

then one of the necessary conditions given in [4] is not satisfied since

$$M\delta\gamma^2 = 0.507089\ldots \le 0.5,$$

so we cannot ensure the convergence of method (9.1.5).

Next, we are going to use our (H) conditions in order to ensure the convergence of the methods. We consider $g_1(x, y) = x - aF(x)$ and $g_2(x, y) = x + bF(x)$, and we define the functions

$$h_1(s, t) = p_0 t + 0.5p,$$

$$h_2(s, t) = p_1 t + 0.5p,$$

$$f_0(s, t) = 0.0579754\ldots(s + t),$$

and

$$f_1(s, t) = 0.0579754\ldots(s + t),$$

where

$$p = \|A_0\|,$$

$$p_0 = \|I - 0.75\delta F[x, x_0] = 1.16988\ldots,$$

and

$$p_1 = \|I + 0.8\delta F[x, x_0] = 1.03601,$$

which satisfy the (C_3) and (C_4) conditions. Moreover, we define

$$\varphi(t) = 0.0579754(2.98245 + 2.20589t),$$

$$\psi(t) = 0.10934 + 0.243838t,$$

and

$$q(t) = \frac{\phi(t)}{1 - \varphi(t)} = \frac{0.10934 + 0.243838t}{1 - 0.0579754(2.98245 + 2.20589t)}.$$

The solutions of equation (9.2.1) are

$$r_1 = 0.598033\ldots \quad \text{and} \quad r_2 = 1.46863\ldots$$

Then, by denoting $r = 0.598033\ldots$ and $\bar{r} = 1.44524\ldots$, it is easy to see that the following conditions are verified:

$$\varphi(r) = 0.24939\ldots < 1,$$

$$h_1(c, 0) = 0.745613\ldots \leq \bar{r}, \quad h_1(r, r) = \bar{r} \ldots \leq \bar{r},$$

and

$$h_2(c, 0) = 0.745613\ldots \leq \bar{r}, \quad h_2(r, r) = \bar{r} \ldots \leq \bar{r}.$$

Finally, to show that condition (C_7) is verified, we choose $r_1 = 0.6$ and obtain

$$\int_0^1 f_0(\theta r + (1 - \theta)r_1, \theta r + (1 - \theta)r_1)d\theta = 0.0695705\ldots < 1.$$

So all the conditions of Theorem 9.2.3 are satisfied, and as a consequence we can ensure the convergence of method (9.1.5).

9.3   LOCAL CONVERGENCE OF THE SECANT METHOD

Next we shall show the local convergence of the method under the set of conditions that will be presented. Conditions (C^*) are as follows:

(C_1^*) F is a Fréchet-differentiable operator defined on a convex subset $\mathcal{D}$ of a Banach space $\mathcal{X}$ with values in a Banach space $\mathcal{Y}$.

(C_2^*) There exists $x^* \in \mathcal{D}$ such that $F(x^*) = 0$ and $F'(x^*)^{-1} \in \mathcal{L}(\mathcal{Y}, \mathcal{X})$.

(C_3^*) There exist continuous and nondecreasing functions $h_1, h_2 : [0, +\infty) \times [0, +\infty) \to [0, +\infty)$, functions $g_1, g_2 : \mathcal{D} \times \mathcal{D} \to \mathcal{D}$ such that for each $u, v \in \mathcal{D}$ there exist $y = g_1(u, v) \in \mathcal{D}$ and $z = g_2(u, v) \in \mathcal{D}$ satisfying

$$\|y - x^*\| \leq h_1(\|u - x^*\|, \|v - x^*\|)$$

and

$$\|z - x^*\| \leq h_2(\|u - x^*\|, \|v - x^*\|).$$

(C_4^*) There exist a mapping $\delta F : \mathcal{D} \times \mathcal{D} \to \mathcal{L}(\mathcal{X}, \mathcal{Y})$, continuous nondecreasing functions $f_1, f_2 : [0, +\infty) \times [0, +\infty) \to [0, +\infty)$ with $f_1(0, 0) = f_2(0, 0) = 0$, $n \geq 0$, and $x_{-1}, x_0, y_0 = g_1(x_{-1}, x_0), z_0 = g_2(x_{-1}, x_0) \in \mathcal{D}$ such that for each $x, y, z \in \mathcal{D}$

$$\|F'(x^*)^{-1}(\delta F(x, y) - F'(x^*))\| \leq f_0(\|x - x^*\|, \|y - x^*\|)$$

and

$$\|F'(x^*)^{-1}(\delta F(x, y) - F'(z))\| \leq f_1(\|x - z\|, \|y - z\|).$$

(C_5^*) Define functions $\varphi, \psi : [0, +\infty) \to [0, +\infty)$ by

$$\varphi(t) = f_0(h_1(t, t), h_2(t, t))$$

and

$$\psi(t) = \int_0^1 f_1(\theta t + h_1(t, t), \theta t + h_2(t, t))d\theta.$$

Moreover, define functions q and p by

$$q(t) = \frac{\psi(t)}{1 - \varphi(t)}$$

and

$$p(t) = \int_0^1 f_1(\theta t + h_1(t, t), \theta t + h_2(t, t))d\theta + f_0(h_1(t, t), h_2(t, t)) - 1.$$

Suppose that equation

$$p(t) = 0 \qquad\qquad (9.3.1)$$

has at least one positive zero and the smallest positive zero, denoted by r, satisfies

$$\varphi(r) < 1,$$

$$h_1(r, r) \leq \bar{r},$$

and

$$h_2(r, r) \leq \bar{r}, \text{ for some } \bar{r} \geq r.$$

Set $q = q(r)$.

(C_6^*) $\bar{U}(x_0, \bar{r}) \subseteq \mathcal{D}$.

(C_7^*) There exists $r_1 \geq r$ such that

$$\int_0^1 f_0((1 - \theta)r_1, (1 - \theta)r_1)d\theta < 1.$$

Theorem 9.3.1. *Suppose that conditions (C^*) hold. Then the sequence $\{x_n\}$ generated for $x_{-1}, x_0 \in U(x^*, r) \setminus \{x^*\}$ by (9.1.2) is well defined, remains in $U(x^*, r)$, and converges to the solution x^* of equation $F(x) = 0$, which is unique in $\bar{U}(x_0, r_1) \cap \mathcal{D}$.*

Proof. The proof is analogous to the proof of Theorem 9.2.1 We shall first show using mathematical induction that sequence $\{x_n\}$ is well defined and belongs in $U(x^*, r)$. We have by the first condition in (C_4^*) that

$$\begin{aligned}
\|F'(x^*)^{-1}(A_0 - F'(x_1))\| &\leq f_0(\|y_0 - x^*\|, \|z_0 - x^*\|) \\
&\leq f_0(h_1(\|x_{-1} - x^*\|, \|x_0 - x^*\|), \\
&\quad h_2(\|x_{-1} - x^*\|, \|x_0 - x^*\|))
\end{aligned}$$

$$\leq \quad f_0(h_1(r, r), h_2(r, r)) \leq \varphi(r) < 1.$$

Hence, $A_0^{-1} \in \mathcal{L}(\mathcal{Y}, \mathcal{X})$ and

$$\|A_0^{-1} F'(x_0)\|$$

$$\leq \frac{1}{1 - f_0(h_1(\|x_{-1} - x^*\|, \|x_0 - x^*\|), h_2(\|x_{-1} - x^*\|, \|x_0 - x^*\|))}$$

$$\leq \frac{1}{1 - \varphi(r)}.$$

That is, x_1 is well defined. Then we have by the second substep of condition (C_4^*) that

$$\|F'(x^*)^{-1}(\int_0^1 F'(x^*) + \theta(x_0 - x^*)) - A_0)d\theta(x_0 - x^*)\|$$

$$\leq \int_0^1 f_1(\|x^* + \theta(x_0 - x^*) - y_0\|, \|x^* + \theta(x_0 - x^*) - z_0\|)d\theta\|x_0 - x^*\|$$

$$\leq \int_0^1 f_1(\theta\|x_0 - x^*\| + \|y_0 - x^*\|, \theta\|x_0 - x^*\| + \|z_0 - x^*\|)d\theta\|x_0 - x^*\|$$

$$\leq \int_0^1 f_1(\theta\|x_0 - x^*\| + h_1(\|x_{-1} - x^*\|, \|x_0 - x^*\|), \theta\|x_0 - x^*\|$$

$$+ h_2(\|x_{-1} - x^*\|, \|x_0 - x^*\|)d\theta\|x_0 - x^*\|$$

$$\leq f_1(\theta r + h_1(r, r), \theta r + h_2(r, r))d\theta\|x_0 - x^*\|.$$

In view of the approximation

$$
\begin{aligned}
x_1 - x^* &= x_0 - x^* - A_0^{-1} F(x_0) \\
&= -A_0^{-1}(F(x_0) - F(x^*) - A_0(x_0 - x^*)) \\
&= -A_0^{-1}\left[\int_0^1 F'(x^* + \theta(x_0 - x^*) - A_0)\right]d\theta(x_0 - x^*),
\end{aligned}
$$

we get that

$$\|x_1 - x^*\|$$

$$\leq \|A_0^{-1} F(x^*)\|\|\int_0^1 F'(x^*)^{-1}(F'(x^* + \theta(x_0 - x^*)) - A_0)d\theta\|x_0 - x^*\|$$

$$\leq q\|x_0 - x^*\| < \|x_0 - x^*\| < r,$$

which shows $x_1 \in U(x^*, r)$. Next we show that $A_1^{-1} \in \mathcal{L}(\mathcal{Y}, \mathcal{X})$. As above, we have

$$\|F'(x^*)^{-1}(A_1 - F'(x^*))\| \leq f_0(h_1(\|x_0 - x^*\|, \|x_1 - x^*\|),$$

$$h_2(\|x_0 - x^*\|, \|x_1 - x^*\|))$$

$$\leq f_0(h_1(r, r), h_2(r, r))\|x_1 - x^*\|$$

$$= \varphi(r) < 1,$$

so

$$\|A_1^{-1}F'(x^*)\| \leq \frac{1}{1 - \varphi(r)}.$$

That is, x_2 is well defined. Then we obtain

$$\|F'(x^*)^{-1}\left[\int_0^1 F'(x^* + \theta(x_1 - x^*)) - A_1\right]d\theta\|x_1 - x^*\|$$

$$\leq f_1(\theta\|x_1 - x^*\| + \|y_1 - x^*\|, \theta\|x_1 - x^*\| + \|z_1 - x^*\|)d\theta\|x_1 - x^*\|$$

$$\leq f_1(\theta r + h_1(r, r), \theta r + h_2(r, r),)d\theta\|x_1 - x^*\|.$$

Using the identity

$$x_2 - x^* = -A_1^{-1}\left[\int_0^1 F'(x^* + \theta(x_1 - x^*)) - A_1\right]d\theta\|x_1 - x^*\|,$$

we get again that

$$\|x_2 - x^*\| \leq q\|x_1 - x^*\| < \|x_1 - x^*\| < r,$$

which shows that $x_2 \in U(x^*, r)$.

Similarly, we have

$$\|x_{k+1} - x_k\| \leq q\|x_k - x^*\| < \|x_k - x^*\| < r,$$

which yields $\lim_{k \to \infty} x_k = x^*$ and $x_{k+1} \in U(x^*, r)$. The uniqueness part has been shown in Theorem 9.2.1. $\qquad\square$

The conditions (H^*) are as follows:

(H_1^*) F is a continuous operator defined on a convex subset $\mathcal{D}$ of a Banach space $\mathcal{X}$ with values in a Banach space $\mathcal{Y}$.

(H_2^*) There exists $x^* \in \mathcal{D}$ such that $F(x^*) = 0$.

(H_3^*) There exist continuous and nondecreasing functions $h_1, h_2 : [0, +\infty) \times [0, +\infty) \to [0, +\infty)$, functions $g_1, g_2 : \mathcal{D} \times \mathcal{D} \to \mathcal{D}$ such that for each $u, v \in \mathcal{D}$ there exist $y = g_1(u, v) \in \mathcal{D}$ and $z = g_2(u, v) \in \mathcal{D}$

$$\|y - x^*\| \leq h_1(\|u - x^*\|, \|v - x^*\|),$$

and

$$\|z - x^*\| \leq h_2(\|u - x^*\|, \|v - x^*\|).$$

(H_4^*) There exist a divided difference of order one δF, a linear operator $M \in \mathcal{L}(\mathcal{X}, \mathcal{Y})$, continuous nondecreasing functions $f_1, f_2 : [0, +\infty) \times [0, +\infty) \to [0, +\infty)$, $n \geq 0$, and $x_{-1}, x_0, y_0 = g_1(x_{-1}, x_0), z_0 = g_2(x_{-1}, x_0) \in \mathcal{D}$ such that $M^{-1} \in \mathcal{L}(\mathcal{Y}, \mathcal{X})$ and for each $x, y, z \in \mathcal{D}$

$$\|M^{-1}(\delta F(x, y) - M)\| \leq f_0(\|x - x^*\|, \|y - x^*\|)$$

and

$$\|M^{-1}(\delta F(x, x^*) - \delta F(u, v))\| \leq f_1(\|x - u\|, \|x^* - v\|).$$

(H_5^*) Define functions $\varphi, \psi : [0, +\infty) \to [0, +\infty)$ by

$$\varphi(t) = f_0(h_1(t, t), h_2(t, t))$$

and

$$\psi(t) = f_1(t + h_1(t, t), t + h_2(t, t)).$$

Moreover, define functions q and p by

$$q(t) = \frac{\psi(t)}{1 - \varphi(t)}$$

and

$$p(t) = f_1(t + h_1(t, t), h_2(t, t)) + f_0(h_1(t, t), h_2(t, t)) - 1.$$

Suppose that the equation

$$p(t) = 0$$

has at least one positive zero and the smallest positive zero, denoted by R, satisfies

$$\varphi(R) < 1,$$

$$h_1(R, R) \leq \bar{R},$$

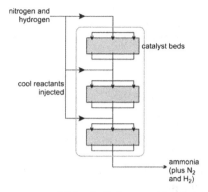

FIGURE 9.1 Taken from www.essentialchemicalindustry.org/chemicals/ammonia.html. Ammonia generation process.

and

$$h_2(R, R) \le \bar{R}, \text{ for some } \bar{R} \ge R.$$

Set $q = q(R)$.

(H_6^*) $\bar{U}(x_0, \bar{R}) \subseteq \mathcal{D}$.

(H_7^*) There exists $R_1 \ge R$ such that

$$f_0(0, R_1) < 1.$$

Theorem 9.3.2. *Suppose that conditions* (H^*) *hold. Then the sequence* $\{x_n\}$ *generated for* $x_{-1}, x_0 \in \bar{U}(x^*, R) \setminus x^*$ *by* (9.1.2) *is well defined, remains in* $U(x^*, R)$, *and converges to the solution* x^* *of equation* $F(x) = 0$, *which is unique in* $\bar{U}(x_0, R_1) \cap \mathcal{D}$.

Remark 9.3.3. *The results in this section hold for Newton-like methods, if we set* $x_{-1} = x_0$ *and* $z_n = y_n = x_n$, *for each* $n = 0, 1, 2 \dots$ *[4–8,10,16,32].*

9.3.1 Local example 1

In order to show the applicability of our theory in a real problem, we are going to consider the following quartic equation that describes the fraction of the nitrogen–hydrogen feed that gets converted to ammonia, called the fractional conversion showed in [18,35]. In Fig. 9.1 the process of ammonia generation is shown.

For 250 atm and 500 °C, this equation takes the form:

$$f(x) = x^4 - 7.79075x^3 + 14.7445x^2 + 2.511x - 1.674.$$

In this case, we take the domain

$$\Omega = B(x_0, R),$$

where

$$x_0 = 0.5$$

and

$$R = 0.3.$$

Choose

$$x_{-1} = 0.4.$$

We are going to use our (C^*) conditions in order to ensure the convergence of the method (9.1.7). Define the functions

$$h_1(s,t) = s + 2t,$$

$$h_2(s,t) = s,$$

$$f_0(s,t) = 0.111304(32.2642s + 20.5458t),$$

and

$$f_1(s,t) = 0.111304(40.9686s + 30.7779t),$$

which satisfy the (C_3^*) and (C_4^*) conditions. Moreover, we define

$$\varphi(t) = 13.0603t,$$

$$\psi(t) = 21.0985t,$$

$$q(t) = \frac{\phi(t)}{1 - \varphi(t)} = \frac{-1 + 34.1587t}{1 - 13.0603t},$$

and

$$p(t) = \int_0^1 f_1(\theta t + h_1(t,t), \theta t + h_2(t,t))d\theta + f_0(h_1(t,t), h_2(t,t)) - 1$$
$$= 34.1587t - 1.$$

The root of $p(t)$ is

$$r_1 = 0.0292751\ldots$$

Then, by denoting

$$r = 0.0292751\ldots$$

and

$$\bar{r} = 3r = 0.0878253\ldots,$$

it is easy to see that the following conditions are verified:

$$\varphi(r) = 0.38234\ldots < 1,$$

$$h_1(r, r) = \bar{r} \leq \bar{r},$$

and

$$h_2(r, r) = r \leq \bar{r}.$$

Finally, it is clear that

$$\bar{U}(x_0, \bar{r}) \subseteq U(x_0, R).$$

Next, to show that condition (C_7^*) is verified, we choose

$$r_1 = 0.1$$

and obtain

$$\int_0^1 f_0(\theta r + (1 - \theta)r_1, \theta r + (1 - \theta)r_1) d\theta = 0.293899\ldots < 1.$$

So, all the conditions of Theorem 9.3.1 are satisfied, and as a consequence we can ensure the convergence of method (9.1.7).

9.3.2 Local example 2

Let

$$X = Y = \mathbb{R}^3,$$

$$D = \bar{U}(0, 1),$$

and

$$x^* = \begin{pmatrix} 0 \\ 0 \\ 0 \end{pmatrix}.$$

Define function F on D for

$$w = \begin{pmatrix} x \\ y \\ z \end{pmatrix}$$

by

$$F(w) = \begin{pmatrix} e^x - 1 \\ \dfrac{e-1}{2}y^2 + y \\ z \end{pmatrix}.$$

Then, the Fréchet-derivative is given by

$$F'(v) = \begin{bmatrix} e^x & 0 & 0 \\ 0 & (e-1)y + 1 & 0 \\ 0 & 0 & 1 \end{bmatrix}.$$

Using the norm of the maximum of the rows, we see that

$$F'(x^*) = \text{diag}\{1, 1, 1\}.$$

In this case, we take the domain

$$\Omega = B(x_0, R),$$

where

$$x_0 = 0.5$$

and

$$R = 0.3.$$

Choosing

$$x_{-1} = 0.4,$$

we will now easily see that our (C^*) conditions are satisfied in order to ensure the convergence of the method (9.1.4). We define the functions

$$h_1(s, t) = t,$$

$$h_2(s, t) = s,$$

$$f_0(s, t) = \frac{e}{2}(s + t),$$

and

$$f_1(s, t) = e(s + t)$$

which satisfy the (H_3^*) and (H_4^*) conditions. Moreover, we define

$$\varphi(t) = et,$$

$$\psi(t) = 3et,$$

$$q(t) = \frac{\phi(t)}{1 - \varphi(t)} = \frac{3et}{1 - et},$$

and

$$p(t) = \int_0^1 f_1(\theta t + h_1(t,t), \theta t + h_2(t,t))d\theta + f_0(h_1(t,t), h_2(t,t)) - 1$$

$$= 10.8731t - 1.$$

The root of $p(t)$ is

$$r_1 = 0.0919699\ldots,$$

so denoting $r = 0.0919699\ldots$ and $\bar{r} = r$, it is easy to see that the following conditions are verified:

$$\varphi(r) = 0.25 < 1,$$

$$h_1(r,r) = \bar{r} \le \bar{r},$$

and

$$h_2(r,r) = r \le \bar{r}.$$

Finally, it is clear that $\bar{U}(x_0, \bar{r}) \subseteq U(x_0, R)$, and, to show that condition (H_7^*) is verified, we choose $r_1 = 0.1$ and obtain

$$\int_0^1 f_0(\theta r + (1-\theta)r_1, \theta r + (1-\theta)r_1)d\theta = 0.135914\ldots < 1.$$

So all the conditions of Theorem 9.3.2 are satisfied, and as a consequence we can ensure the convergence of method (9.1.4).

REFERENCES

[1] S. Amat, S. Busquier, J.M. Gutiérrez, On the local convergence of secant-type methods, Int. J. Comput. Math. 81 (8) (2004) 1153–1161.

[2] S. Amat, S. Busquier, Á.A. Magreñán, Reducing chaos and bifurcations in Newton-type methods, Abst. Appl. Anal. 2013 (2013) 726701, https://doi.org/10.1155/2013/726701.

[3] S. Amat, Á.A. Magreñán, N. Romero, On a family of two step Newton-type methods, Appl. Math. Comput. 219 (4) (2013) 11341–11347.

[4] S. Amat, J.A. Ezquerro, M.A. Hernández-Verón, On a Steffensen-like method for solving equations, Calcolo (2) (2016), https://doi.org/10.1007/s10092-015-0142-3.

[5] I.K. Argyros, A unifying local–semilocal convergence analysis and applications for two-point Newton-like methods in Banach space, J. Math. Anal. Appl. 298 (2004) 374–397.

[6] I.K. Argyros, Computational theory of iterative methods, in: C.K. Chui, L. Wuytack (Eds.), Stud. Comput. Math., Elsevier Science, San Diego USA, 2007.

[7] I.K. Argyros, A semilocal convergence analysis for directional Newton methods, Math. Comput. AMS 80 (2011) 327–343.

[8] I.K. Argyros, Weaker conditions for the convergence of Newton's method, J. Comp. 28 (2012) 364–387.

[9] I.K. Argyros, A. Cordero, Á.A. Magreñán, J.R. Torregrosa, On the convergence of a damped secant method with modified right-hand side vector, Appl. Math. Comput. 252 (2015) 315–323.

[10] I.K. Argyros, D. González, Á.A. Magreñán, A semilocal convergence for a uniparametric family of efficient secant-like methods, J. Funct. Spaces 2014 (2014) 467980, https://doi.org/10.1155/2014/467980.

[11] I.K. Argyros, J.A. Ezquerro, M.A. Hernández-Verón, S. Hilout, Á.A. Magreñán, Enlarging the convergence domain of secant-like methods for equations, Taiwanese J. Math. 19 (2) (2015) 629–652.

[12] I.K. Argyros, Á.A. Magreñán, Relaxed secant-type methods, Nonlinear Stud. 21 (3) (2014) 485–503.

[13] I.K. Argyros, Á.A. Magreñán, A unified convergence analysis for secant-type methods, J. Korean Math. Soc. 51 (6) (2014) 1155–1175.

[14] I.K. Argyros, Á.A. Magreñán, Expanding the applicability of the secant method under weaker conditions, Appl. Math. Comput. 266 (2015) 1000–1012.

[15] I.K. Argyros, Á.A. Magreñán, Extending the convergence domain of the secant and Moser method in Banach space, J. Comput. Appl. Math. 290 (2015) 114–124.

[16] J.E. Dennis, Toward a unified convergence theory for Newton-like methods, in: L.B. Rall (Ed.), Nonlinear Funct. Anal. App., Academic Press, New York, 1971, pp. 425–472.

[17] J.A. Ezquerro, M.A. Hernández-Verón, A.I. Velasco, An analysis of the semilocal convergence for secant-like methods, Appl. Math. Comput. 226 (C) (September 2015) 883–892.

[18] V.B. Gopalan, J.D. Seader, Application of interval Newton's method to chemical engineering problems, Reliab. Comput. 1 (3) (1995) 215–223.

[19] J.M. Gutiérrez, Á.A. Magreñán, N. Romero, On the semilocal convergence of Newton–Kantorovich method under center-Lipschitz conditions, Appl. Math. Comput. 221 (2013) 79–88.

[20] M.A. Hernández, M.J. Rubio, J.A. Ezquerro, Solving a special case of conservative problems by secant-like method, Appl. Math. Comput. 169 (2005) 926–942.

[21] M.A. Hernández, M.J. Rubio, J.A. Ezquerro, Secant-like methods for solving nonlinear integral equations of the Hammerstein type, J. Comput. Appl. Math. 115 (2000) 245–254.

[22] V.A. Kurchatov, On a method of linear interpolation for the solution of functional equations, Dokl. Akad. Nauk SSSR 198 (3) (1971) 524–526 (in Russian). Translation in: Soviet Math. Dokl. 12 (1974) 835–838.

[23] P. Laarsonen, Ein überquadratisch konvergenter iterativer Algorithmus, Ann. Acad. Sci. Fenn. Ser. I 450 (1969) 1–10.

[24] Á.A. Magreñán, Estudio de la dinámica del método de Newton, amortiguado (PhD thesis), Servicio de Publicaciones, Universidad de La Rioja, 2013, http://dialnet.unirioja.es/servlet/tesis?codigo=38821.

[25] Á.A. Magreñán, A new tool to study real dynamics: the convergence plane, Appl. Math. Comput. 248 (2014) 215–224.

[26] Á.A. Magreñán, I.K. Argyros, New semilocal and local convergence analysis for the secant method, Appl. Math. Comput. 262 (2015) 298–307.

[27] Á.A. Magreñán, I.K. Argyros, Expanding the applicability of secant method with applications, Bull. Korean Math. Soc. 52 (3) (2015) 865–880.

[28] Á.A. Magreñán, I.K. Argyros, New improved convergence analysis for the secant method, Math. Comput. Simulation 119 (2016) 161–170.

[29] Á.A. Magreñán, I.K. Argyros, Optimizing the applicability of a theorem by F. Potra for Newton-like methods, Appl. Math. Comput. 242 (2014) 612–623.

[30] F.A. Potra, An error analysis for the secant method, Numer. Math. 38 (1982) 427–445.

[31] F.A. Potra, Sharp error bounds for a class of Newton-like methods, Lib. Math. 5 (1985) 71–84.

[32] F.A. Potra, V. Pták, Nondiscrete Induction and Iterative Processes, Pitman, New York, 1984.

[33] J.W. Schmidt, Untere Fehlerschranken für Regula–Falsi Verhafren, Period. Hungar. 9 (1978) 241–247.

[34] A.S. Sergeev, On the method of chords, Sibirsk. Math. J. 11 (1961) 282–289 (in Russian).

[35] M. Shacham, An improved memory method for the solution of a nonlinear equation, Chem. Eng. Sci. 44 (1989) 1495–1501.

[36] S.M. Shakhno, On a Kurchatov's method of linear interpolation for solving nonlinear equations, PAMM-Proc. Appl. Math. Mech. 4 (2004) 650–651.

[37] S. Ulm, Majorant principle and the method of chords, Izv. Akad. Nauk Eston. SSR, Ser. Fiz.-Mat. 13 (1964) 217–227 (in Russian).

[38] T. Yamamoto, A convergence theorem for Newton-like methods in Banach spaces, Numer. Math. 51 (1987) 545–557.

Chapter 10

Robust convergence of Newton's method for cone inclusion problem

10.1 INTRODUCTION

In this chapter we consider the problem of approximately solving nonlinear inclusion problem of the form

$$\text{find } x \text{ such that } F(x) \in C, \qquad (10.1.1)$$

where C is a nonempty closed convex cone in a Banach space $\mathbb{Y}$ and $F : \mathbb{X} \longrightarrow \mathbb{Y}$ is a nonlinear function between the Banach spaces $\mathbb{X}$ and $\mathbb{Y}$. The non-linear equations in Banach spaces are solved widely using Newton's method and its variants; see [4–8,12–17,21–28,31–33] for details. Robinson in [31] generalized Newton's method for solving (10.1.1) in the special case in which C is the degenerate cone $\{0\} \subset \mathbb{Y}$. Lipschitz continuity or Lipschitz-like continuity of the derivative of the nonlinear operator in question is usually used in convergence analysis of Newton's type method.

There are plethora of works, written by relevant authors, dealing with the convergence of analysis of Newton-like methods by relaxing the assumption of Lipschitz type continuity of the derivative of the operator involved [4–8,12–17, 21–28,31–33]. Moreover, the authors of this chapter have presented different works related to the convergence of Newton-like methods [1–3,9–11,18–20,29, 30].

This chapter uses the idea of restricted convergence domains to present a convergence analysis of Newton's method for solving a nonlinear inclusion problems of the form (10.1.1). This analysis relaxes the Lipschitz type continuity of the derivative of the operator involved. The main aim of the analysis is to find a larger convergence domain for the Newton's method for solving (10.1.1). A finer convergence analysis is obtained using the restricted convergence domains, with the following advantages, under the same computational cost:

- The error estimates on the distances involved are tighter.
- The information on the location of the solution is at least as precise.

A Contemporary Study of Iterative Methods. DOI: 10.1016/B978-0-12-809214-9.00010-3

135

10.2 SEMILOCAL ANALYSIS FOR NEWTON'S METHOD

In this section we are concerned with the problem of solving the equation

$$F(x) \in C, \qquad (10.2.1)$$

where $F : D \subseteq \mathbb{X} \to \mathbb{Y}$ is a nonlinear function which is continuously differentiable, D is an open set, and $C \subset \mathbb{Y}$ a nonempty closed convex cone. Recall that [31] a nonlinear continuously Fréchet differentiable function $F : D \to \mathbb{Y}$ satisfies Robinson's condition at $\bar{x} \in D$ if

$$rge\, T_{\bar{x}} = \mathbb{Y}, \qquad (10.2.2)$$

where $T_{\bar{x}} : \mathbb{X} \rightrightarrows \mathbb{Y}$ is a sublinear mapping as defined in [31]. Let $R > 0$ be a scalar constant. Set $\rho^* := \sup\{t \in [0, R) : U(x_0, t) \subseteq D\}$. A continuously differentiable function $f_0 : [0, R) \to \mathbb{R}$ is a center-majorant function at a point $\bar{x} \in D$ for a continuously differentiable function $F : D \to Y$ if

$$\|T_{\bar{x}}^{-1}[F'(x) - F'(\bar{x})]\| \leq f_0'(\|x - \bar{x}\|) - f_0'(0) \qquad (10.2.3)$$

for each $x \in U(\bar{x}, R)$ and satisfies the following conditions:

(h_1^0) $f_0(0) > 0$, $f_0'(0) = -1$;
(h_2^0) $f'(0)$ is convex and strictly increasing;
(h_3^0) $f_0(t) = 0$ for some $t \in (0, R)$.

Then sequence $\{t_n^0\}$ generated by $t_0^0 = 0$,

$$t_{k+1}^0 = t_k^0 - \frac{f_0(t_k)}{f_0'(t_k)}, \quad k = 0, 1, \ldots \qquad (10.2.4)$$

is well defined, strictly increasing, remains in $(0, t_0^*)$, and converges to t_0^*, where t_0^* is the smallest zero of function f_0 in $(0, R)$. Set $D_1 := \bar{U}(\bar{x}, \rho^*) \cap U(\bar{x}, t_0^*)$. Suppose that there exists $f_1 : [0, \rho_1) \to \mathbb{R}$, $\rho_1 = \min\{\rho^*, t_0^*\}$ such that

$$\|T_{\bar{x}}^{-1}[F'(y) - F'(x)]\| \leq f_1'(\|y - x\| + \|x - \bar{x}\|) - f_1'(\|x - \bar{x}\|) \qquad (10.2.5)$$

for each $x, y \in D_1$ and satisfying

(h_1) $f_1(0) > 0$, $f_1'(0) = -1$;
(h_2) f_1' is convex and strictly increasing;
(h_3) $f_1(t) = 0$ for some $t \in (0, R)$;
(h_4) $f_0(t) \leq f_1(t)$ and $f_0'(t) \leq f_1'(t)$ for each $t \in [0, \rho_1)$.

Throughout this chapter we assume that the above "h" conditions hold. Moreover, we impose the following condition on the majorant condition f, since it is required, which is considered valid only when explicitly stated.

(h_5) $f_1(t) < 0$ for some $t \in (0, R)$.

Remark 10.2.1. *Since $f_1(0) > 0$ and f_1 is continuous, (h_4) implies condition (h_3).*

We can state now the following theorem.

Theorem 10.2.2. *Let $\mathbb{X}$, $\mathbb{Y}$ be Banach spaces, with $\mathbb{X}$ being reflexive, $D \subseteq \mathbb{X}$ an open set, $F : D \to \mathbb{Y}$ a continuously Fréchet differentiable function, $C \subset \mathbb{Y}$ a nonempty closed convex cone. Suppose $\bar{x} \in D$, F satisfies Robinson's condition at $\bar{x}$, f_0 is a center-majorant function, f_1 is a majorant function for F at $\bar{x}$, and*

$$\|T_{\bar{x}}^{-1}(-F(\bar{x}))\| \le f_1(0). \tag{10.2.6}$$

Then f_1 has the smallest zero $t_* \in (0, R)$, the sequences generated by Newton's method for solving the inclusion $F(x) \in C$ and equation $f(t) = 0$, with starting points $x_0 = \bar{x}$ and $t_0 = 0$, respectively,

$$x_{k+1} \in x_k + \text{argmin}\{\|d\| : F(x_k) + F'(x_k)d \in C\},$$

$$t_{k+1} = t_k - \frac{f_1(t_k)}{f_1'(t_k)}, \quad k = 0, 1, \ldots, \tag{10.2.7}$$

are well defined, $\{x_k\}$ is a constrained in $B(\bar{x}, t_*)$, $\{t_k\}$ is strictly increasing, constrained in $[0, t_*)$, converges to t_* and together satisfy the inequalities:

$$\|x_{k+1} - x_k\| \le -\frac{f_1(t_k) - f_1(t_{k-1}) - f_1'(t_{k-1})(t_k - t_{k-1})}{f_0'(t_k)}\left(\frac{\|x_k - x_{k-1}\|}{t_k - t_{k-1}}\right)^2$$

$$\le -\frac{f_1(t_k)}{f_1'(t_k)}\left(\frac{\|x_k - x_{k-1}\|}{t_k - t_{k-1}}\right)^2,$$

$$\|x_{k+1} - x_k\| \le t_{k+1} - t_k,$$

$$\|x_{k+1} - x_k\| \le \frac{t_{k+1} - t_k}{(t_k - t_{k-1})^2}\|x_k - x_{k-1}\|^2, \tag{10.2.8}$$

for all $k = 0, 1, \ldots$, and $k = 1, 2, \ldots$, respectively. Moreover, $\{x_k\}$ converges to $x_* \in B[\bar{x}, t_*]$ such that $F(x_*) \in C$,

$$\|x_* - x_k\| \le t_* - t_k, \ t_* - t_{k+1} \le \frac{1}{2}(t_* - t_k), \quad k = 0, 1, \ldots, \tag{10.2.9}$$

and therefore $\{t_k\}$ converges Q-linearly to t_* and $\{x_k\}$ converges R-linearly to x_*. If, additionally, f_1 satisfies (h_4) then the following inequalities hold:

$$\|x_{k+1} - x_k\} \le \frac{D^- f_1'(t^*)}{-2f_0'(t^*)}\|x_k - x_{k-1}\|^2,$$

$$t_{k+1} - t_k \le \frac{D^- f_1'(t^*)}{-2f_0'(t^*)}(t_k - t_{k-1})^2, \tag{10.2.10}$$

$$k = 1, 2, \ldots,$$

and, consequently, $\{x_k\}$ and $\{t_k\}$ converge Q-quadratically to x_* and t_*, respectively, as follows:

$$\limsup_{k \to \infty} \frac{\|x_* - x_{k+1}\|}{\|x_* - x_k\|^2} \le \frac{D^- f_1'(t^*)}{-2 f_0'(t_*)},$$

$$t_* - t_{k+1} \le \frac{D^- f_1'(t_*)}{-2 f_0'(t_*)} (t_* - t_k)^2, \tag{10.2.11}$$

$$k = 1, 2, \ldots$$

We will use the above result to prove a robust semilocal affine invariant theorem for Newton's method for solving a nonlinear inclusion of the form as in [27].

Now we can state our main result of the chapter.

Theorem 10.2.3. *Let $\mathbb{X}$, $\mathbb{Y}$ be Banach spaces, with $\mathbb{X}$ being reflexive, $D \subseteq \mathbb{X}$ an open set, $F : D \to \mathbb{Y}$ a continuously Fréchet differentiable function, $C \subset \mathbb{Y}$ a nonempty closed convex cone, $R > 0$ and $f : [0, R) \to R$ a continuously differentiable function. Suppose $\bar{x} \in D$, F satisfies Robinson's condition at $\bar{x}$, f_0 is a center-majorant function, f_1 is a majorant function for F at $\bar{x}$ satisfying (h_5), and*

$$\|T_{\bar{x}}^{-1}(-F(\bar{x}))\| \le f_1(0). \tag{10.2.12}$$

Define $\beta := \sup\{-f(t) : t \in [0, R]\}$. Let $0 < \rho < \beta/2$ and consider $g : [0, R - \rho) \to \mathbb{R}$, defined by

$$g(t) = \frac{-1}{f_0'(\rho)} [f(t + \rho) + 2\rho]. \tag{10.2.13}$$

Then g has a smallest zero $t_{*,\rho} \in (0, R - \rho)$, the sequences generated by Newton's method for solving the inclusion $F(x) \in C$ and equation $g(t) = 0$, with starting points $x_0 \in B(\bar{x}, \rho)$ and $t_0 = 0$, respectively,

$$x_{k+1} \in x_k + \operatorname{argmin}\{\|d\| : F(x_k) + F'(x_k)d \in C\},$$

$$t_{k+1} = t_k - \frac{g(t_k)}{g'(t_k)}, \quad k = 0, 1, \ldots, \tag{10.2.14}$$

are well defined, $\{x_k\}$ is constrained in $B(\bar{x}, t_{*,\rho})$, $\{t_k\}$ is strictly increasing, is contained in $[0, t_{*,\rho})$, and converges to $t_{*,\rho}$, and together they satisfy the inequalities:

$$\|x_{k+1} - x_k\| \le t_{k+1} - t_k, \quad k = 0, 1, \ldots, \tag{10.2.15}$$

$$\|x_{k+1} - x_k\| \leq \frac{t_{k+1} - t_k}{t_k - t_{k-1}^2} \|x_k - x_{k-1}\|^2$$

$$\leq \frac{D^- g'(t_{*,\rho})}{-2g'(t_{*,\rho})} \|x_k - x_{k-1}\|^2, \quad k = 1, 2, \ldots$$

(10.2.16)

Moreover, $\{x_k\}$ converges to $x_* \in B[\bar{x}, t_{*,\rho})$ such that $F(x_*) \in C$, satisfying the inequalities

$$\|x_* - x_k\| \leq t_{*,\rho} - t_k, \quad t_{*,\rho} \leq \frac{1}{2}(t_{*,\rho} - t_k), \quad k = 0, 1, \ldots, \quad (10.2.17)$$

and the convergence of $\{x_k\}$ and $\{t_k\}$ to x_* and $t_{*,\rho}$, respectively, is quadratic as follows:

$$\limsup_{k \to \infty} \frac{\|x_* - x_{k+1}\|}{\|x_* - x_k\|^2} \leq \frac{D^- g'(t_{*,\rho})}{-2g'(t_{*,\rho})},$$

$$t_{*,\rho} - t_{k+1} \leq \frac{D^- g'(t_{*,\rho})}{-2g'(t_{*,\rho})}(t_{*,\rho} - t_k)^2, \ k = 0, 1, \ldots$$

(10.2.18)

Based on the main theorem, we can now present the following remarks.

Remark 10.2.4. *It is easy to see that the inequalities (10.2.3) and (10.2.12) are well defined.*

Remark 10.2.5. *The definition of the sequence $\{x_k\}$ in (10.2.7) is equivalent to the conditions*

$$x_{k+1} - x_k \in T_{x_k}^{-1}(-F(x_k)) \text{ and } \|x_{k+1} - x_k\| = \|T_{x_k}^{-1}(-F(x_k))\|, \ k = 0, 1, \ldots$$

Remark 10.2.6. *Theorems 10.2.2 and 10.2.3 are affine invariant in the following sense. Letting $A : \mathbb{Y} \to \mathbb{Y}$ be an invertible continuous linear mapping, $\bar{F} := A \circ F$ and the set $\bar{C} := A(C)$, the corresponding inclusion problem (10.2.1) is given by*

$$\bar{F}(x) \in \bar{C},$$

and the convex process associated is denoted by $\bar{T}_{\bar{x}}d = \bar{F}'(\bar{x})d - \bar{C}$. Then $\bar{T}_{\bar{x}} = A \circ T_{\bar{x}}$ and $\bar{T}_{\bar{x}}^{-1} = T_{\bar{x}}^{-1} \circ A^{-1}$. Moreover, we have the conditions $rge \ \bar{T}_{\bar{x}} = Y$,

$$\|\bar{T}_{\bar{x}}^{-1}(-\bar{F}(\bar{x}))\| \leq f_1(0),$$

and the affine majorant condition (Lipschitz-like condition)

$$\|\bar{T}_{\bar{x}}^{-1}[\bar{F}'(y) - \bar{F}'(x)]\| \leq f_1'(\|y - x\| + \|x - \bar{x}\|) - f_1'(\|x - \bar{x}\|),$$

for $x, y \in B(\bar{x}, R)$ such that $\|x - \bar{x}\| + \|y - x\| < R$. Therefore the assumptions in Theorems 10.2.2 and 10.2.3 are insensitive with respect to invertible continuous linear mappings. Note that such property does not hold in [31].

Remark 10.2.7. (a) *Suppose that*

$$\|T_{\bar{x}}^{-1}[F'(y) - F'(x)]\| \leq f'(\|y - x\| + \|x - \bar{x}) - f'(\|x - \bar{x})$$

(10.2.19)

holds for each $x, y \in \bar{U}(\bar{x} - \rho^*)$ *and some continuously differentiable function* $f : [0, R] \to \mathbb{R}$ *satisfying conditions* (h_1)–(h_3) *and* (h_5). *Then if* $f_0(t) = f(t) = f_1(t)$ *for each* $t \in [0, R)$, *Theorems 10.2.2 and 10.2.3 reduce to Theorems 7 and 8, respectively, in [27]. Notice, however, that for each* $t \in [0, R)$,

$$f_0' \leq f'(t),$$

(10.2.20)

$$f_1'(t) \leq f'(t),$$

(10.2.21)

hold in general. Therefore, if a strict inequality holds in (10.2.20) *or* (10.2.21) *then the advantages as stated in the abstract of this chapter hold.*
(b) *Let* $t^* := \sup\{t \in [0, R) : f_0'(t) < 0\}$. *Define* $D_1^* = \bar{U}(\bar{x}, \rho^*) \cap U(\bar{x}, \bar{t})$. *Then the conclusion of the two proceeding theorems hold with* D_1^*, t^* *replacing* D_1, t_0^*, *respectively. In this case we do not have to define sequence* $\{t_k^0\}$. *Notice, however, that the function* f_1^* *satisfying* (10.2.5) *will be different in general from* f_1.

Related to the proofs of the preceding theorems, we need the following two auxiliary results:

Lemma 10.2.8. *If* $\|x - \bar{x}\| \leq t < \bar{t}$ *then* $\text{dom}\,(T_x^{-1}F'(\bar{x})) = \mathbb{X}$ *and*

$$\|T_x^{-1}F'(\bar{x})\| \leq -1/f_0'(t).$$

(10.2.22)

Consequently, $T_x = \mathbb{Y}$.

Proof. Take $0 \leq t < \bar{t}$ and $x \in B[\bar{x}, t]$. Since f_0 is a center-majorant function for F at $\bar{x}$, using (10.2.3), (h_2), and (h_1), we obtain

$$\|T_{\bar{x}}^{-1}[F'(x) - F'(\bar{x})]\| \leq f_0'(\|x - \bar{x}\|) - f'(0)$$

(10.2.23)

$$\leq f_0'(t) - f_0'(0)$$

(10.2.24)

$$\leq f_0'(t) + 1 < 1.$$

(10.2.25)

To simplify the notation, define $S = T_{\bar{x}}^{-1}[F'(x) - F'(\bar{x})]$. Since $[F'(x) - F'(\bar{x})]$ is a continuous linear mapping and $T_{\bar{x}}^{-1}$ is a sublinear mapping with closed graph, it is easy to see that S is a sublinear mapping with closed graph. Moreover, by assumption $rge T_{\bar{x}} = \mathbb{Y}$, we have $\text{dom}\,S = \mathbb{X}$. Because S is a closed graph, it is easy to see that $(S + I)(x)$ also has a closed graph for all $x \in X$, where I is the identity mapping on $\mathbb{X}$. We conclude that $rge(T_{\bar{x}}^{-1}[F'(x) - F'(\bar{x})] + I) = \mathbb{X}$ and

$$\|(T_{\bar{x}}^{-1}[F'(x) - F'(\bar{x})] + I)^{-1}\| \leq \frac{1}{1 - (f_0'(t) + 1)} = \frac{1}{-f_0'(t)}. \qquad (10.2.26)$$

The rest of the proof is similar to Proposition 12 in [27]. $\qquad\qquad\square$

Remark 10.2.9. *Newton's method at a point $x \in D$ is a solution of the linearizion of the inclusion $F(y) \in C$ at such a point, namely, a solution of the linear inclusion $F(x) + F'(x)(x - y) \in C$. So we study the linearization error of F at a point in D*

$$E(x, y) := F(y) - [F(x) + F'(x)(y - x)], \quad y, x \in D. \qquad (10.2.27)$$

We will bound this error by the error in the linearization of the majorant function f,

$$e(t, s) := f(s) - [f(t) + f'(t)(s - t)], \quad t, s \in [0, R). \qquad (10.2.28)$$

Lemma 10.2.10. *Let $R > 0$ and let $f : [0, R) \to R$ be a continuously differentiable function. Suppose that $\bar{x} \in D$, f is a majorant function for F at $\bar{x}$ and satisfies (h_4). If $0 \leq \rho \leq \beta/2$, where $\beta := \sup\{-f_1(t) : t \in [0, R)\}$ then for any $\hat{x} \in B(\bar{x}, \rho)$ the scalar function $g : [0, R - \rho) \to \mathbb{R}$,*

$$g(t) = \frac{-1}{f_0'(\rho)}[f_1(t + \rho) + 2\rho],$$

is a majorant function for F at $\hat{x}$ and also satisfies condition (h_4).

Proof. Since the domain of f_1 is $[0, R)$ and $f_1'(\rho) \neq 0$, we conclude that g is well defined. First we will prove that function g satisfies conditions (h_1), (h_2), (h_3), and (h_4). We trivially have that $g'(0) = -1$. Since f_1 is convex, combining this with (h_1), we have $f_1(t) + t \geq f_1(0) > 0$, for all $0 \leq t < R$, which implies $g(0) = -[f_1(\rho) + 2\rho]/f_0'(\rho) > 0$, hence g satisfies (h_1) and (h_2). Now, as $\rho < \beta/2$, we have

$$\lim_{t \to \bar{t} - \rho} = \frac{-1}{f_0'(\rho)}(2\rho - \beta) < 0, \qquad (10.2.29)$$

which implies that g satisfies (h_4) and, as g is continuous and $g(0) > 0$, it also satisfies (h_3). It remains only to prove that g satisfies (h_2). First of all, note that for any $\hat{x} \in B(\bar{x}, \rho)$, we have $\|\hat{x} - \bar{x}\| < \rho < \bar{t}$ and, by using Lemma 10.2.8, we get

$$\|T_{\bar{x}}^{-1} F'(\bar{x})\| \leq \frac{-1}{f_0'(\rho)}. \qquad (10.2.30)$$

Because $B(\bar{x}, R) \subseteq D$, for any $\hat{x} \in B(\bar{x}, \rho)$ we trivially have $B(\hat{x}, R - \rho) \subset D$. Now take $x, y \in \mathbb{X}$ such that $x, y \in B(\bar{x}, R - \rho)$, $\|x - \hat{x}\| + \|y - x\| < R - \rho$. Hence $x, y \in B(\bar{x}, R)$ and $\|x - \bar{x}\| + |y - x| < R$. Thus, we have

$$\|T_{\hat{x}}^{-1}[F'(y) - F'(x)]\| \leq \|T_{\hat{x}}^{-1}F'(\bar{x})T_{\bar{x}}^{-1}[F'(y) - F'(x)]\|$$
$$\leq \|T_{\hat{x}}^{-1}F'(x)\| \|T_{\bar{x}}^{-1}[F'(y) - F'(x)]\|$$
$$\leq \frac{-1}{f_0'(\rho)}[f'(\|y - x\| + \|x - \bar{x}\|) - f'(\|x - \bar{x}\|)].$$

On the other hand, since f' is convex, the function $s \mapsto f_1'(t + s) - f_1'(s)$ is increasing for $t \geq 0$ and, as $\|x - \bar{x}\| \leq \|x - \hat{x}\| + \rho$, we conclude that

$$f_1'(\|y - x\| + \|x - \bar{x}\|) - f_1'(\|x - \bar{x}\|)$$
$$\leq f_1'(\|y - x\| + \|x - \hat{x}\| + \rho) - f_1'(\|x - \hat{x}\| + \rho).$$

Now, using the two above inequalities, together with the definition of the function g, we get

$$\|T_{\hat{x}}^{-1}[F'(y) - F'(x)]\| \leq g'(\|y - x\| + \|x - \hat{x}\|) - g'(\|x - \hat{x}\|),$$

implying that the function g satisfies (10.2.5). ☐

Remark 10.2.11. *If $f_0(t) = f(t) = f_1(t)$ for each $t \in [0, R)$, then the last two lemmas reduce to Propositions 12 and 17, respectively, in [27]. Otherwise, in view of (10.2.20) and (10.2.21), the new lemmas constitute an improvement.*

Proof of Theorems. Simply notice that the iterates $\{x_k\}$ lie in D_1 which is a more precise location than $\bar{U}(\bar{x}, \rho)$ used in [27], since $D_1 \subseteq \bar{U}(\bar{x}, \rho^*)$. Then, follow the proofs in [27] using f_1, Lemma 10.2.9, Lemma 10.2.11 instead of f, Proposition 12, and Proposition 17, respectively. ☐

10.3   SPECIAL CASES AND A NUMERICAL EXAMPLE

We present a robust semilocal convergence result in Newton's method for solving nonlinear inclusion problem using Lipschitz-like condition. In particular, Theorems 10.2.2 and 10.2.3, reduce respectively to:

Theorem 10.3.1. *Let $\mathbb{X}$, $\mathbb{Y}$ be Banach spaces, with $\mathbb{X}$ being reflexive, and $D \subseteq \mathbb{X}$ be an open set. Let also $F : D \to \mathbb{Y}$ be a continuously differentiable operator and $C \subset \mathbb{Y}$ a nonempty closed convex cone. Let $\bar{x} \in D$, $L_0 > 0$, $L_1 > 0$, and $R > 0$. Suppose that F satisfies Robinson's condition at $\bar{x}$,*

$$\|T_{\bar{x}}^{-1}[F'(x) - F'(\bar{x})]\| \leq L_0\|x - \bar{x}\| \quad \text{for each } x \in U(\bar{x}, R), \quad (10.3.1)$$
$$\|T_{\bar{x}}^{-1}[F'(x) - F'(\bar{x})]\| \leq L_1\|x - y\| \quad \text{for each } x, y \in D_1, \quad (10.3.2)$$
$$\|T_{\bar{x}}^{-1}(-F(\bar{x}))\| \leq \eta,$$

and

$$h_1 = 2L_1\eta \le 1. \tag{10.3.3}$$

Define $f_1 : [0, +\infty) \to \mathbb{R}$ by $f_1(t) := \frac{L_1}{2}t^2 - t + n$ and $t_* := \frac{1-\sqrt{1-h_1}}{L_1}$. Then the sequences generated by Newton's method for solving the inclusion $F(x) \in C$ and equation $f_1(t) = 0$, with starting points $x_0 = \bar{x}$ and t_0, respectively,

$$x_{k+1} \in x_k + \operatorname{argmin}\{\|d\| : F(x_k) + F'(x_k)d \in C\},$$

$$t_{k+1} = t_k - \frac{f_1(t_k)}{f_1'(t_k)},$$

are well defined, $\{x_k\}$ is contained in $B(\bar{x}, t^*)$, $\{t_k\}$ is strictly increasing, is contained in $[0, t_*)$, and converges to t_*. Moreover, the following estimates hold:

$$\|x_{k+1} - x_k\| \le \frac{L_1\|x_k - x_{k-1}\|^2}{2(1 - L_0\|x_k - \bar{x}\|)}. \tag{10.3.4}$$

Furthermore, $\{x_k\}$ converges to $x_* \in B[\bar{x}, t_*, \rho)$ such that $F(x_*) \in C$.

Proof. It is easy to see that function f_1 is a majorant function for F at $\bar{x}$ and function $f_0(t) = \frac{L_0}{2}t^2 - t + n$ is a center-majorant function for F at $\bar{x}$. The rest follows from the proof of Theorem 10.2.2. $\square$

Theorem 10.3.2. *Under the hypotheses of Theorem 10.3.1, further suppose that*

$$0 \le \rho < B := \frac{(1 - 2L_1n)}{4L_1}.$$

Define $g : [0, +\infty) \to \mathbb{R}$ by

$$g(t) = \frac{-1}{L_0\rho - 1}\left[\frac{L_1}{2}(t + \rho)^2 - (t + \rho) + n + 2\rho\right].$$

Denote by $t_{*,\rho}$ the smallest zero of g. Then the sequences generated by Newton's method for solving the inclusion $F(x) \in C$ and equation $g(t) = 0$ with starting point $x_0 = \bar{x}$ and for any $\hat{x} \in B(\bar{x}, \rho)$ and $t_0 = 0$, respectively, defined by

$$x_{k+1} = x_k + \operatorname{argmin}\{\|d\| : F(x_k) + F'(x_k)d \in C\},$$

$$t_{k+1} = t_k - \frac{g(t_k)}{g'(t_k)},$$

are well defined, $\{x_k\}$ is contained in $B(\bar{x}, t_{*,\rho})$, converges to $x_* \in B[\bar{x}, t_{*,\rho})$ with $F(x_*) \in C$, while $\{t_k\}$ is strictly increasing, is contained in $[0, t_{*,\rho})$, and converges to $t_{*,\rho}$.

Proof. The proof follows from the proof of Theorem 10.2.2 using the special choices of the functions f_0 and f_1. $\square$

Remark 10.3.3. *Let* $f(t) = \frac{L}{2}t^2 - t + n$. *Then the corresponding Kantorovich condition to* (10.3.3) *given in [27] is* $h_k = 2L\eta \leq 1$.

However, we have that

$$L_1 \leq L, \tag{10.3.5}$$

so

$$h_k \leq 1 \Rightarrow h \leq 1, \tag{10.3.6}$$

but not necessarily vice versa, unless $L_1 = L$. It follows from the proof of Theorem 10.3.1 that the following sequence $\{s_n\}$ defined by

$$s_0 = 0, \quad s_1 = n,$$

$$s_2 = s_1 + \frac{L_0(s_1 - s_0)^2}{2(1 - L_0 s_1)}, \tag{10.3.7}$$

$$s_{n+2} = s_{n+1} + \frac{L_1(s_{n+1} - s_n)^2}{2(1 - L_0 s_{n+1})}$$

is also majorizing for the sequence $\{x_n\}$ which certainly converges if (10.3.3) holds. Moreover, we have that

$$t_0 = 0, \quad t_1 = n,$$

$$t_{n+1} = t_n + \frac{L_1(t_n - t_{n-1})^2}{2(1 - L_0 t_n)}, \tag{10.3.8}$$

and

$$u_0 = 0, \quad u_1 = n,$$

$$u_{n+1} = u_n + \frac{L(u_n - u_{n-1})^2}{2(1 - L u_n)}. \tag{10.3.9}$$

Sequence $\{u_n\}$ was used in [27]. Then we have for $L_0 < L_1 < L$ that

$$s_n < t_n < u_n, \quad n = 2, 3, \ldots, \tag{10.3.10}$$

$$s_{n+1} - s_n < t_{n+1} - t_n < u_{n+1} - u_n, \quad n = 1, 2, \ldots, \tag{10.3.11}$$

and

$$s_* = \lim_{n \to \infty} s_n \leq t_* = \lim_{n \to \infty} t_n \leq u_* = \lim_{n \to \infty} u_n. \tag{10.3.12}$$

Furthermore, we have that sequence $\{s_k\}$ converges, provided that

$$h_1 = \bar{L}\eta \leq 1, \tag{10.3.13}$$

where

$$\bar{L} = \frac{1}{4}\left(4L_0 + \sqrt{L_1 L_0} + \sqrt{L_1 L_0 + 8L_0^2}\right).$$

It follows from (10.3.3) and (10.3.12) that

$$h_k \leq 1 \Rightarrow h \leq 1 \Rightarrow h_1 \leq 1. \tag{10.3.14}$$

Estimates (10.3.10)–(10.3.12) and (10.3.13) justify the advantages of our new approach over that in [27]. Notice also that these advantages are obtained under the same computational cost, since in practice the computation of the function f requires the computation of f_0 or f_1 as special cases. Next we present an academic example to show that $L_0 < L_1 < L$, so that the aforementioned advantages will hold.

Example 10.3.4. *Let*

$$X = Y = \mathbb{R},$$

$$\bar{x} = 1,$$

$$D = U(1, 1 - q),$$

$$q \in \left[0, \frac{1}{2}\right)$$

and define function F on D by

$$F(x) = x^3 - q.$$

Then we get that

$$\eta = \frac{1 - q}{3},$$

$$L_0 = 3 - q,$$

$$L = 2(2 - q),$$

and

$$L_1 = 2(1 + \frac{1}{L_0}).$$

So

$$L_0 < L_1 < L$$

holds for each

$$q \in \left[0, \frac{1}{2}\right).$$

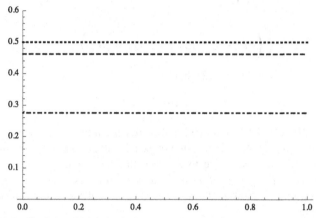

FIGURE 10.1 Condition (10.3.3) holds for each q between the dashed and the dotted line and condition (10.3.13) holds for each q between the dot-dashed and the dotted line.

However, the old Kantorovich convergence condition is not satisfied, since $h_k > 1$ *for each*

$$q \in \left[0, \frac{1}{2}\right).$$

However, conditions (10.3.3) and (10.3.13) hold respectively for

$$q \in [0.4620, 0.5]$$

and

$$q \in [0.2757, 0.5].$$

Hence, our results can apply, see Fig. 10.1.

Hence there is no guarantee that sequence $\{x_k\}$ converges according to Theorem 18 in [27]. We leave the details to the motivated readers. Our results can also improve along the same lines the corresponding ones given in [32] concerning Smale's α-theory and Wang's theory [33].

REFERENCES

[1] S. Amat, S. Busquier, C. Bermúdez, Á.A. Magreñán, Expanding the applicability of a third order Newton-type method free of bilinear operators, Algorithms 8 (3) (2015) 669–679.

[2] S. Amat, S. Busquier, C. Bermúdez, Á.A. Magreñán, On the election of the damped parameter of a two-step relaxed Newton-type method, Nonlinear Dynam. 84 (1) (2016) 9–18.

[3] S. Amat, Á.A. Magreñán, N. Romero, On a two-step relaxed Newton-type method, Appl. Math. Comput. 219 (24) (2013) 11341–11347.

[4] I.K. Argyros, Computational theory of iterative methods, in: K. Chui, L. Wuytack (Eds.), Stud. Comput. Math., vol. 15, Elsevier, New York, U.S.A., 2007.

[5] I.K. Argyros, Concerning the convergence of Newton's method and quadratic majorants, J. Appl. Math. Comput. 29 (2009) 391–400.

[6] I.K. Argyros, A semilocal convergence analysis for directional Newton methods, Math. Comp. 80 (2011) 327–343.

[7] I.K. Argyros, Y.J. Cho, S. Hilout, Numerical Methods for Equations and Its Applications, CRC Press/Taylor and Francis Group, New York, 2012.

[8] I.K. Argyros, D. González, Local convergence analysis of inexact Gauss–Newton method for singular systems of equations under majorant and center-majorant condition, SeMA J. 69 (1) (2015) 37–51.

[9] I.K. Argyros, J.A. Ezquerro, M.A. Hernández-Verón, Á.A. Magreñán, Convergence of Newton's method under Vertgeim conditions: new extensions using restricted convergence domains, J. Math. Chem. 55 (7) (2017) 1392–1406.

[10] I.K. Argyros, S. González, Á.A. Magreñán, Majorizing sequences for Newton's method under centred conditions for the derivative, Int. J. Comput. Math. 91 (12) (2014) 2568–2583.

[11] I.K. Argyros, J.M. Gutiérrez, Á.A. Magreñán, N. Romero, Convergence of the relaxed Newton's method, J. Korean Math. Soc. 51 (1) (2014) 137–162.

[12] I.K. Argyros, S. Hilout, On the Gauss–Newton method, J. Appl. Math. (2010) 1–14.

[13] I.K. Argyros, S. Hilout, Extending the applicability of the Gauss–Newton method under average Lipschitz-conditions, Numer. Alg. 58 (2011) 23–52.

[14] I.K. Argyros, S. Hilout, On the solution of systems of equations with constant rank derivatives, Numer. Alg. 57 (2011) 235–253.

[15] I.K. Argyros, S. Hilout, Improved local convergence of Newton's method under weaker majorant condition, J. Comput. Appl. Math. 236 (7) (2012) 1892–1902.

[16] I.K. Argyros, S. Hilout, Weaker conditions for the convergence of Newton's method, J. Complexity 28 (2012) 364–387.

[17] I.K. Argyros, S. Hilout, Computational Methods in Nonlinear Analysis, World Scientific Publ. Comp., New Jersey, 2013.

[18] I.K. Argyros, Á.A. Magreñán, Extended convergence results for the Newton–Kantorovich iteration, J. Comput. Appl. Math. 286 (2015) 54–67.

[19] I.K. Argyros, Á.A. Magreñán, L. Orcos, J.A. Sicilia, Local convergence of a relaxed two-step Newton like method with applications, J. Math. Chem. 55 (7) (2017) 1427–1442, https://www.scopus.com/inward/record.uri?eid=2-.

[20] I.K. Argyros, Á.A. Magreñán, J.A. Sicilia, Improving the domain of parameters for Newton's method with applications, J. Comput. Appl. Math. 318 (2017) 124–135.

[21] L. Blum, F. Cucker, M. Shub, S. Smale, Complexity and Real Computation, Springer-Verlag, New York, 1997.

[22] J.E. Dennis Jr., R.B. Schnabel, Numerical Methods for Unconstrained Optimization and Nonlinear Equations, Classics Appl. Math., SIAM, Philadelphia, 1996.

[23] P. Deuflhard, G. Heindl, Affine invariant convergence for Newton's method and extensions to related methods, SIAM J. Numer. Anal. 16 (1) (1979) 1–10.

[24] R.S. Dembo, S.C. Eisenstat, T. Steihaug, Inexact Newton methods, SIAM J. Numer. Anal. 19 (1982) 400–408.

[25] A.L. Dontchev, R.T. Rockafellar, Implicit Functions and Solution Mappings. A View From Variational Analysis, Springer Monogr. Math., Springer, Dordrecht, 2009.

[26] O.P. Ferreira, M.L.N. Goncalves, P.R. Oliveira, Convergence of the Gauss–Newton method for convex composite optimization under a majorant condition, SIAM J. Optim. 23 (3) (2013) 1757–1783.

[27] O.P. Ferreira, A robust semilocal convergence analysis of Newton's method for cone inclusion problems in Banach spaces under affine invariant majorant condition, J. Comput. Appl. Math. 279 (2015) 318–335.

[28] C. Li, K.F. Ng, Convergence analysis of the Gauss–Newton method for convex inclusion and convex-composite optimization problems, J. Math. Anal. Appl. 389 (1) (2012) 469–485.

[29] Á.A. Magreñán, I.K. Argyros, Optimizing the applicability of a theorem by F. Potra for Newton-like methods, Appl. Math. Comput. 242 (2014) 612–623.

[30] Á.A. Magreñán, I.K. Argyros, J.A. Sicilia, New improved convergence analysis for Newton-like methods with applications, J. Math. Chem. 55 (7) (2017) 1505–1520.

[31] S.M. Robinson, Normed convex processes, Trans. Amer. Math. Soc. 174 (1972) 127–140.

[32] S. Smale, Newton method estimates from data at one point, in: R. Ewing, K. Gross, C. Martin (Eds.), The Merging of Disciplines: New Directions in Pure, Applied and Computational Mathematics, Springer-Verlag, New York, 1986, pp. 185–196.

[33] X. Wang, Convergence of Newton's method and uniqueness of the solution of equation in Banach space, IMA J. Numer. Anal. 20 (2000) 123–134.

Chapter 11

Gauss–Newton method for convex composite optimization

11.1 INTRODUCTION

Let $F : \mathbb{R}^i \to \mathbb{R}^j$ be a continuously differentiable operator and $h : \mathbb{R}^i \to \mathbb{R}$ a real-valued convex functional. We consider the convex composite optimization problem

$$\min h(F(x)). \tag{11.1.1}$$

The study of (11.1.1) is related to the convex inclusion problem

$$F(x) \in C = \arg \min h \tag{11.1.2}$$

because if $x^* \in \mathbb{R}^i$ satisfies the convex inclusion (11.1.2), then x^* is the solution of (11.1.1), but if $x^* \in \mathbb{R}^i$ is the solution of (11.1.1), it does not necessarily satisfy the convex inclusion (11.1.2).

Due to the wide range of application in the mathematical programming problems, e.g., penalization method, minimax, and goal programming, etc., the study of (11.1.1) is an important problem in mathematics [1–15]. We consider the algorithm (see Algorithm) considered in [10] for solving (11.1.1). But our semilocal convergence analysis is based on the restricted majorant condition. Using the idea of restricted majorant condition, we extended the applicability of the Algorithm.

Throughout this chapter $\| \cdot \|$ stands for a norm in $\mathbb{R}^i$, $U(x, r)$ and $\overline{U(x, r)}$ respectively stand for open and closed ball with center $x \in \mathbb{R}^i$ and radius r. The set $W^o := \{z \in \mathbb{R}^i : \langle z, w \rangle \leq 0, w \in W\}$ denotes the polar of a closed convex $W \subset \mathbb{R}^i$. The distance from a point x to a set $W \subset \mathbb{R}^i$ is given by $d(x, W) := \inf\{\|x - w\| : w \in W\}$, $P(\mathbb{R}^i)$ denotes the set of all subsets of $\mathbb{R}^i$ and $Ker(A)$ represents the kernel of the linear map A. $\dim(S)$ denotes the dimension the vector subspace S in $\mathbb{R}^i$. If $v \in \mathbb{R}^i$, then $v^\perp = \{u \in \mathbb{R}^i : \langle u, v \rangle = 0\}$. The sum of a point $x \in \mathbb{R}^i$ with a set $X \in P(\mathbb{R}^i)$ is the set given by $y + X = \{y + x : x \in X\}$. The vector space of the $i \times i$ matrices with the Frobenius norm is denoted by $\mathbb{M}_i$. Finally, $C^l(\mathbb{R}^i, \mathbb{R}^j)$ is the set of l-times continuously differentiable functions from $\mathbb{R}^i$ to $\mathbb{R}^j$ and, in the case $i = j$, we use the short notation $C^l(\mathbb{R}^i)$.

A Contemporary Study of Iterative Methods. DOI: 10.1016/B978-0-12-809214-9.00011-5

In Section 11.2, we present the Gauss–Newton algorithm, and the semilocal convergence of the algorithm is presented in Section 11.3. Special cases and examples are given in the concluding Section 11.4 of the chapter.

11.2   THE GAUSS–NEWTON ALGORITHM

For $F \in C^1(\mathbb{R}^i, \mathbb{R}^j)$, $\Delta \in (0, +\infty]$, and $x \in \mathbb{R}^i$, let

$$D_\Delta(x) := \operatorname{argmin}\{h(F(x) + F'(x)d) : d \in \mathbb{R}^i, \|d\| \leq \Delta\}. \qquad (11.2.1)$$

It follows that $D_\Delta(x)$ is the solution set for the following problem:

$$\min\{h(F(x) + F'(x)d) : d \in \mathbb{R}^i, \|d\| \leq \Delta\}. \qquad (11.2.2)$$

Choose $\Delta \in (0, +\infty]$, $\eta \in (1, +\infty]$, and a point $x_0 \in \mathbb{R}^i$. Then, the *Gauss–Newton* algorithm associated with (Δ, η, x_0) as defined in [7] (see also, [10–12]) is as follows:

Algorithm

INITIALIZATION. Choose $\Delta \in (0, +\infty]$, $\eta \in (1, +\infty]$, and $x_0 \in \mathbb{R}^i$. Set $k = 0$.

STOP CRITERION. Compute $D_\Delta(x_k)$. If $0 \in D_\Delta(x_k)$, STOP. Otherwise.

ITERATIVE STEP. Compute d_k satisfying $d_k \in D_\Delta(x_k)$, $\|d_k\| \leq \eta d(0, D_\Delta(x_k))$, set $x_{k+1} = x_k + d_k$, $k = k + 1$, and go to Stop Criterion.

The sequence $\{x_k\}$ generated by Algorithm is well defined as it can be seen in [10].

11.2.1   Regularity

Let C be as in (11.1.2), then for $F \in C^1(\mathbb{R}^i, \mathbb{R}^j)$ and $x \in \mathbb{R}^i$, we define the set $D_C(x)$ as

$$D_C(x) := \{d \in \mathbb{R}^i : F(x) + F'(x)d \in C\}.$$

The next proposition gives the relation between the sets $D_\Delta(x)$ and $D_C(x)$.

Proposition 11.2.1. *(see [10]) Let $x \in \mathbb{R}^i$. If $D_C(x) \neq 0$ and $d(0, D_C(x)) \leq \Delta$, then*

(i) $D_\Delta(x) = \{d \in \mathbb{R}^i : \|d\| \leq \Delta, F(x) + F'(x)d \in C\} \subset D_\Delta(x)$.
(ii) $d(0, D_\Delta(x)) = d(0, D_C(x))$.

Proof. By definition of C and $D_\Delta(x)$, we have

$$\{d \in \mathbb{R}^i : \|d\| \leq \Delta, F(x) + F'(x)d \in C\} \subset D_\Delta(x).$$

Let $d \in D_\Delta(x)$. Note that $D_C(x) \neq 0$ and $d(0, D_C(x)) \leq \Delta$, so there is $\bar{d} \in D_C(x)$ such that $\|\bar{d}\| \leq \Delta$ and $F(x) + F'(x)\bar{d} \in C$. Hence, by (11.1.2) and (11.2.1), we obtain $\bar{d} \in D_\Delta(x)$. Therefore, as $d, \bar{d} \in D_\Delta(x)$, using (11.2.1), we have

$$h(F(x) + F'(x)d) = h(F(x) + F'(x)\bar{d}).$$

Again using $F(x) + F'(x)\bar{d} \in C$, the last equality and definition of C, we obtain $F(x) + F'(x)d \in C$, which proves (i). To prove (ii), it is enough to prove that $D_\Delta(x) \subset D_C(x)$, which follows by definition of $D_C(x)$. To conclude the proof, first note that the inclusion $D_\Delta(x) \subset D_C(x)$ implies that

$$d(0, D_\Delta(x)) \geq d(0, D_C(x)). \tag{11.2.3}$$

Since $D_C(x) \neq 0$ and $d(0, D_C(x)) \leq \Delta$, there exists $\bar{d} \in D_C(x)$ such that

$$\|\bar{d}\| = d(0, D_C(x)) \leq \Delta.$$

Using (11.1.2) and (11.2.1), we conclude that $\bar{d} \in D_\Delta(x)$ and hence

$$d(0, D_\Delta(x)) \leq \|\bar{d}\| = d(0, D_C(x)). \tag{11.2.4}$$

The result now follows from (11.2.3) and (11.2.4). □

We use the following definition, which was introduced in [11] for studying the Gauss–Newton method (see also [10], [12]).

Definition 11.2.2. *(see [10–12]) Let $F \in C^1(\mathbb{R}^i, \mathbb{R}^j)$ and let $h : \mathbb{R}^i \to \mathbb{R}$ be a real-valued convex function. A point $x_0 \in \mathbb{R}^i$ is called a quasi-regular point of the inclusion (11.1.2), that is, of the inclusion*

$$F(x) \in C = \arg\min h := \{z \in \mathbb{R}^i : h(z) \leq h(x), x \in \mathbb{R}^j\},$$

if $r \in (0, +\infty)$ exists, and there is an increasing positive-valued function β : $[0, r) \to (0, +\infty)$ such that

$$D_C(x) \neq 0, \quad d(0, D_C(x)) \leq \beta(\|x - x_0\|)d(F(x), C), \quad x \in U(x_0, r). \tag{11.2.5}$$

Let $x_0 \in \mathbb{R}^i$ be a quasi-regular point of the inclusion (11.1.2). We denote by r_{x_0} the supremum of r such that (11.2.5) holds for some increasing positive-valued function β on $[0, r)$, that is,

$$r_{x_0} := \sup\{r : \exists \beta : [0, r) \to (0, +\infty) \text{ satisfying } (11.2.5)\}. \tag{11.2.6}$$

Let $r \in [0, r_{x_0})$. The set $U_r(x_0)$ denotes the set of all increasing positive-valued functions β on $[0, r)$ such that (11.2.5) holds, that is,

$$U_r(x_0) := \{\beta : [0, r) \to (0, +\infty) : \beta \text{ satisfying } (11.2.5)\}.$$

Define

$$\beta_{x_0}(t) := \inf\{\beta(t) : \beta \in U_{r_{x_0}}(x_0), \ t \in [0, r_{x_0})\}. \tag{11.2.7}$$

The number r_{x_0} and the function β_{x_0} are called respectively the quasi-regular radius and the quasi-regular bound function of the quasi-regular point x_0.

11.3   SEMILOCAL CONVERGENCE

In order to make the chapter as self-contained as possible, in the first part of this section and until the end of Theorem 11.3.5, we restate the results from [10]. Then, in the second part we introduce the improvements.

We shall use the concept of the majorant f for a mapping F.

Definition 11.3.1. *Let $F \in C^1(\mathbb{R}^i, \mathbb{R}^j)$, $R > 0$, and $x_0 \in \mathbb{R}^i$. A twice differentiable function $f : [0, R) \to \mathbb{R}$ is a majorant function for the mapping F on $U(x_0, R)$ if it satisfies*

$$\|F'(y) - F'(x)\| \le f'(\|y - x\| + \|x - x_0\|) - f'(\|x - x_0\|) \tag{11.3.1}$$

for any $x, y \in U(x_0, R)$, $\|x - x_0\| + \|y - x\| < R$,

(h_1) $f(0) = 0$, $f'(0) = -1$;
(h_2) f' is convex and strictly increasing.

We also consider the following condition.

(h_3) There exists $t^* \in (0, R)$ such that $f_{\xi,\alpha}(t) > 0$, for all $t \in (0, t^*)$ and $f_{\xi,\alpha}(t^*) = 0$;
(h_4) $f'_{\xi,\alpha}(t^*) < 0$.

From now on, (h_1)–(h_4), (11.2.6), and (11.2.7) shall be called the (h) conditions. Then the following semilocal convergence result was presented in [10] under the (h) conditions.

Theorem 11.3.2. *(see [10]) Suppose that the (h) conditions are satisfied. Choose the constants $\alpha > 0$, $\xi > 0$ and consider the function $f_{\xi,\alpha} : [0, R) \to \mathbb{R}$, defined by*

$$f_{\xi,\alpha}(t) := \xi + (\alpha - 1)t + \alpha f(t). \tag{11.3.2}$$

Then the sequence $\{t_k\}$ defined by

$$t_0 = 0, \quad t_{k+1} = t_k - f'_{\xi,\alpha}(t_k)^{-1} f_{\xi,\alpha}(t_k), \quad k = 0, 1, \ldots, \tag{11.3.3}$$

is well defined, $\{t_k\}$ is strictly increasing, contained in $[0, t_*)$, and it converges Q-linearly to t_*, the solution of $f_{\xi,\alpha}(t) = 0$. Let $\eta \in [1, \infty)$, $\Delta \in (0, \infty]$, and let $h : \mathbb{R}^j \to \mathbb{R}$ be a real-valued convex function with nonempty minimizer set C. Further, suppose that $x_0 \in \mathbb{R}^i$ is a quasi-regular point of the inclusion $F(x) \in C$

with the quasi-regular radius r_{x_0} and the quasi-regular bound function β_{x_0} as in (11.2.6) and (11.2.7), respectively. If $d(F(x_0), C) > 0, t_* \leq r_{x_0}$,

$$\triangle \geq \xi \geq \eta\beta_{x_0}(0)d(F(x_0), C), \quad \alpha \geq \sup\left\{\frac{\eta\beta_{x_0}(t)}{\eta\beta_{x_0}(t)[f'(t) + 1] + 1} : \xi \leq t < t_*\right\},$$

$$(11.3.4)$$

then the sequence generated by Algorithm is contained in $U(x_0, t_*)$,

$$F(x_k) + F'(x_k)(x_{k+1} - x_k) \in C, \quad k = 0, 1, \ldots, \quad (11.3.5)$$

it satisfies the inequalities

$$\|x_k - x_{k-1}\| \leq t_k - t_{k-1}, \quad \|x_{k+1} - x_k\| \leq \frac{t_{k+1} - t_k}{(t_k - t_{k-1})^2}\|x_k - x_{k-1}\|^2,$$

$$(11.3.6)$$

for all $k = 1, 2, \ldots$, converges to a point $x_* \in U[x_0, t_*]$ such that $F(x_*) \in C$,

$$\|x_* - x_k\| \leq t_* - t_k, \quad k = 0, 1, \ldots, \quad (11.3.7)$$

and the convergence is R-linear. If, additionally, $f_{\xi,\alpha}$ satisfies (h4), then the following inequalities hold:

$$\|x_{k+1} - x_k\| \leq \frac{f''_{\xi,\alpha}(t_*)}{-2f'_{\xi,\alpha}(t_*)}\|x_k - x_{k-1}\|^2, \quad t_{k+1} - t_k \leq \frac{f''_{\xi,\alpha}(t_*)}{-2f'_{\xi,\alpha}(t_*)}(t_k - t_{k-1})^2,$$

$$(11.3.8)$$

for all $k = 1, 2, \ldots$. Moreover, the sequences $\{x_k\}$ and $\{t_k\}$ converge Q-quadratically to x_* and t_*, respectively, as follows:

$$\limsup_{k \to \infty} \frac{\|x_* - x_{k+1}\|}{\|x_* - x_k\|^2} \leq \frac{f''_{\xi,\alpha}(t_*)}{-2f'_{\xi,\alpha}(t_*)}, \quad t_* - t_{k+1} \leq \frac{f''_{\xi,\alpha}(t_*)}{-2f'_{\xi,\alpha}(t_*)}(t_* - t_k)^2,$$

$$(11.3.9)$$

for all $k = 0, 1, \ldots$.

Next we present the improvements of Theorem 11.3.2 by introducing tighter than f majorant functions for mapping F.

Remark 11.3.3. *In view of (11.3.1), there exists a twice-differentiable function $f_0 : [0, R) \to \mathbb{R}$, which is a center-majorant function for the mapping F on $U(x_0, \mathbb{R})$. That is, f satisfies*

$$\|F'(x) - F'(x_0)\| \leq f'_0(\|x - x_0\|) - f'_0(0) \quad (11.3.10)$$

for any $x \in U(x_0, R), \|x - x_0\| < R$.

(h'_1) $f_0(0) = 0$, $f'_0(0) = -1$.
(h'_2) f'_0 is convex and strictly increasing.
Define parameter ρ_0 by

$$\rho_0 := \sup\{t \geq 0 : f'_0(t) < 0\} \qquad (11.3.11)$$

and set

$$\rho = \min\{\rho_0, R\}. \qquad (11.3.12)$$

We also suppose that
(h'_3) *There exists $t^*_0 \in (0, R)$ such that for*

$$f_{0,\xi,\alpha}(t) = \xi + (\alpha - 1)t + \alpha f_0(t), \qquad (11.3.13)$$

$f_{0,\xi,\alpha}(t) > 0$ *for all $t \in (0, R)$ and $f_{0,\xi,\alpha}(t^*_0) = 0$;*
(h'_4) $f'_{0,\xi,\alpha}(t^*_0) < 0.$
Then, again in view of (11.3.1), there exists a restricted majorant function f_1 : $[0, \rho) \to \mathbb{R}$ for the mapping F on $U(x_0, \rho)$ satisfying

$$\|F'(y) - F'(x)\| \leq f'_1(\|y - x\| + \|x - x_0\|) - f'_1(\|x - x_0\|) \qquad (11.3.14)$$

for any $x, y \in U(x_0, \rho)$, $\|x - x_0\| + \|y - x\| < \rho$.
(h''_1) $f_1(0) = 0$, $f'_1(0) = -1$.
(h''_2) f'_1 is convex and strictly increasing.
We also suppose that
(h''_3) *There exists $t^*_1 \in (0, \rho)$ such that for*

$$f_{1,\xi,\alpha}(t) = \xi + (\alpha - 1)t + \alpha f_1(t), \qquad (11.3.15)$$

$f_{1,\xi,\alpha}(t) > 0$ *for all $t \in (0, t^*_1)$ and $f_{1,\xi,\alpha}(t^*_1) = 0$;*
(h''_4) $f'_{1,\xi,\alpha}(t^*_1) < 0.$

Remark 11.3.4. *It follows from (11.3.1), (11.3.10), and (11.3.14) that*

$$f_0(t) \leq f(t), \qquad (11.3.16)$$
$$f_1(t) \leq f(t), \qquad (11.3.17)$$
$$f'_0(t) \leq f'(t), \qquad (11.3.18)$$
$$f'_1(t) \leq f'(t), \qquad (11.3.19)$$
$$f_{0,\xi,\alpha}(t) \leq f_{\xi,\alpha}(t), \qquad (11.3.20)$$
$$f'_{0,\xi,\alpha}(t) \leq f'_{\xi,\alpha}(t), \qquad (11.3.21)$$
$$f_{1,\xi,\alpha}(t) \leq f_{\xi,\alpha}(t), \qquad (11.3.22)$$

$$f'_{1,\xi,\alpha}(t) \le f'_{\xi,\alpha}(t), \tag{11.3.23}$$

$$t^* \le t_0^*, \tag{11.3.24}$$

$$t^* \le t_1^*. \tag{11.3.25}$$

It is worth noticing that in practice the computation of function f requires the computation of functions f_0 and f_1 as special cases. Therefore, (11.3.10) and (11.3.14) are not additional to (11.3.1) hypotheses. Notice also that in view of (11.3.16)–(11.3.25) and (11.3.5)–(11.3.9), the new semilocal sufficient convergence criterion is always at least as weak and the error bounds on the distances $\|x_n - x^*\|$, $\|x_{n+1} - x_n\|$ are at least as tight. Moreover, we also have the following favorable comparisons: The estimates in [10] give

$$\|Tx^{-1}\| \le \frac{\beta_0}{1 - \beta_0[f'(\|x - x_0\|) - f'(0)]}, \tag{11.3.26}$$

$$\alpha \ge \sup \left\{ \frac{n\beta_{x_0}(t)}{n\beta_{x_0}(t)(1 + f'(t)) + 1}, \xi \le t \le t^* \right\}, \tag{11.3.27}$$

and the new ones are

$$\|Tx^{-1}\| \le \frac{\beta_0}{1 - \beta_0[g'(\|x - x_0\|) - g'(0)]}, \tag{11.3.28}$$

$$\alpha \ge \sup \left\{ \frac{n\beta_{x_0}(t)}{n\beta_{x_0}(t)(1 + g'(t)) + 1}, \xi \le t \le \rho \right\}, \tag{11.3.29}$$

where

$$T_x d = F'(x)d - C = T_{x_0}d + [F'(x) - F'(x_0)]d, \ d \in \mathbb{R}^i, \tag{11.3.30}$$

g is f_0 or f_1 and $\rho = \begin{cases} t_0^*, & \text{if } g' = f'_0, \\ t_1^*, & \text{if } g' = f'_1. \end{cases}$ Then, in view of (11.3.16)–(11.3.25) and (11.3.26)–(11.3.29), the new estimates on $\|T_x^{-1}\|$ and α are better.

We shall also suppose that

(h_5) $f_0(t) \le f_1(t)$ and $f'_0(t) \le f'_1(t)$ for each $t \in [0, R]$.

From now on we denote (h'_1), (h'_2), (h'_3), (h'_4), (h''_1), (h''_2), (h''_3), (h''_4), and (h_5) as the conditions (C). Notice that

$$(h) \Rightarrow (C) \tag{11.3.31}$$

but not necessarily vice versa, unless $f(t) = f_0(t) = f'(t)$ for $t \in [0, R]$. Next we present the new semilocal convergence result for Algorithm. The proof is obtained from the proof of Theorem 11.3.2, if we simply replace in the corresponding estimates function f by f_1 or f_0 (see also (11.3.26)–(11.3.30)).

Theorem 11.3.5. *Suppose that the (C) conditions are satisfied. Choose the constants $\alpha > 0$, $\xi > 0$ and define functions $f_{1,\xi,\alpha} : [0, R) \to \mathbb{R}$, $f_{0,\xi,\alpha} : [0, R) \to \mathbb{R}$ by*

$$f_{1,\xi,\alpha}(t) := \xi + (\alpha - 1)t + \alpha f(t), \quad f_{0,\xi,\alpha}(t) := \xi + (\alpha - 1)t + \alpha f_0(t).$$
$$(11.3.32)$$

Then the sequences generated for $k = 0, 1, 2, \dots$ by

$$t_0 = 0, \ t_{k+1} = t_k - f'_{1,\xi,\alpha}(t_k)^{-1} f_{1,\xi,\alpha}(t_k) \ and$$
$$s_0 = 0, \ s_{k+1} = s_k - f'_{0,\xi,\alpha}(s_k)^{-1} f_{0,\xi,\alpha}(s_k)$$
$$(11.3.33)$$

are well defined, $\{t_k\}$ and $\{s_k\}$ are strictly increasing, contained in $[0, t_1^*)$ and $[0, t_0^*)$, respectively, and converge Q-linearly to the solution t_1^* and t_0^* of $f_{1,\xi,\alpha}(t) = 0$, and $f_{0,\xi,\alpha}(t) = 0$. Let $\eta \in [1, \infty)$, $\Delta \in (0, \infty]$, and let $h : \mathbb{R}^j \to \mathbb{R}$ be a real-valued convex function with a nonempty minimizer set C. Further suppose that $x_0 \in \mathbb{R}^i$ is a quasi-regular point of the inclusion $F(x) \in C$ with the quasi-regular radius r_{x_0} and the quasi-regular bound function β_{x_0} as defined in (11.2.6) and (11.2.7), respectively. If $d(F(x_0), C) > 0$, $t_1^* \leq r_{x_0}$, $t_0^* \leq r_{x_0}$, and

$$\Delta \geq \xi \geq \eta \beta_{x_0}(0) d(F(x_0), C), \quad \alpha \geq \sup \left\{ \frac{\eta \beta_{x_0}(t)}{\eta \beta_{x_0}(t)[g'(t) + 1] + 1} : \xi \leq t < \rho \right\},$$
$$(11.3.34)$$

then the sequence generated by Algorithm is contained in $U(x_0, t_*)$,

$$F(x_k) + F'(x_k)(x_{k+1} - x_k) \in C, \quad k = 0, 1, \dots,$$
$$(11.3.35)$$

satisfies the inequalities

$$\|x_k - x_{k-1}\| \leq t_k - t_{k-1}, \ \|x_{k+1} - x_k\| \leq \frac{t_{k+1} - t_k}{(t_k - t_{k-1})^2} \|x_k - x_{k-1}\|^2,$$
$$(11.3.36)$$

for all $k = 1, 2, \dots$, and converges to a point $x_* \in U[x_0, t_*]$ such that $F(x_*) \in C$,

$$\|x_* - x_k\| \leq t_* - t_k, \ k = 0, 1, \dots,$$
$$(11.3.37)$$

and convergence is R-linear. Moreover, if $f_{1,\xi,\alpha}$ satisfies (h_4), then the following inequalities hold:

$$\|x_{k+1} - x_k\| \le \frac{f''_{1,\xi,\alpha}(t_*)}{-2f'_{1,\xi,\alpha}(t_*)}\|x_k - x_{k-1}\|^2,$$

$$t_{k+1} - t_k \le \frac{f''_{1,\xi,\alpha}(t_*)}{-2f'_{1,\xi,\alpha}(t_*)}(t_k - t_{k-1})^2, \tag{11.3.38}$$

for all $k = 1, 2, \ldots$. Furthermore, the sequences $\{x_k\}$ and $\{t_k\}$ converge Q-quadratically to x_* and t_*, respectively, as follows:

$$\limsup_{k\to\infty} \frac{\|x_* - x_{k+1}\|}{\|x_* - x_k\|^2} \le \frac{f''_{1,\xi,\alpha}(t_*)}{-2f'_{1,\xi,\alpha}(t_*)}, \quad t_* - t_{k+1} \le \frac{f''_{1,\xi,\alpha}(t_*)}{-2f'_{1,\xi,\alpha}(t_*)}(t_* - t_k)^2, \tag{11.3.39}$$

for all $k = 0, 1, \ldots$.

Remark 11.3.6. *If $f_0(t) = f_1(t) = f(t)$ for $t \in [0, R]$, then Theorem 11.3.5, reduces to Theorem 11.3.2. Otherwise it constitutes an improvement (see Remark 11.3.4).*

11.4 SPECIAL CASES AND EXAMPLES

We specialize Theorem 11.3.5 in this section, when x_0 is a regular point of the inclusion (11.1.2), under Lipschitz condition [1,2,5] and under Robinson's condition [10,11,13].

Theorem 11.4.1. *Suppose that the (C) conditions are satisfied. Then the sequence generated for each $k = 0, 1, 2, \ldots$ by*

$$t_0 = 0, \ t_{k+1} = t_k - f'_{1,\xi,\alpha}(t_k)^{-1} f_{1,\xi,\alpha}(t_k), \ k = 0, 1, \ldots,$$

is well defined, strictly increasing, contained in $[0, t_1^*)$, and converges Q-linearly to the solution t_1^* of the equation $f_{1,\xi,\alpha}(t) = 0$. Let $\eta \in [1, \infty)$, $\Delta \in (0, \infty]$, and let $h : \mathbb{R}^j \to \mathbb{R}$ be a real-valued convex function with a nonempty minimizer set C. Further, suppose that $x_0 \in \mathbb{R}^i$ is a quasi-regular point of the inclusion $F(x) \in C$ with associated constants $r > 0$ and $\beta > 0$. If $d(F(x_0), C) > 0$, $t_1^* \le r$,

$$\Delta \ge \xi \ge \eta\beta d(F(x_0), C), \quad \alpha \ge \left\{ \frac{\eta\beta}{\eta\beta[f'_{1,\xi,\alpha}(\xi) + 1] + 1} \right\},$$

then the sequence generated by Algorithm is contained in $U(x_0, t_1^*)$,

$$F(x_k) + F'(x_k)(x_{k+1} - x_k) \in C, \quad k = 0, 1, \ldots,$$

satisfies the inequalities

$$\|x_k - x_{k-1}\| \le t_k - t_{k-1}, \ \|x_{k+1} - x_k\| \le \frac{t_{k+1} - t_k}{(t_k - t_{k-1})^2}\|x_{k+1} - x_k\|^2,$$

for all $k = 1, 2, \ldots,$ *and converges to a point* $x_* \in U[x_0, t_1^*]$ *such that* $F(x_*) \in C,$

$$\|x_* - x_k\| \leq t_1^* - t_k, \ k = 0, 1, \ldots,$$

and the convergence is R-linear. Moreover, if $f_{1,\xi,\alpha}$ *satisfies* (h_4''), *then the following inequalities hold:*

$$\|x_{k+1} - x_k\| \leq \frac{f_{1,\xi,\alpha}''(t_1^*)}{-2f_{1,\xi,\alpha}'(t_*)} \|x_k - x_{k-1}\|^2,$$

$$t_{k+1} - t_k \leq \frac{f_{1,\xi,\alpha}''(t_1^*)}{-2f_{1,\xi,\alpha}'(t_1^*)} (t_k - t_{k-1})^2,$$

for all $k = 1, 2, \ldots.$ *Furthermore, the sequences* $\{x_k\}$ *and* $\{t_k\}$ *converge Q-quadratically to* x_* *and* t_1^*, *respectively, as follows:*

$$\limsup_{k \to \infty} \frac{\|x_* - x_{k+1}\|}{\|x_* - x_k\|^2} \leq \frac{f_{1,\xi,\alpha}''(t_1^*)}{-2f_{1,\xi,\alpha}'(t_1^*)}, \ t_1^* - t_{k+1} \leq \frac{f_{1,\xi,\alpha}''(t_1^*)}{-2f_{1,\xi,\alpha}'(t_*)} (t_1^* - t_k)^2,$$

for all $k = 0, 1, \ldots.$

Under the Lipschitz condition, Theorem 11.4.1 becomes the following.

Theorem 11.4.2. *Let* $F \in C^1(\mathbb{R}^j, \mathbb{R}^i)$. *Assume that* $x_0 \in \mathbb{R}^i$, $R > 0$, $K_0 > 0$, *and* $K_1 > 0$ *such that* $\|F'(x) - F'(x_0)\| \leq K_0 \|x - x_0\|$, *for each* $x \in U(x_0, R)$ *and* $\|F'(y) - F'(x)\| \leq K_1 \|y - x\|$, *for each* $x, y \in D_0 := U(x_0, R) \cap U(x_0, \frac{1}{K_0})$. *Choose the constants* $\alpha > 0$, $\xi > 0$ *and define function* $f_{1,\xi,\alpha} : [0, R) \to \mathbb{R}$,

$$f_{1,\xi,\alpha}(t) = \xi - t + (\alpha K_1 t^2)/2.$$

If $2\alpha K_1 \xi \leq 1$, *then* $t_1^* = (1 - \sqrt{1 - 2\alpha K_1 \xi})/(\alpha K_1)$ *is the smallest zero of* $f_{1,\xi,\alpha}$, *the sequence generated with starting point* $t_0 = 0$,

$$t_{k+1} = t_k - f_{1,\xi,\alpha}'(t_k)^{-1} f_{1,\xi,\alpha}(t_k), \quad k = 0, 1, \ldots,$$

is well defined, strictly increasing, contained in $[0, t_1^*)$, *and converges Q-linearly to the solution* t_1^* *of* $f_{1,\xi,\alpha}(t) = 0$. *Let* $\eta \in [1, \infty)$, $\Delta \in (0, \infty]$, *and let* $h : \mathbb{R}^j \to \mathbb{R}$ *be a real-valued convex function with a nonempty minimizer set* C. *Further suppose that* $x_0 \in \mathbb{R}^i$ *is a quasi-regular point of the inclusion* $F(x) \in C$ *with associated constants* $r > 0$ *and* $\beta > 0$. *If* $d(F(x_0), C) > 0$, $t_1^* \leq r$,

$$\Delta \geq \xi \geq \eta \beta d(F(x_0), C), \quad \alpha \geq \left\{ \frac{\eta \beta}{\eta \beta \xi K_1 + 1} \right\},$$

then the sequence generated by Algorithm is contained in $U(x_0, t_1^*)$,

$$F(x_k) + F'(x_k)(x_{k+1} - x_k) \in C, \quad k = 0, 1, \dots,$$

satisfies the inequalities

$$\|x_k - x_{k-1}\| \leq t_k - t_{k-1}, \quad \|x_{k+1} - x_k\| \leq \frac{t_{k+1} - t_k}{(t_k - t_{k-1})^2} \|x_k - x_{k-1}\|^2,$$

for all $k = 1, 2, \dots$, *converging to a point* $x_* \in U[x_0, t_1^*]$ *such that* $F(x_*) \in C$,

$$\|x_* - x_k\| \leq t_1^* - t_k, \quad k = 0, 1, \dots,$$

and the convergence is R-linear. Moreover, if $2K_1 \xi \alpha < 1$ *then the following inequalities hold:*

$$\|x_{k+1} - x_k\| \leq \frac{\alpha K_1}{2\sqrt{1 - 2\alpha K_1 \xi}} \|x_k - x_{k-1}\|^2,$$

$$t_{k+1} - t_k \leq \frac{\alpha K_1}{2\sqrt{1 - 2\alpha K_1 \xi}} (t_k - t_{k-1})^2,$$

for all $k = 1, 2, \dots$ *Furthermore, the sequences* $\{x_k\}$ *and* $\{t_k\}$ *converge Q-quadratically to* x_* *and* t_1^*, *respectively, as follows:*

$$\limsup_{k \to \infty} \frac{\|x_* - x_{k+1}\|}{\|x_* - x_k\|^2} \leq \frac{\alpha K_1}{2\sqrt{1 - 2\alpha K_1 \xi}}, \quad t_1^* - t_{k+1} \leq \frac{\alpha K_1}{2\sqrt{1 - 2\alpha K_1 \xi}} (t_1^* - t_k)^2,$$

for all $k = 0, 1, \dots$

Theorem 11.4.3. *Suppose the (C) conditions are satisfied. Then, the sequence generated with starting point* $t_0 = 0$,

$$t_{k+1} = t_k - f'_{1,\xi,\alpha}(t_k)^{-1} f_{1,\xi,\alpha}(t_k), \quad k = 0, 1, \dots,$$

is well defined, strictly increasing, contained in $[0, t_1^*)$, *and converges Q-linearly to the solution* t_1^* *of* $f_{1,\xi,\alpha}(t) = 0$. *Let* $\eta \in [1, \infty)$, $\Delta \in (0, \infty]$, *and let* $h : \mathbb{R}^j \to \mathbb{R}$ *be a real-valued convex function with a nonempty minimizer set* C. *Further suppose that* C *is a cone and* $x_0 \in \mathbb{R}^i$ *satisfies the Robinson condition with respect to* C *and* F. *Let* $\beta_0 = \|T_{x_0}^{-1}\|$. *If* $d(F(x_0), C) > 0$, $t_1^* \leq r_{\beta_0} := \sup\{t \in [0, R) : \beta_0 - 1 + \beta_0 f'(t) < 0\}$,

$$\Delta \geq \xi \geq \eta \beta_0 d(F(x_0), C), \quad \alpha \geq \left\{ \frac{\eta \beta_0}{1 + (\eta - 1)\beta_0 [f'(\xi) + 1]} \right\},$$

then the sequence generated by Algorithm is contained in $U(x_0, t_1^*)$,

$$F(x_k) + F'(x_k)(x_{k+1} - x_k) \in C, \quad k = 0, 1, \dots,$$

satisfies the inequalities

$$\|x_k - x_{k-1}\| \le t_k - t_{k-1}, \quad \|x_{k+1} - x_k\| \le \frac{t_{k+1} - t_k}{(t_k - t_{k-1})^2}\|x_k - x_{k-1}\|^2,$$

for all $k = 1, 2, \ldots,$ converging to a point $x_* \in U[x_0, t_1^*]$ such that $F(x_*) \in C,$

$$\|x_* - x_k\| \le t_1^* - t_k, \quad k = 0, 1, \ldots,$$

and the convergence is R-linear. Moreover, if $f_{1,xi\alpha}$ satisfies (h_4''), then the following inequalities hold:

$$\|x_{k+1} - x_k\| \le \frac{f_{1,\xi,\alpha}''(t_1^*)}{-2\alpha f_{1,\xi,\alpha}'(t_1^*)}\|x_k - x_{k-1}\|^2,$$

$$t_{k+1} - t_k \le \frac{f_{1,\xi,\alpha}''(t_1^*)}{-2\alpha f_{1,\xi,\alpha}'(t_1^*)}(t_k - t_{k-1})^2,$$

for all $k = 1, 2, \ldots$. Furthermore, the sequences $\{x_k\}$ and $\{t_k\}$ converge Q-quadratically to x_* and t_1^*, respectively, as follows:

$$\limsup_{k \to \infty} \frac{\|x_* - x_{k+1}\|}{\|x_* - x_k\|^2} \le \frac{f_{1,\xi,\alpha}''(t_1^*)}{-2\alpha f_{1,\xi,\alpha}'(t_1^*)},$$

$$t_1^* - t_{k+1} \le \frac{f_{1\xi,\alpha}''(t_1^*)}{-2\alpha f_{1,\xi,\alpha}'(t_1^*)}(t_1^* - t_k)^2,$$

for all $k = 0, 1, \ldots$

Remark 11.4.4. **(a)** *If in Theorem 11.4.3 we suppose that there exist $R > 0$, $K_0 > 0$, and $K_1 > 0$ such that $\|F'(x) - F'(x_0)\| \le K_0\|x - x_0\|$, for each $x \in U(x_0, R)$ and $\|F'(y) - F'(x)\| \le K_1\|y - x\|$, for each $x, y \in D_0 := U(x_0, R) \cap U(x_0, \frac{1}{K_0})$, then $f : [0, R) \to \mathbb{R}$ defined by $f(t) = K_1 t^2/2 - t$ is a restricted majorant function for F on D_0 and (h_3''), (h_4''), and t_1^* become*

$$2\alpha K_1 \xi < 1,$$

$$2\alpha K_1 \xi < 1,$$

$$t_1^* = (1 - \sqrt{1 - 2\alpha K_1 \xi})/(\alpha K_1),$$

while α satisfies

$$\alpha \ge \frac{\eta \beta_0}{1 + (\eta - 1)K_1 \beta_0 \xi}.$$

(b) *In particular, if $C = \{0\}$ and $i = j$, the Robinson condition is equivalent to the condition that $F'(x_0)^{-1}$ is nonsingular. Hence, for $\eta = 1$, Theorem 11.4.3 becomes a semilocal convergence result for the Newton's method under the Lipschitz condition, which improves the Newton–Kantrorovich theorem for solving nonlinear equations using Newton's method [1,2].*

Remark 11.4.5. *The results of this section reduce to the corresponding ones in [10, Section 4] if $f_0(t) = f_1(t) = f(t)$ for all $t \in [0, R]$ (as special cases of Theorem 11.3.5). Otherwise they constitute an improvement (see also Remark 11.3.4).*

Let us complete this section by a concrete example. Other examples where $K_0 < K$ or $K_1 < K$ can be found in [3–6].

Example 11.4.6. *Let*

$$\alpha = 1,$$
$$\eta = 1,$$
$$i = j = 1,$$
$$R = 1 - q,$$
$$x_0 = 1,$$
$$C = \{0\},$$

and define a function on

$$U(x_0, 1 - q)$$

for

$$q \in [0, \frac{1}{2})$$

by

$$F(x) = x^3 - q.$$

Then we have that

$$\xi = \frac{1}{3}(1 - q),$$
$$K_0 = 3 - q,$$
$$K_1 = 2(1 + \frac{1}{3 - q}),$$

$$K = 2(2 - q),$$

and clearly,

$$K_0 < K_1 < K.$$

The Kantorovich hypothesis

$$2K\xi \leq 1$$

is violated, since

$$2K\xi > 1$$

for each

$$q \in [0, \frac{1}{2}).$$

Hence, there is no guarantee under the old hypothesis that Newton's method converges to

$$x^* = \sqrt[3]{q}.$$

However, our hypothesis becomes

$$2K_1\xi = 4(1 + \frac{1}{3 - q})\frac{1}{3}(1 - q) \leq 1,$$

which is true for

$$q \in [0, 0.4619].$$

Hence our result guarantees the convergence of Newton's method.

REFERENCES

[1] I.K. Argyros, Computational theory of iterative methods, in: C.K. Chui, L. Wuytack (Eds.), Stud. Comput. Math., vol. 15, Elsevier Publ. Co., New York, USA, 2007.

[2] I.K. Argyros, Convergence and Application of Newton-Type Iterations, Springer, 2008.

[3] I.K. Argyros, Local convergence analysis of proximal Gauss–Newton method for penalized least squares problems, Appl. Math. Comput. 241 (2014) 401–408.

[4] I.K. Argyros, Local convergence analysis of the Gauss-Newton method for injective overdetermined systems under the majorant condition, J. Kor. Math. Soc. 51 (5) (2014) 955–970.

[5] I.K. Argyros, Expanding the applicability of the Gauss–Newton method for convex optimization under a majorant condition, SeMA J. 65 (1) (2014) 1–11.

[6] I.K. Argyros, Expanding the applicability of the Gauss–Newton method for convex optimization on Riemannian manifolds, Appl. Math. Comput. 249 (2014) 453–467.

[7] J.V. Burke, M.C. Ferris, A Gauss–Newton method for convex composite optimization, Math. Program. 71 (1995) 179–194.

[8] J.E. Dennis Jr., R.B. Schnabel, Numerical Methods for Unconstrained Optimization and Non-linear Equations, Classics in Appl. Math., vol. 16, SIAM, Philadelphia, 1996.

[9] O.P. Ferreira, M.L.N. Goncalves, P.R. Oliveira, Local convergence analysis of inexact Gauss–Newton like methods under majorant condition, J. Comput. Appl. Math. 236 (2012) 2487–2498.

[10] O.P. Ferreira, M.L.N. Goncalves, P.R. Oliveira, Convergence of the Gauss–Newton for convex composite optimization under a majorant condition, SIAM J. Optim. 23 (3) (2013) 1757–1783.

[11] C. Li, K.F. Ng, Majorizing functions and convergence of the Gauss–Newton method for convex composite optimization, SIAM J. Optim. 18 (2007) 613–642.

[12] C. Li, K.F. Ng, Convergence analysis of the Gauss–Newton method for convex inclusion and convex composite optimization problems, J. Math. Anal. Appl. 389 (2012) 469–485.

[13] S.M. Robinson, Extension of Newton's method to nonlinear functions with vales in a cone, Numer. Math. 19 (1972) 341–347.

[14] R.T. Rockafellar, Monotone Processes of Convex and Concave Type, Mem. Amer. Math. Soc., vol. 77, AMS, Providence, RI, 1967.

[15] R.T. Rockafellar, Convex Analysis, Princeton Math. Ser., vol. 28, Princeton University Press, Princeton, NJ, 1970.

Chapter 12

Domain of parameters

12.1 INTRODUCTION

In this chapter we are concerned with the problem of approximating a locally unique solution x^* of equation

$$F(x) = 0, \tag{12.1.1}$$

where F is a Fréchet-differentiable operator defined on a convex subset D of a Banach space X with values in a Banach space Y.

Many problems in applied sciences including engineering can be solved by means of finding the solutions of equations in a form like (12.1.1) using mathematical modeling [4,6,10–12,16,17,19–21,24,26], and for solving that equation the vast majority of methods are iterative since it is very complicated to find the solution in a closed form.

Convergence analysis of iterative methods is usually divided into two categories: semilocal and local convergence analysis. The semilocal convergence matter is, based on the information around an initial point, to give criteria ensuring the convergence of iteration procedures; while the local one is, based on the information about a solution, to find estimates of the radii of the convergence balls. A very important problem in the study of iterative procedures is the convergence domain. In general, the convergence domain is small. Therefore it is important to enlarge the convergence domain without additional hypotheses. Another important problem is to find more precise error estimates on the distances $\|x_{n+1} - x_n\|$ and $\|x_n - x^*\|$.

Newton's method defined for each $n = 0, 1, 2, \ldots$ by

$$x_{n+1} = x_n - F'(x_n)^{-1} F(x_n), \tag{12.1.2}$$

where x_0 is an initial guess, is the most popular, well-known and used method for generating a sequence $\{x_n\}$ approximating x^*.

Throughout this chapter we will denote by $U(z, \varrho)$ and $\bar{U}(z, \varrho)$ the open and closed ball in X with center $z \in X$ and radius $\varrho > 0$. Let also $L(X, Y)$ stand for the space of bounded linear operators from X into Y.

The most famous semilocal convergence result for Newton's method is the Newton–Kantorovich theorem [6,10,11,18] which is based on the following hypotheses:

A Contemporary Study of Iterative Methods. DOI: 10.1016/B978-0-12-809214-9.00012-7
Copyright © 2018 Elsevier Inc. All rights reserved.

(H_1) There exist $x_0 \in D$ such that $F'(x_0)^{-1} \in L(Y, X)$ and a parameter $\eta \geq 0$ such that

$$\|F'(x_0)^{-1}F(x_0)\| \leq \eta;$$

(H_2) There exists a parameter $L > 0$ such that for each $x, y \in D$,

$$\|F'(x_0)^{-1}(F'(x) - F'(y))\| \leq L\|x - y\|;$$

and

(H_3) $\bar{U}(x_0, R) \subseteq D$ for some $R > 0$.

The sufficient semilocal convergence condition of Newton's method is given by the Kantorovich hypothesis

$$h = 2L\eta \leq 1. \tag{12.1.3}$$

There are simple examples in the literature to show that hypothesis (12.1.3) is not satisfied but Newton's method converges starting at x_0. Moreover, the convergence domain of Newton's method depending on the parameters L and η is in general small. Therefore, it is important to enlarge the convergence domain by using the same constants L and η applying techniques as those in [17,18, 24]. Different authors in a series of works [6,10–12] presented weaker sufficient convergence conditions for Newton's method by using more precise majorizing sequences than before [18]. Moreover, the authors of this chapter have presented different works related to the convergence of Newton-like methods [2,3,5,7–9, 13–15,22,23].

These conditions are the following:

$$h_1 = 2A_1\eta \leq 1, \tag{12.1.4}$$

$$h_2 = 2A_2\eta \leq 1, \tag{12.1.5}$$

$$h_3 = 2A_3\eta \leq 1, \tag{12.1.6}$$

$$h_4 = 2A_4\eta_0 \leq 1, \tag{12.1.7}$$

where

$$A_1 = \frac{L_0 + L}{2}, \quad A_2 = \frac{1}{8}\left(L + 4L_0 + \sqrt{L^2 + 8L_0L}\right),$$

$$A_3 = \frac{1}{8}\left(4L_0 + \sqrt{L_0L + 8L_0^2} + \sqrt{L_0L}\right), \quad A_4 = \frac{1}{\eta_0},$$

η_0 is the small positive root of a quadratic polynomial, and $L_0 > 0$ is the center-Lipschitz constant such that

$$\|F'(x_0)^{-1}(F'(x) - F'(x_0))\| \leq L_0\|x - x_0\| \quad \text{for each } x \in D. \tag{12.1.8}$$

The existence of L_0 is always implied by (H_2).

We have that

$$L_0 \leq L \tag{12.1.9}$$

holds in general and $\frac{L}{L_0}$ can be arbitrarily large [6]. Notice also that (12.1.8) is not an additional hypothesis to (H_2), since in practice the computation of parameter L involves the computation of L_0 as a special case. Notice that if $L_0 = L$, conditions (12.1.4)–(12.1.7) reduce to condition (12.1.3). However, if $L_0 < L$, then we have [10,12]

$$h \leq 1 \Rightarrow h_1 \leq 1 \Rightarrow h_2 \leq 1 \Rightarrow h_3 \leq 1 \Rightarrow h_4 \leq 1, \tag{12.1.10}$$

$$\frac{h_1}{h} \to \frac{1}{2}, \quad \frac{h_2}{h} \to \frac{1}{4}, \quad \frac{h_2}{h_1} \to \frac{1}{2}, \quad \frac{h_3}{h} \to 0,$$
and
$$\frac{h_3}{h_2} \to 0, \quad \frac{h_3}{h_1} \to 0, \quad \text{as } \frac{L_0}{L} \to 0. \tag{12.1.11}$$

Estimates (12.1.11) show by how many times (at most) a condition is improving the previous one.

Notice also that the error bounds on the distances involved, as well as the location on the solution x^*, are also improved under these weaker conditions [6,10–12]. In the present chapter the main goal is to further improve conditions (12.1.3)–(12.1.7) by using parameters smaller than L_0 and L and by restricting the domain D. Similar ideas are used to improve the error bounds and enlarge convergence radii in the local convergence case.

12.2   CONVERGENCE ANALYSIS

We present first the semilocal convergence analysis of Newton's method. Next we state the following version of the Newton–Kantorovich theorem [6,10,11, 18].

Theorem 12.2.1. *Let $F : \mathcal{D} \subset X \to Y$ be a Fréchet-differentiable operator. Suppose that (12.1.3) and conditions (H_1)–(H_3) hold, where*

$$R = \frac{1 - \sqrt{1-h}}{L}.$$

Then the sequence $\{x_n\}$ generated by Newton's method is well defined, remains in $U(x_0, R)$ for each $n = 0, 1, 2, \ldots$, and converges to a unique solution $x^* \in \bar{U}(x_0, R)$ of equation (12.1.1).

Let us consider an academic example, where the Newton–Kantorovich hypothesis (12.1.3) is not satisfied.

Example 12.2.2. Let $X = Y = \mathbb{R}$, $x_0 = 1$, $D = U(1, 1 - p)$ for $p \in (0, \frac{1}{2})$ and define function F on D by

$$F(x) = x^3 - p.$$

We have that $\eta = \frac{1-p}{3}$ and $L = 2(2 - p)$. Then hypothesis (12.1.3) is not satisfied, since

$$h = \frac{4}{3}(2 - p)(1 - p) > 1 \quad \text{for each } p \in (0, \frac{1}{2}).$$

Hence there is no guarantee under the hypotheses of Theorem 12.2.1 that sequence $\{x_n\}$ starting from $x_0 = 1$ converges to $x^* = \sqrt[3]{p}$.

Next we present a semilocal convergence result that extends the applicability of Theorem 12.2.1.

Theorem 12.2.3. Let $F : \mathcal{D} \subset X \to Y$ be a Fréchet-differentiable operator. Suppose that there exist $x_0 \in D$, $\eta \geq 0$, $\gamma > 1$, and $L_\gamma > 0$ such that

$$F'(x_0) \in L(Y, X),$$

$$\|F'(x_0)^{-1} F(x_0)\| \leq \eta,$$

$$D_\gamma = U(x_0, \gamma \eta) \subseteq D,$$

$$\|F'(x_0)^{-1}(F'(x) - F'(y))\| \leq L_\gamma \|x - y\| \quad \text{for each } x, y \in D_\gamma,$$

$$h_\gamma = 2L_\gamma \eta \leq 1,$$

and

$$R_\gamma \leq \gamma \eta,$$

where

$$R_\gamma = \frac{1 - \sqrt{1 - h_\gamma}}{L_\gamma}.$$

Then the sequence $\{x_n\}$ generated by Newton's method is well defined, remains in $U(x_0, R_\gamma)$ for each $n = 0, 1, 2, \ldots$, and converges to a unique solution $x^* \in \bar{U}(x_0, R_\gamma)$ of equation (12.1.1).

Proof. The hypotheses of Theorem 12.2.1 on D_γ are satisfied. $\qquad\Box$

Example 12.2.4. Let $D = U(x_0, 1)$, $x^* = \sqrt[3]{2}$ and define function F on D by

$$F(x) = x^3 - 2. \tag{12.2.1}$$

We are going to consider such initial points for which previous conditions cannot be satisfied but our new conditions are satisfied. That is, the improvement that we get is the new weaker conditions.

The sufficient condition in Theorem 12.2.1 is

$$h = 2L\eta \le 1.$$

So we want to get the values of x_0 and γ for which condition $h \le 1$ is not satisfied but the conditions of Theorem 12.2.3 hold.

We get that

$$\eta = \frac{1}{3}\left|\frac{-2+x_0^3}{x_0^2}\right|,$$

$$L = \left|\frac{2(1+x_0)}{x_0^2}\right|,$$

$$h = \frac{4|1+x_0|\left|\frac{-2+x_0^3}{x_0^2}\right|}{3x_0^2},$$

$$R = \frac{x_0^2\left(1-\sqrt{1-\frac{4|1+x_0|\left|\frac{-2+x_0^3}{x_0^2}\right|}{3x_0^2}}\right)}{2|1+x_0|},$$

$$L_\gamma = \frac{2\left(x_0 + \frac{1}{3}\gamma\left|\frac{-2+x_0^3}{x_0^2}\right|\right)}{x_0^2},$$

$$h_\gamma = \frac{4\left|\frac{-2+x_0^3}{x_0^2}\right|\left(x_0 + \frac{1}{3}\gamma\left|\frac{-2+x_0^3}{x_0^2}\right|\right)}{3x_0^2},$$

and

$$R_\gamma = \frac{x_0^2\left(1-\sqrt{1-\frac{4\left|\frac{-2+x_0^3}{x_0^2}\right|\left(x_0+\frac{1}{3}\gamma\left|\frac{-2+x_0^3}{x_0^2}\right|\right)}{3x_0^2}}\right)}{2\left(x_0 + \frac{1}{3}\gamma\left|\frac{-2+x_0^3}{x_0^2}\right|\right)}.$$

Imposing the following conditions, and recalling (SL) conditions:

- $h > 1$,
- $h_\gamma \le 1$,
- $R_\gamma \le \gamma\eta$,
- $\gamma\eta \le 1$,

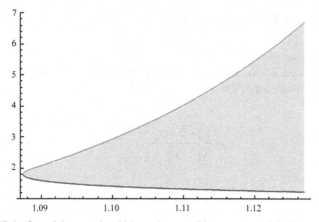

FIGURE 12.1   One of the cases in which previous conditions are not satisfied but conditions of Theorem 12.2.3 hold.

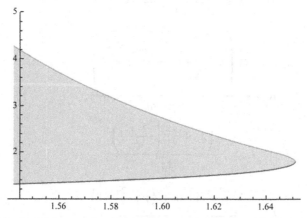

FIGURE 12.2   One of the cases in which previous conditions are not satisfied but conditions of Theorem 12.2.3 hold.

we obtain that the colored zone corresponds to the cases for which previous conditions cannot guarantee the convergence to the solution but our new weaker criteria can. In order to obtain the graphics, we associate the pair (x_0, γ) of the xy-plane, where $x = x_0$ and $y = \gamma$. Moreover, if we consider the set of points

$$V = \{(x_0, \gamma) \in \mathbb{R}^2 : (SL) \ conditions \ are \ satisfied\},$$

we can observe that every point x_0 chosen such that the associated pair (x_0, γ) belongs to V cannot be chosen as a starting point with the old condition but can be chosen using the conditions of Theorem 12.2.3. Two examples of these regions in which Theorem 12.2.3 can guarantee the convergence but previous results can't are shown in Figs. 12.1 and 12.2.

Next we present a semilocal result given in [12] involving condition (12.1.6).

Theorem 12.2.5. *Let $F : \mathcal{D} \subset X \to Y$ be a continuously Fréchet-differentiable operator. Suppose that (12.1.6) and conditions (H_1)–(H_3) hold, where*

$$r_3 = \eta + \frac{L_0 \eta^2}{2(1 - \alpha)(1 - L_0 \eta)},$$

and

$$\alpha = \frac{2L}{L + \sqrt{L^2 + 8L_0 L}}.$$

Then the sequence $\{x_n\}$ generated by Newton's method is well defined, remains in $U(x_0, r_3)$ for each $n = 0, 1, 2, \ldots$, and converges to a unique solution $x^* \in \bar{U}(x_0, r_3)$ of equation (12.1.1).

Next we present an improvement of Theorem 12.2.5.

Theorem 12.2.6. *Let $F : \mathcal{D} \subset X \to Y$ be a Fréchet-differentiable operator. Suppose that there exist $x_0 \in D$, $\eta \geq 0$, $\gamma > 1$, $L_\gamma > 0$, $L_{0,\gamma} \geq 0$ such that*

$$F'(x_0) \in L(Y, X),$$

$$\|F'(x_0)^{-1} F(x_0)\| \leq \eta,$$

$$D_\gamma \subseteq D,$$

$$\|F'(x_0)^{-1}(F'(x) - F'(x_0))\| \leq L_{0,\gamma} \|x - x_0\| \quad \text{for each } x \in D_\gamma,$$

$$\|F'(x_0)^{-1}(F'(x) - F'(y))\| \leq L_\gamma \|x - y\| \quad \text{for each } x, y \in D_\gamma,$$

$$h_{3,\gamma} = 2L_{3,\gamma} \eta \leq 1,$$

and

$$r_{3,\gamma} \leq \gamma \eta,$$

where the set D_γ is given in Theorem 12.2.3,

$$L_{3,\gamma} = \frac{1}{4}(4L_{0,\gamma} + \sqrt{L_\gamma L_{0,\gamma} + 8L_{0,\gamma}^2} + \sqrt{L_{0,\gamma} L_\gamma}),$$

$$r_{3,\gamma} = \eta + \frac{L_{0,\gamma} \eta^2}{2(1 - \alpha_\gamma)(1 - L_{0,\gamma} \eta)},$$

and

$$\alpha_\gamma = \frac{2L_\gamma}{L_\gamma + \sqrt{L_\gamma^2 + 8L_{0,\gamma} L_\gamma}}.$$

Then the sequence $\{x_n\}$ generated by Newton's method is well defined, remains in $U(x_0, r_{3,\gamma})$ for each $n = 0, 1, 2, \ldots$, and converges to a unique solution $x^* \in \bar{U}(x_0, r_{3,\gamma})$ of equation (12.1.1).

Proof. The hypotheses of Theorem 12.2.5 on D_γ are satisfied.   □

Example 12.2.7. *Let $X = Y = C[0, 1]$, the space of continuous functions defined on $[0, 1]$, be equipped with the max-norm. Let $\Omega = \{x \in C[0, 1]; \|x\| \leq R\}$ for $R > 0$. Define operator F on Ω by*

$$F(x)(s) = x(s) - f(s) - \lambda \int_0^1 G(s, t)x(t)^3 \, dt, \quad x \in C[0, 1], \ s \in [0, 1],$$

where $f \in C[0, 1]$ is a given function, λ is a real constant, and the kernel G is the Green's function defined by

$$G(s, t) = \begin{cases} (1 - s)t, & t \leq s, \\ s(1 - t), & s \leq t. \end{cases}$$

In this case, for each $x \in \Omega$, $F'(x)$ is a linear operator defined on Ω by the following expression:

$$[F'(x)(v)](s) = v(s) - 3\lambda \int_0^1 G(s, t)x(t)^2 v(t) \, dt, \quad v \in C[0, 1], \ s \in [0, 1].$$

If we choose $x_0(s) = f(s) = 1$, it follows that $\|I - F'(x_0)\| \leq 3|\lambda|/8$. Thus if $|\lambda| < 8/3$, $F'(x_0)^{-1}$ is defined and

$$\|F'(x_0)^{-1}\| \leq \frac{8}{8 - 3|\lambda|}.$$

Moreover,

$$\|F(x_0)\| \leq \frac{|\lambda|}{8},$$

$$\eta = \|F'(x_0)^{-1}F(x_0)\| \leq \frac{|\lambda|}{8 - 3|\lambda|}.$$

On the other hand, for each $x, y \in \Omega$, we have

$$\|F'(x) - F'(y)\| \leq \|x - y\|\frac{1 + 3|\lambda|(\|x + y\|)}{8} \leq \|x - y\|\frac{1 + 6R|\lambda|}{8}$$

and

$$\|F'(x) - F'(1)\| \leq \|x - 1\|\frac{1 + 3|\lambda|(\|x\| + 1)}{8} \leq \|x - 1\|\frac{1 + 3(1 + R)|\lambda|}{8}.$$

Choosing $\lambda = 1$ and $R = 2$, we have

$$\eta = 0.2,$$
$$L = 2.6,$$
$$L_0 = 2,$$
$$L_3 = 22.8624,$$
$$\alpha = 0.544267\ldots,$$

and

$$r_3 = 0.346284\ldots$$

Condition (12.1.3) is not satisfied, since

$$h = 1.04 > 1.$$

However, choosing $\gamma = 6$, conditions of Theorem 12.2.6 are satisfied, since

$$\eta = 0.2,$$
$$L_\gamma = 1.64,$$
$$L_{0,\gamma} = 1.52,$$
$$L_{3,\gamma} = 12.3667\ldots,$$
$$\alpha_\gamma = 0.512715\ldots,$$
$$r_{3,\gamma} = 0.289636\ldots \le \gamma\eta = 1.2,$$

and

$$2L_{3,\gamma}\eta = 0.656 \le 1.$$

The convergence of Newton's method is ensured by Theorem 12.2.6.

We present a semilocal convergence result of Newton's method involving condition (12.1.7) [12].

Theorem 12.2.8. *Let $F : \mathcal{D} \subset X \to Y$ be a Fréchet-differentiable operator. Suppose that conditions (H_1)–(H_3) and $D_\gamma \subset D$ hold. Moreover, suppose that there exist $K_0 > 0$, $K_1 > 0$ such that*

$$\|F'(x_0)^{-1}(F'(x_1) - F'(x_0))\| \le K_0\|x_1 - x_0\|,$$

$$\|F'(x_0)^{-1}(F'(x_0 + \theta(x_1 - x_0)) - F'(x_0))\| \le K_1\theta\|x_1 - x_0\|$$
for each $\theta \in [0, 1]$,

$$h_4 = 2L_4\eta_0 \le 1,$$

where

$$x_1 = x_0 - F'(x_0)^{-1}F(x_0),$$

$$L_4 = \frac{1}{2\eta_0} \leq 1,$$

$$\alpha_0 = \frac{L(t_2 - t_1)}{2(1 - L_0 t_2)},$$

$$t_1 = \eta, \quad t_2 = \eta + \frac{K_1 \eta^2}{2(1 - K_0 \eta)},$$

$$r_4 = \eta + \left(1 + \frac{\alpha_0}{1 - \alpha_\gamma}\right) \frac{K \eta^2}{2(1 - K_0 \eta)},$$

η_0 *is defined by*

$$\eta_0 = \begin{cases} \dfrac{1}{L_0 + K_0}, & \text{if } B = LK + 2\alpha_\gamma(L_0(K - 2K_0)) = 0, \\ \text{positive root of } p, & \text{if } B > 0, \\ \text{small positive root of } p, & \text{if } B < 0, \end{cases}$$

and

$$p(t) = (LK + 2\alpha_\gamma L_0(K - 2K_0))t^2 + 4\alpha_\gamma(L_0 + K_0)t - 4\alpha_\gamma.$$

Then the sequence $\{x_n\}$ generated by Newton's method is well defined, remains in $U(x_0, r_4)$ for each $n = 0, 1, 2, \ldots$, and converges to a unique solution $x^* \in \bar{U}(x_0, r_4)$ of equation (12.1.1).

Notice that if $B = 0$, then $p(t) = 0$ for $t = \eta_0$. The discriminant of the quadratic polynomial p is positive, since

$$16\alpha_\gamma^2(L_0 + K_0)^2 + 16\alpha_\gamma(LK + 2\alpha_\gamma L_0(K - 2K_0))$$

$$= 16\alpha_\gamma[\alpha_\gamma(L_0 - K_0)^2 + 2\alpha_\gamma L_0(K_0 + K) + KL].$$

If $B > 0$, then p has a unique positive root by the Descarte's rule of signs. But if $B < 0$, by the Vietta relations of the roots of the quadratic polynomial, the product of the roots equals

$$-\frac{4\alpha_\gamma}{B} > 0$$

and the sum of the roots equals

$$\frac{-4\alpha_\gamma(L_0 + K_0)}{B} > 0.$$

Therefore, p has two positive roots.

An improvement of Theorem 12.2.8 is

Theorem 12.2.9. *Let $F : D \subset X \to Y$ be a Fréchet-differentiable operator. Suppose that there exist $x_0 \in D$, $\eta \geq 0$, $\gamma > 1$, $K_{0,\gamma} > 0$, $K_\gamma > 0$, $L_\gamma > 0$, and $L_{0,\gamma} \geq 0$ such that*

$$F'(x_0) \in L(Y, X),$$

$$\|F'(x_0)^{-1} F(x_0)\| \leq \eta,$$

$$D_\gamma \subseteq D,$$

$$\|F'(x_0)^{-1}(F'(x_1) - F'(x_0))\| \leq K_{0,\gamma} \|x_1 - x_0\|,$$

$$\|F'(x_0)^{-1}(F'(x_0 + \theta(x_1 - x_0)) - F'(x_0))\| \leq K_\gamma \theta \|x_1 - x_0\|$$
$$\text{for each } \theta \in [0, 1],$$

$$\|F'(x_0)^{-1}(F'(x) - F'(x_0))\| \leq L_{0,\gamma} \|x - x_0\| \quad \text{for each } x \in D_\gamma,$$

$$\|F'(x_0)^{-1}(F'(x) - F'(y))\| \leq L_\gamma \|x - y\| \quad \text{for each } x, y \in D_\gamma,$$

$$h_{4,\gamma} = 2L_{4,\gamma} \eta_{0,\gamma} \leq 1,$$

and

$$r_{4,\gamma} \leq \gamma \eta,$$

where

$$x_1 = x_0 - F'(x_0)^{-1} F(x_0),$$

$$\alpha_{0,\gamma} = \frac{L_\gamma (t_2 - t_1)}{2(1 - L_{0,\gamma} t_2)},$$

$$t_1 = \eta, \quad t_2 = \eta + \frac{K_\gamma \eta^2}{2(1 - K_{0,\gamma} \eta)},$$

$$r_{4,\gamma} = \eta + \left(1 + \frac{\alpha_{0,\gamma}}{1 - \alpha_\gamma}\right) \frac{K_\gamma \eta^2}{2(1 - K_{0,\gamma} \eta)},$$

$\eta_{0,\gamma}$ is defined by

$$\eta_{0,\gamma} = \begin{cases} \dfrac{1}{L_{0,\gamma} + K_{0,\gamma}}, & \text{if } B = \begin{aligned} & L_\gamma K_\gamma + 2\alpha_\gamma (L_{0,\gamma}(K_\gamma \\ & -2K_{0,\gamma})) = 0, \end{aligned} \\[2mm] \text{positive root of } p_\gamma, & \text{if } L_\gamma K_\gamma + 2\alpha_\gamma L_{0,\gamma}(K_\gamma - 2K_{0,\gamma}) > 0, \\ \text{small positive root of } p_\gamma, & \text{if } L_\gamma K_\gamma + 2\alpha_\gamma L_{0,\gamma}(K_\gamma - 2K_{0,\gamma}) < 0, \end{cases}$$

and

$$p_\gamma(t) = (L_\gamma K_\gamma + 2\alpha_\gamma L_{0,\gamma}(K_\gamma - 2K_{0,\gamma}))t^2 + 4\alpha_\gamma(L_{0,\gamma} + K_{0,\gamma})t - 4\alpha_\gamma.$$

Then the sequence $\{x_n\}$ generated by Newton's method is well defined, remains in $U(x_0, r_{4,\gamma})$ for each $n = 0, 1, 2, \ldots$, and converges to a unique solution $x^* \in \bar{U}(x_0, r_{4,\gamma})$ of equation (12.1.1).

Proof. The hypotheses of Theorem 12.2.9 on D_γ are satisfied. □

Next we present the local convergence results.

Theorem 12.2.10. *[11,12] Let $F : \mathcal{D} \subset X \to Y$ be a Fréchet-differentiable operator. Suppose that there exist $x^* \in D$, $l_0 > 0$, and $l > 0$ such that*

$$F(x^*) = 0,$$
$$F'(x^*) \in L(Y, X),$$
$$\|F'(x^*)^{-1}(F'(x) - F'(x^*))\| \le l_0\|x - x^*\|, \quad \text{for each} \quad x \in D,$$
$$\|F'(x^*)^{-1}(F'(x) - F'(y))\| \le l\|x - y\|, \quad \text{for each} \quad x, y \in D,$$

and

$$\bar{U}(x^*, \varrho) \subset D,$$

where

$$\varrho = \frac{2}{2l_0 + l}.$$

Then the sequence $\{x_n\}$ generated for $x_0 \in U(x^*, \varrho) \setminus \{x^*\}$ by Newton's method is well defined, remains in $U(x^*, \varrho)$ for each $n = 0, 1, 2, \ldots$, and converges to a x^*. Moreover, for $T \in [\varrho, \frac{2}{l_0})$, the limit point x^* is the only solution of equation (12.1.1) in $\bar{U}(x^*, T) \cap D$.

Then an improvement of Theorem 12.2.10 is given by

Theorem 12.2.11. *Let $F : \mathcal{D} \subset X \to Y$ be a Fréchet-differentiable operator. Suppose that there exist $x^* \in D$, $\delta \ge 1$, $l_{0,\delta} > 0$, and $l_\delta > 0$ such that*

$$F(x^*) = 0, \quad F'(x^*)^{-1} \in L(Y, X),$$
$$\|F'(x^*)^{-1}(F'(x) - F'(x^*))\| \le l_{0,\delta}\|x - x^*\| \quad \text{for each } x \in D_\delta,$$
$$\|F'(x^*)^{-1}(F'(x) - F'(y))\| \le l_\delta\|x - y\| \quad \text{for each } x, y \in D_\delta,$$
$$D_\delta = \bar{U}(x^*, \delta\|x_0 - x^*\|) \subseteq D \quad \text{for } x_0 \in D,$$

and

$$\varrho_\delta \le \delta\|x_0 - x^*\|,$$

where

$$\varrho_\delta = \frac{2}{2l_{0,\delta} + l_\delta}.$$

Then the sequence $\{x_n\}$ generated for $x_0 \in U(x^*, \varrho_\delta) \setminus \{x^*\}$ by Newton's method is well defined, remains in $U(x^*, \varrho_\delta)$ for each $n = 0, 1, 2, \ldots$, and converges to x^*. Moreover, for $T \in [\varrho_\delta, \frac{2}{l_{0,\delta}})$, the limit point x^* is the only solution of equation (12.1.1) in $\bar{U}(x^*, T) \cap D$.

Proof. The hypotheses of Theorem 12.2.11 are satisfied on the domain D_δ. □

Remark 12.2.12. *(a) If $D = U(x_0, \xi)$ for some $\xi > \eta$, then $\gamma \in [1, \frac{\xi}{\eta}]$, provided $\eta \neq 0$.*

(b) If we set $\gamma = \dfrac{2}{1 + \sqrt{1 - 2L_\gamma \eta}}$, then condition $R_\gamma \leq \gamma\eta$ is satisfied as an equality. Another choice for γ is given by $\gamma = 2$. Then again $R_\gamma \leq \gamma\eta$, since we have that $R_\gamma \leq 2\eta = \gamma\eta$.

(c) Clearly, we have

$$L_{0,\gamma} \leq L,$$
$$L_\gamma \leq L,$$
$$K_{0,\gamma} \leq K_0,$$

and

$$K_\gamma \leq K.$$

Therefore, we get

$$h \leq 1 \Rightarrow h_\gamma \leq 1,$$
$$h_3 \leq 1 \Rightarrow h_{3,\gamma} \leq 1,$$

and

$$h_4 \leq 1 \Rightarrow h_{4,\gamma} \leq 1,$$

but not necessarily vice versa, unless $L_{0,\gamma} = L$, $L_\gamma = L$, $K_{0,\gamma} = K_0$, and $K_\gamma = K$.
Notice also that the new majorizing sequences are more precise that the corresponding older ones. As an example, the majorizing sequences $\{t_n\}$, $\{\bar{t}_n\}$ for Newton's method corresponding to conditions $h \leq 1$ and $h_\gamma \leq 1$ are:

$$t_0 = 0, \quad t_1 = \eta, \quad t_{n+1} = t_n + \frac{L(t_n - t_{n-1})^2}{2(1 - Lt_n)},$$

$$\bar{t}_0 = 0, \quad \bar{t}_1 = \eta, \quad \bar{t}_{n+1} = \bar{t}_n + \frac{L_\gamma (\bar{t}_n - \bar{t}_{n-1})^2}{2(1 - L_{0,\gamma} \bar{t}_n)}.$$

Then, a simple induction argument shows that

$$\bar{t}_n \leq t_n,$$

$$0 \leq \bar{t}_{n+1} - \bar{t}_n \leq t_{n+1} - t_n,$$

and

$$R_\gamma \leq R.$$

If $L_\gamma < L$, then the first inequality is strict for $n \geq 2$ and the second inequality is strict for $n \geq 1$. Moreover, we have that $R_\gamma < R$. Hence in this case the information on the location of the solution x^* is more precise under the new approach. Similar comments can be made for the majorizing sequences corresponding to the other "h" and corresponding "h_γ" conditions. Finally, notice that the majorizing sequences corresponding to conditions (12.1.5)–(12.1.7) have already been shown to be more precise than sequence $\{t_n\}$ which corresponds to condition (12.1.4) [10,12].

(d) If $D = U(x^*, \xi)$ for some $\xi > \|x_0 - x^*\|$, then δ can be chosen so that $\delta \in [1, \frac{\xi}{\|x_0 - x^*\|})$ for $x_0 \neq x^*$.

(e) We have

$$l_{0,\delta} \leq l_0$$

and

$$l_\delta \leq l.$$

Therefore we get

$$\varrho \leq \varrho_\delta.$$

Moreover, if $l_{0,\delta} < l_0$ or $l_\delta < l$, then $\varrho < \varrho_\delta$. The corresponding error bounds are also improved, since we have

$$\|x_{n+1} - x^*\| \leq \frac{l\|x_n - x^*\|^2}{2(1 - l_0\|x_n - x^*\|)}.$$

Notice that, if $l_0 = l$, then Theorem 12.2.9 reduces to the corresponding one due to Rheinboldt [25] and Traub [26]. The radius found independently by these authors is given by

$$\bar{\varrho} = \frac{2}{3l}.$$

However, if $l_0 < l$, then our radius is such that

$$\bar{\varrho} < \varrho < \varrho_\delta$$

and

$$\frac{\bar{\varrho}}{\varrho} \to \frac{1}{3} \ as \ \frac{l_0}{l} \to 0.$$

Hence our radius of convergence ϱ can be at most three times larger than $\bar{\varrho}$.

12.3 NUMERICAL EXAMPLES

Example 12.3.1. *Let $\mathcal{X} = \mathcal{Y} = C[0, 1]$ be equipped with the max-norm. Consider the following nonlinear boundary value problem:*

$$\begin{cases} u'' = -u^3 - \alpha \, u^2, \\ u(0) = 0, \quad u(1) = 1. \end{cases}$$

It is well known that this problem can be formulated as the integral equation

$$u(s) = s + \int_0^1 G(s, t) \, (u^3(t) + \alpha \, u^2(t)) \, dt \qquad (12.3.1)$$

where G is the Green's function defined by

$$G(s, t) = \begin{cases} t \, (1 - s), & t \leq s, \\ s \, (1 - t), & s < t. \end{cases}$$

We observe that

$$\max_{0 \leq s \leq 1} \int_0^1 |G(s, t)| \, dt = \frac{1}{8}.$$

Then problem (12.3.1) is in the form (12.1.1), where $F : \mathcal{D} \longrightarrow \mathcal{Y}$ is defined as

$$[F(x)] \, (s) = x(s) - s - \int_0^1 G(s, t) \, (x^3(t) + \alpha \, x^2(t)) \, dt.$$

Set $u_0(s) = s$ and $\mathcal{D} = U(u_0, 1)$. It is easy to verify that $\|u_0\| = 1$. If $2\alpha < 5$, the operator F' satisfies the invertibility conditions. Choosing $\alpha = 0.5$ and $\gamma = 1.5$, we obtain

$$\eta = 0.375,$$

$$L_\gamma = 0.59375,$$

TABLE 12.1 Radius of convergence.

δ	R_{TR}	ϱ	ϱ_δ
3	0.245253	0.324947	0.448563
4	0.245253	0.324947	0.418223
5	0.245253	0.324947	0.324947

TABLE 12.2 Corresponding error bounds for Traub's condition, i.e., $l = l_0 = e$.

n	$\|x_n - x^*\|$
0	0.000955501
1	3.94473×10^{-7}
2	8.41053×10^{-14}
3	3.84158×10^{-27}

$$h_\gamma = 0.429688\ldots,$$

and

$$R_\gamma = 0.411354 \le \gamma\eta = 0.5625.$$

So we can ensure the convergence of $\{x_n\}$.

Example 12.3.2. *Let $X = Y = \mathbb{R}^3$, $D = \overline{U}(0,1)$. Define F on D for $v = (x, y, z)^T$ by*

$$F(v) = (e^x - 1, \frac{e-1}{2}y^2 + y, z)^T. \qquad (12.3.2)$$

Then, the Fréchet-derivative is given by

$$F'(v) = \begin{bmatrix} e^x & 0 & 0 \\ 0 & (e-1)y+1 & 0 \\ 0 & 0 & 1 \end{bmatrix}.$$

Notice that $x^* = (0,0,0)^T$, $F'(x^*) = F'(x^*)^{-1} = diag\{1,1,1\}$, $l_0 = e - 1 < l = e$. Choosing $x_0 = (0.2, 0.2, 0.2)^T$, in Table 12.1 we see the radius found by Traub, our old one, and the new one presented in this chapter. Notice that our radius ϱ_δ is larger than the older ϱ and that given by Traub, r_{TR} (see also Table 12.2). Moreover, the corresponding error bounds are shown in Tables 12.3–12.5.

Example 12.3.3. *We consider the following Planck's radiation law [1] problem which calculates the energy density within an isothermal blackbody and is given*

TABLE 12.3 Corresponding error bounds for our old condition, i.e., $l = e$, $l_0 = e - 1$.	
n	$\|x_n - x^*\|$
0	0.00092994
1	3.9426×10^{-7}
2	8.41053×10^{-14}
3	3.84158×10^{-27}

TABLE 12.4 Corresponding error bounds for our new condition with $\delta = 4$.	
n	$\|x_n - x^*\|$
0	0.000584918
1	2.49195×10^{-7}
2	5.31647×10^{-14}
3	2.42834×10^{-27}

TABLE 12.5 Corresponding error bounds for our new condition with $\delta = 3$.	
n	$\|x_n - x^*\|$
0	0.000582411
1	2.49173×10^{-7}
2	5.31647×10^{-14}
3	2.42834×10^{-27}

by:

$$\vartheta(\lambda) = \frac{8\pi c P \lambda^{-5}}{e^{\frac{cP}{\lambda BT}} - 1}, \tag{12.3.3}$$

where λ is the wavelength of the radiation, T is the absolute temperature of the blackbody, B is the Boltzmann constant, P is the Planck constant, and c is the speed of light. We are interested in determining the wavelength λ which corresponds to the maximum energy density $\vartheta(\lambda)$.

From (12.3.3), we obtain

$$\vartheta'(\lambda) = \left(\frac{8\pi c P \lambda^{-6}}{e^{\frac{cP}{\lambda BT}} - 1} \right) \left(\frac{\frac{cP}{\lambda Bt} e^{\frac{cP}{\lambda BT}}}{e^{\frac{cP}{\lambda BT}} - 1} - 5 \right), \tag{12.3.4}$$

so that the maximum of ϑ occurs when

$$\frac{\frac{cP}{\lambda Bt}e^{\frac{cP}{\lambda BT}}}{e^{\frac{cP}{\lambda BT}}-1}=5. \tag{12.3.5}$$

Furthermore, if $x = \frac{cP}{\lambda BT}$, then (12.3.5) is satisfied if

$$F(x)=e^{-x}+\frac{x}{5}-1=0. \tag{12.3.6}$$

Therefore the solutions of $F(x)=0$ give the maximum wavelength of radiation λ by means of the following formula:

$$\lambda \approx \frac{cP}{x^*BT}, \tag{12.3.7}$$

where x^* is a solution of (12.3.6).

Function (12.3.6) is continuous and such that $F(1) = -0.432121\ldots$ and $F(9) = 0.800123\ldots$ By the intermediate value theorem, function $F(x)$ has a zero in the interval $[1,9]$. So we consider $\mathcal{D} = U(5,4)$ and $x^* = 4.965114\ldots$ We choose $x_0 = 3$, $\delta = 1.5$ and obtain:

$$l_{0,\delta}=0.2214851\ldots,$$
$$l_\delta = 0.6890122\ldots,$$

and

$$\rho_\delta = 1.76681\ldots \le 2.94767\ldots = \delta\|x_0 - x^*\|.$$

As a consequence, conditions of Theorem 12.2.11 are satisfied and we can ensure the convergence of Newton's method.

REFERENCES

[1] M. Abramowitz, I.S. Stegun, Handbook of Mathematical Functions With Formulas, Graphs and Mathematical Tables, Applied Math. Ser., vol. 55, United States Department of Commerce, National Bureau of Standards, Washington DC, 1964.

[2] S. Amat, S. Busquier, C. Bermúdez, Á.A. Magreñán, Expanding the applicability of a third order Newton-type method free of bilinear operators, Algorithms 8 (3) (2015) 669–679.

[3] S. Amat, S. Busquier, C. Bermúdez, Á.A. Magreñán, On the election of the damped parameter of a two-step relaxed Newton-type method, Nonlinear Dynam. 84 (1) (2016) 9–18.

[4] S. Amat, S. Busquier, S. Plaza, Chaotic dynamics of a third-order Newton-type method, J. Math. Anal. Appl. 366 (1) (2010) 24–32.

[5] S. Amat, Á.A. Magreñán, N. Romero, On a two-step relaxed Newton-type method, Appl. Math. Comput. 219 (24) (2013) 11341–11347.

[6] I.K. Argyros, Convergence and Application of Newton-Type Iterations, Springer, 2008.

[7] I.K. Argyros, J.A. Ezquerro, M.A. Hernández-Verón, Á.A. Magreñán, Convergence of Newton's method under Vertgeim conditions: new extensions using restricted convergence domains, J. Math. Chem. 55 (7) (2017) 1392–1406.

[8] I.K. Argyros, S. González, Á.A. Magreñán, Majorizing sequences for Newton's method under centred conditions for the derivative, Int. J. Comput. Math. 91 (12) (2014) 2568–2583.

[9] I.K. Argyros, J.M. Gutiérrez, Á.A. Magreñán, N. Romero, Convergence of the relaxed Newton's method, J. Korean Math. Soc. 51 (1) (2014) 137–162.

[10] I.K. Argyros, S. Hilout, Weaker conditions for the convergence of Newton's method, J. Complexity 28 (2012) 364–387.

[11] I.K. Argyros, S. Hilout, Numerical Methods in Nonlinear Analysis, World Scientific Publ. Comp., NJ, 2013.

[12] I.K. Argyros, S. Hilout, On an improved convergence analysis of Newton's method, Appl. Math. Comput. 225 (2013) 372–386.

[13] I.K. Argyros, Á.A. Magreñán, Extended convergence results for the Newton–Kantorovich iteration, J. Comput. Appl. Math. 286 (2015) 54–67.

[14] I.K. Argyros, Á.A. Magreñán, L. Orcos, J.A. Sicilia, Local convergence of a relaxed two-step Newton like method with applications, J. Math. Chem. 55 (7) (2017) 1427–1442, https://www.scopus.com/inward/record.uri?eid=2-.

[15] I.K. Argyros, Á.A. Magreñán, J.A. Sicilia, Improving the domain of parameters for Newton's method with applications, J. Comput. Appl. Math. 318 (2017) 124–135.

[16] F. Chicharro, A. Cordero, J.R. Torregrosa, Drawing dynamical and parameters planes of iterative families and methods, Sci. World J. 2013 (2013) 780153.

[17] J.A. Ezquerro, M.A. Hernández, How to improve the domain of parameters for Newton's method, Appl. Math. Let. 48 (2015) 91–101.

[18] L.V. Kantorovich, G.P. Akilov, Functional Analysis, Pergamon Press, Oxford, 1982.

[19] Á.A. Magreñán, I.K. Argyros, On the local convergence and the dynamics of Chebyshev–Halley methods with six and eight order of convergence, J. Comput. Appl. Math. 298 (2016) 236–251.

[20] Á.A. Magreñán, Different anomalies in a Jarratt family of iterative root-finding methods, Appl. Math. Comput. 233 (2014) 29–38.

[21] Á.A. Magreñán, A new tool to study real dynamics: the convergence plane, Appl. Math. Comput. 248 (2014) 215–224.

[22] Á.A. Magreñán, I.K. Argyros, Optimizing the applicability of a theorem by F. Potra for Newton-like methods, Appl. Math. Comput. 242 (2014) 612–623.

[23] Á.A. Magreñán, I.K. Argyros, J.A. Sicilia, New improved convergence analysis for Newton-like methods with applications, J. Math. Chem. 55 (7) (2017) 1505–1520.

[24] H. Ramos, J. Vigo-Aguiar, The application of Newton's method in vector form for solving nonlinear scalar equations where the classical Newton method fails, J. Comput. Appl. Math. 275 (2015) 228–237.

[25] W.C. Rheinboldt, An adaptive continuation process for solving systems of nonlinear equations, in: Banach Ctr. Publ., vol. 3, Polish Academy of Science, 1978, pp. 129–142.

[26] J.F. Traub, Iterative Methods for the Solution of Equations, Prentice-Hall Ser. Automat. Comput., Prentice-Hall, Englewood Cliffs, NJ, 1964.

Chapter 13

Newton's method for solving optimal shape design problems

13.1 INTRODUCTION

In this chapter we are concerned with the following problem

$$\min_{u \in U} F(u) \tag{13.1.1}$$

where $F(u) = J(u, S(u), z(u)) + \frac{\varepsilon}{2}\|u - u_T\|^2$, $\varepsilon \in \mathbb{R}$, functions u_T, S, z, and J are defined on a function space (Banach or Hilbert) U with values in another function space V.

Many optimal shape design problems can be formulated as in (13.1.1) [6]. In the elegant work by W. Laumen [6] the mesh independence principle [1] (see also [2]) was transferred to the minimization problem by the necessary first order condition

$$F'(u) = 0 \text{ in } U. \tag{13.1.2}$$

The most well-known and used method for solving (13.1.2) is Newton's method, which is given by

$$F''(u_{n-1})(w)(v) = -F'(u_{n-1})(v),$$
$$u_n = u_{n-1} + w,$$

for $n \in \mathbb{N}$, where F, F', and F'' also depend on functions defined on the infinite-dimensional Hilbert space V. The discretization of this method is obtained by replacing the infinite-dimensional spaces V and U with the finite-dimensional subspaces V^M, U^M and the discretized Newton's method

$$F''_N\left(u_{n-1}^M\right)\left(w^M\right)\left(v^M\right) = -F'_N\left(u_{n-1}^M\right)\left(v^M\right),$$
$$u_n^M = u_{n-1}^M + w^M.$$

In this chapter we show that under the same hypotheses and computational cost, a finer mesh independence principle can be given.

A Contemporary Study of Iterative Methods. DOI: 10.1016/B978-0-12-809214-9.00013-9

185

13.2 THE MESH INDEPENDENCE PRINCIPLE

Let u_0 be chosen in the closed ball

$$U_* = U(u_*, r_*) = \{u \in U : \|u - u_*\| \le r_*, r_* > 0\}$$

in order to guarantee convergence to the solution u_*. The assumptions concerning the cost function F_N, which are assumed to hold on a possibly smaller ball $\hat{U}_* = U(u_*, \hat{r}_*)$ with $\hat{r}_* \le r_*$, are stated below.

- **Assumption C1.** There exist positive constants δ, L_0, and L such that

$$\left\| F''(u_*)^{-1} \right\| \le \delta,$$

$$\left\| F''(u) - F''(u_*) \right\| \le L_0 \|u - u_*\| \text{ for all } u \in U_*,$$

$$\left\| F''(u) - F''(v) \right\| \le L \|u - v\| \text{ for all } u, v \in U_x^0 = U_* \cap U(u_*, \frac{1}{L_0}).$$

- **Assumption C2.** There exist uniformly bounded Lipschitz constants $L_N^{(i)}$, $i = 1, 2$, such that

$$\left\| F_N'(u) - F_N'(v) \right\| \le L_N^{(1)} \|u - v\|, \quad \text{for all } u, v \in \hat{U}_*, N \in \mathbb{N},$$

$$\left\| F_N''(u) - F_N''(v) \right\| \le L_N^{(2)} \|u - v\|, \quad \text{for all } u, v \in \hat{U}_*, N \in \mathbb{N}.$$

Without loss of generality, we assume $L_N^{(i)} \le L$, $i = 1, 2$, for all N.

- **Assumption C3.** There exists a sequence $z_N^{(1)}$, with $z_N^{(1)} \to 0$ as $N \to \infty$, such that

$$\left\| F_N'(u) - F'(u) \right\| \le z_N^{(1)}, \quad \text{for all } u \in \hat{U}_*, N \in \mathbb{N},$$

$$\left\| F_N''(u) - F''(u) \right\| \le z_N^{(2)}, \quad \text{for all } u \in \hat{U}_*, N \in \mathbb{N}.$$

- **Assumption C4.** There exists a sequence $z_N^{(2)}$, with $z_N^{(2)} \to 0$ as $N \to \infty$, such that for all $N \in \mathbb{N}$ there exists a $\hat{u}^N \in U^N \times \hat{U}_*$ such that

$$\left\| \hat{u}^N - u_* \right\| \le z_N^{(2)}.$$

- **Assumption C5.** F_N' and F_N'' correspond to the derivatives of F_N.
  The cost function F is assumed to be twice continuously Fréchet differentiable. As a consequence, its first derivative is also Lipschitz continuous:

$$\left\| F'(u) - F'(v) \right\| \le \hat{L} \|u - v\|, \quad \text{for all } u, v \in U_*.$$

Without loss of generality, we assume $\hat{L} \le L$. Furthermore, we assume that $L_0 \le L$. Otherwise, i.e., if $L \le L_0$, the results that follow hold with L_0 replacing L.

Remark 13.2.1. *In general,*

$$L_0 \le L_1, \; L \le L_1 \tag{13.2.1}$$

hold and $\dfrac{L_1}{L_0}$ *can be arbitrarily large [3], where L_1 is the third Lipschitz constant in Assumption C_1 but for $u, v \in U_*$. Notice that $U_*^0 \subseteq U_*$. If $L_0 = L = L_1$, our Assumptions C1–C5 coincide with those in [6, p. 1074].*

Otherwise, our assumptions are finer and preserve the same computational cost since in practice the evaluation of L requires the evaluation of L_0. This modification of the assumptions in [6] will result in larger convergence balls U_* and $\hat{U}_*$, which in turn implies a wider choice of initial guesses for Newton's method and finer bounds on the distances involved. This observation is important in computational mathematics.

Notice that

$$\delta \left\| F''(u_*) - F_N''\left(\hat{u}^M\right) \right\| \le \delta \left[\left\| F''(u_*) - F''\left(\hat{u}^M\right) \right\| \right.$$
$$+ \left. \left\| F''\left(\hat{u}^M\right) - F_N''\left(\hat{u}^M\right) \right\| \right]$$
$$\le \delta \left[L_0 z_M^{(2)} + z_N^{(1)} \right]$$
$$\le \delta \hat{z} < 1$$

holds for a constant $\hat{z}$ if M and N are sufficiently large. It also follows by the Banach lemma on invertible operators that $F''\left(\hat{u}^M\right)^{-1}$ exists and

$$\left\| F''\left(\hat{u}^M\right)^{-1} \right\| \le \frac{\delta}{1 - \delta \hat{z}} = \hat{\delta}.$$

We showed in [3] that if

$$r_* \le \frac{2}{(2L_0 + L)\delta} < \frac{1}{\delta L_0}, \tag{13.2.2}$$

then the estimates

$$\delta \left\| F''(u_i) - F''(u_*) \right\| \le \delta L_0 \|u_i - u_*\| \le \delta L_0 r_* < 1$$

hold, which again also imply the existence of $F''(u_i)$ with

$$\left\| F''(u_i)^{-1} \right\| \le \frac{\delta}{1 - \delta L_0 r_*} = \hat{\delta}. \tag{13.2.3}$$

Hence by redefining δ by $\hat{\delta}$ if necessary, we assume that

$$\left\| F''(u_i)^{-1} \right\| \le \delta, \quad \text{for all } i \in \mathbb{N} \tag{13.2.4}$$

$$\left\| F_N'' \left(\hat{u}^M \right)^{-1} \right\| \le \delta, \quad \text{for all } \hat{u}^M \in U^M, N \in \mathbb{N}, \tag{13.2.5}$$

for M and N satisfying

$$\delta \left[L_0 z_M^{(2)} + z_N^{(1)} \right] \le \delta \hat{z} < 1. \tag{13.2.6}$$

The next result is a refinement of Theorem 2.1 in [6, p. 1075], which also presents sufficient conditions for the existence of a solution of the problem

$$\min_{u \in U^M} F_N \left(u^M \right) \tag{13.2.7}$$

and shows the convergence of Newton's method for $M, N \to \infty$.

Theorem 13.2.2. *Let Assumptions C1–C5 hold and parameters M, N satisfy*

$$z_{MN} = 2\delta \left[\max\{1, L_0\} + \frac{1}{2\delta} \right] \left(z_N^{(1)} + z_M^{(2)} \right) \le \min \left\{ r_*, \frac{1}{\delta L} \right\}. \tag{13.2.8}$$

Then the discretized Newton's method has a local solution $u_*^M \in \hat{U}_*$ satisfying

$$\left\| u_*^M - u_* \right\| < z_{MN}. \tag{13.2.9}$$

Proof. We apply the Newton–Kantorovich theorem [5] to Newton's method starting at $u_0^M = \hat{u}^M$ to obtain the existence of a solution u_*^M of the infinite-dimensional minimization problem. Using Assumptions C2–C4, (13.2.5), and (13.2.8), we obtain in turn

$$2h = 2\delta L \left\| F_N'' \left(\hat{u}^M \right)^{-1} F_N' \left(\hat{u}^M \right) \right\|$$

$$\le 2\delta L \left\| F_N'' \left(\hat{u}^M \right)^{-1} \right\| \left\| F_N' \left(\hat{u}^M \right) \right\|$$

$$\le 2\delta^2 L \left(\left\| F_N' \left(\hat{u}^M \right) - F' \left(\hat{u}^M \right) \right\| + \left\| F' \left(\hat{u}^M \right) - F' \left(u_* \right) \right\| \right)$$

$$\le 2\delta^2 L \left(z_N^{(1)} + L_0 \left\| \hat{u}^M - u_* \right\| \right)$$

$$\le 2\delta^2 L \left(\max\{1, L_0\} + \frac{1}{2\delta} \right) \left(z_N^{(1)} + z_M^{(2)} \right)$$

$$\le \delta L z_{MN} \le 1, \tag{13.2.10}$$

which implies the required assumption $2h < 1$ (for the quadratic convergence).

We also need to show $U \left(\hat{u}^M, r(h) \right) \subset U \left(u_*, \hat{r}_* \right)$. By Assumption C4, is suffices to show

$$r(h) = \frac{1}{\delta L} \left(1 - \sqrt{1 - 2h} \right) \le \hat{r}_* - z_M^{(2)}. \tag{13.2.11}$$

But by (13.2.8) and the definition of $r(h)$, we get

$$r(h) = 2\delta \max\{1, L_0\} \left(z_N^{(1)} + z_M^{(2)}\right)$$

$$< 2\delta \left(\max\{1, L_0\} + \frac{1}{2\delta}\right)\left(z_N^{(1)} + z_M^{(2)}\right) - z_M^{(2)}$$

$$\leq z_{MN} - z_M^{(2)}$$

$$\leq \hat{r}_* - z_M^{(2)}, \tag{13.2.12}$$

which is (13.2.11).

Hence, there exists a solution $u_*^M \in U\left(\hat{u}^M, r(h)\right)$ such that

$$\left\|u_*^M - u_*\right\| \leq \left\|u_*^M - \hat{u}^M\right\| + \left\|\hat{u}^M - u_*\right\|$$

$$< z_{MN} - z_M^{(2)} + z_M^{(2)} = z_{MN}. \quad \square \tag{13.2.13}$$

Remark 13.2.3. *If equalities hold in (13.2.5) then Theorem 13.2.2 reduces to Theorem 2.1 in [6]. Otherwise, it is an improvement (and under the same computational cost) since $\hat{z}, M, N, z_{MN}$ are smaller and $\hat{\delta}$ (i.e., δ), $r_*, \hat{r}_*$ are larger than the corresponding ones in [6, p. 1075] and our condition (13.2.10) is weaker than the corresponding (2.5) in [6] (i.e., set $L_0 = L$ in (13.2.10)). That is, the claims made in Remark 13.2.1 are justified and our theorem extends the applicability of the mesh independence principle.*

Remark 13.2.4. *(a) In practice we want $\min\left\{r_*, \dfrac{1}{\delta L}\right\}$ to be as large as possible. It then immediately follows from (13.2.2), (13.2.10), and (13.2.12) that the conclusions of Theorem 13.2.2 hold if z_{MN} given in (13.2.8) is replaced by*

$$z_{MN}^1 = \frac{2\delta L}{L_0}\left[\max\{1, L_0\} + \frac{1}{2\delta}\right]\left(z_N^{(1)} + z_M^{(2)}\right) \leq \min\left\{r_*, \frac{1}{\delta L_0}\right\}. \tag{13.2.14}$$

(b) Another way is to rely on our Theorem 2.1 in [4] using the weaker (than (13.2.10)) Newton–Kantorovich-type hypothesis:

$$2\,h_0 = 2\,\overline{L}\,\delta\left\|F_N''\left(\hat{u}^M\right)^{-1} F_N'\left(\hat{u}^M\right)\right\| \leq 1, \tag{13.2.15}$$

where

$$\frac{1}{2}\overline{L}^{-1} = \begin{cases} \frac{1}{L_0+H}, & \text{if } b = LK + \alpha L_0(K - 2H) = 0, \\ 2\frac{-\alpha(L_0+H)+\sqrt{\alpha^2(L_0+H)^2+\alpha b}}{b}, & \text{if } b > 0, \\ -2\frac{\alpha(L_0+H)+\sqrt{\alpha^2(L_0+H)^2+\alpha b}}{b}, & \text{if } b < 0, \end{cases}$$

$$\alpha = \frac{2L}{L+\sqrt{L^2+8L_0L}}, \text{ and } H, K \text{ satisfy}$$

$$\|F'(u_1) - F'(u_0)\| \leq H\|u_1 - u_0\|,$$

$$\|F'(u_0 + \theta(u_1 - u_0)) - F'(u_0)\| \leq K\theta\|u_1 - u_0\| \text{ for all } \theta \in [0, 1].$$

As in (13.2.10), for (13.2.15) to hold we must have

$$2\,h_0 \leq 2\,\overline{L}\,\delta^2 \left[\max\{1, L_0\} + \frac{1}{2\delta}\right] \left(z_N^{(1)} + z_M^{(2)}\right)$$

$$\leq \delta L_0 z_{MN}^0 \leq 1,$$

provided that

$$z_{MN}^0 = \frac{2\,\overline{L}}{L_0}\,\delta \left[\max\{1, L_0\} + \frac{1}{2\delta}\right] \left(z_N^{(1)} + z_M^{(2)}\right)$$

$$\leq \min\left\{r_*, \frac{1}{\delta L_0}\right\}. \tag{13.2.16}$$

The other hypothesis for the application of our Theorem 2.1 in [4] is

$$U\left(\hat{u}^M, r^1(h)\right) \subseteq U\left(u_*, \hat{r}_*\right),$$

where

$$r^1(h) = 2\delta \max\{1, L_0\} \left(z_N^{(1)} + z_M^{(2)}\right).$$

As a consequence, we can state the following result.

Theorem 13.2.5. *Under the hypotheses of Theorem 13.2.2 with z_{MN} replaced by z_{MN}^0 (given in (13.2.13)) the conclusions of this theorem hold.*

Theorem 13.2.5 improves the earlier result given by Theorem 5 in [6], since L_1 was used there and $H \leq K \leq L_0 \leq L \leq L_1$.

Note also that this improvement is obtained under the same computational cost, since in practice the computation of L_1 requires the computation of the other constants as special cases. Moreover, we showed that a solution

$$u_*^M \in U\left(\hat{u}^M, r(h)\right) \text{ (or } \left(\hat{u}^M, r^1(h)\right)) \subset U\left(u_*, \hat{r}_*\right)$$

of the discretized minimization problem exists.

Furthermore, in the main results of this study we show two different ways of improving the corresponding Theorem 2.2 in [6, p. 1076], where it was shown that the discretized Newton's method converges to the solution u_*^M for any $u_0^M \in U(u_*, r_1)$ for sufficiently small r_1.

In order to further motivate the reader, let us provide a simple numerical example.

TABLE 13.1 Comparison table.

M	z_{MN} (13.2.20)	r_*^L (13.2.18)	z_{MN} (13.2.8)	z_{MN} (13.2.14)	z_{MN} (13.2.16)	r_* (13.2.19)
27	0.238391246	0.24525296	0.164317172	0.259945939	0.194620625	0.3826919122323857
22				0.319024562		
19			0.23350335			
18					0.291930937	

Example 13.2.6. *Let $U = \mathbb{R}$, $U_* = U(0, 1)$ and define the real function F on U_* by*

$$F(x) = e^x - 1. \tag{13.2.17}$$

Then we obtain using (13.2.17) that $L = e^{\frac{1}{L_0}}$, $L_0 = e - 1$, $L_1 = e$, and $\delta = 1$. We let $z_N^{(1)} = 0$, $z_M^{(2)} = \frac{1}{M}$. Set $r^* = \hat{r}_*$ and $N = 0$.
The convergence radius given in [6, p. 1075] is

$$r_*^L = \hat{r}_*^L = \frac{2}{3\delta L_1} = 0.24525296, \tag{13.2.18}$$

whereas by (13.2.2) our radius is given by

$$r^* = \hat{r}_* = \frac{2}{(2L_0 + L)\,\delta} = 0.3826919122323857. \tag{13.2.19}$$

That is, (13.2.1) holds as a strict inequality and

$$r_*^L = \hat{r}_*^L < r^* = \hat{r}_*.$$

The condition (2.4) used in [6, p. 1075] corresponding to our (13.2.8) is given by

$$z_{MN}^L = 2\delta \left[\max\{1, L\} + \frac{1}{2\delta} \right] \left(z_N^{(1)} + z_M^{(2)} \right) \leq \min \left\{ \hat{r}_*^L, \frac{1}{\delta L} \right\}. \tag{13.2.20}$$

We can tabulate the following results containing the minimum M, for which conditions (13.2.6), (13.2.8), (13.2.14), and (13.2.16) are satisfied.
Table 13.1 indicates the superiority of our results over those in [6, p. 1075].

We can now present a finer version of Theorem 2.2 given in [6] of the mesh independence principle.

Theorem 13.2.7. *Suppose Assumptions C1–C5 are satisfied and there exist discretization parameters M and N such that*

$$z_{MN} \leq \frac{1}{6} \min \left\{ \frac{\hat{r}_*}{4}, \frac{1}{(2L_0 + 3L)\delta + 1} \right\}. \tag{13.2.21}$$

Then the discretized Newton's method converges to u_*^M for all starting points $u_0^M \in U(u_*, r_1)$, where

$$r_1 = \frac{3}{4} \min \left\{ \frac{1}{(2L_0 + L)\delta}, \frac{\hat{r}_*}{2} \right\}.$$
(13.2.22)

Moreover, if

$$\left\| u_0^M - u_0 \right\| \leq \tau,$$

where

$$\tau = \frac{2 \left(\frac{1}{2} + \| u_0 - u_* \| \right) z_{MN}}{b^2 + \sqrt{b^2 - 6L\delta \left(\frac{1}{2} + \| u_0 - u_* \| \right) z_{MN}}}$$
(13.2.23)

and

$$b = 1 + \frac{1}{2} z_{MN} - 2\delta L \left\| u_0^M - u_* \right\|,$$

the following estimates hold for $c_i \in \mathbb{R}$, $i = 1, 2, 3, 4$, $n \in \mathbb{N}$:

$$\left\| u_{n+1}^M - u_*^M \right\| \leq c_1 \left\| u_n^M - u_*^M \right\|^2,$$
(13.2.24)

$$\left\| u_n^M - u_n \right\| \leq c_2 z_{MN},$$

$$\left\| F_N' \left(u_n^M \right) - F'(u_n) \right\| \leq c_3 z_{MN},$$

and

$$\left\| u_n^M - u_*^M \right\| \leq \| u_n - u_* \| + c_4 z_{MN}.$$

Proof. We first show the convergence of the discretized Newton's method for all u_0^M in a suitable ball around u_*. Since the assumptions of Theorem 13.2.2 are satisfied, the existence of a solution $u_*^M \in \hat{U}_*$ is guaranteed. We shall show that the discretized Newton's method converges to u_*^M if $u_0^M \in U(u_*, r_2)$, where $r_2 = \min \left\{ \frac{1}{(2L_0 + L)\delta}, \frac{\hat{r}_*}{2} \right\}$.

The estimates

$$\left\| u_*^M - u_* \right\| + \left\| u_0^M - u_*^M \right\| \leq 2 \left\| u_*^M - u_* \right\| + \left\| u_0^M - u_* \right\|$$

$$\leq 2 z_{MN} + r_2 \leq \hat{r}_*$$

imply $U \left(u_*^M, \left\| u_0^M - u_*^M \right\| \right) \subset \hat{U}_*$. Hence, Assumptions C1–C5 hold in $U \left(u_*^M, \left\| u_0^M - u_*^M \right\| \right)$.

We also have

$$\left\| F''\left(u_*\right)^{-1}\right\| \left\| F_N''\left(u_*^M\right) - F''\left(u_*\right)\right\| \leq \delta \left[\left\| F_N''\left(u_*^M\right) - F_N''\left(u_*\right)\right\|\right.$$
$$\left. + \left\| F_N''\left(u_*\right) - F''\left(u_*\right)\right\|\right]$$
$$\leq \delta \left[L \left\| u_*^M - u_*\right\| + z_{MN}\right]$$
$$\leq (\delta L + 1) z_{MN}$$
$$\leq \frac{\delta L + \frac{1}{2}}{(2_0 + 3L)\delta + 1} < 1, \qquad (13.2.25)$$

where we used the fact that $z_n^{(1)} \leq \frac{1}{2\delta} z_{MN}$ (by (13.2.8)).

It follows by (13.2.25) and the Banach lemma on invertible operators that $F_N''\left(u_*^M\right)^{-1}$ exists and

$$\left\| F_N''\left(u_*^M\right)^{-1}\right\| \leq \frac{\delta}{1 - \left(\delta L + \frac{1}{2}\right) z_{MN}}.$$

By the theorem on quadratic convergence of Newton's method and since all assumptions hold, the convergence to u_*^M has been established.

Using a refined formulation of this theorem given in [3], the convergence is guaranteed for all $u_0^M \in U\left(u_*^M, r_3\right)$, where

$$r_3 = \frac{2}{(2L_0 + L)\left\| F_N''\left(u_*^M\right)^{-1}\right\|}.$$

Therefore we should show

$$U(u_*, r_2) \subset U\left(u_*^M, r_3\right),$$

or, equivalently,

$$\left\| u_0^M - u_*^M\right\| \leq \left\| u_0^M - u_*\right\| + \left\| u_* - u_*^M\right\|$$
$$\leq r_2 + z_{MN}$$
$$\leq \frac{1}{(2L_0 + L)\delta} + z_{MN}$$
$$= \frac{1 + (2L_0 + L)\delta}{(2L_0 + L)\delta}$$
$$\leq \frac{2\left(1 - L\delta z_{MN} - \frac{1}{2}z_{MN}\right)}{(2L_0 + L)\delta}$$
$$\leq \frac{2}{(2L_0 + L)\left\| F_N''\left(u_*^M\right)^{-1}\right\|} = r_3.$$

Hence the discretized Newton's method converges to u_*^M for all $u_0^M \in U(u_*, r_2)$ such that (13.2.24) holds for $c_1 = \delta L$. Next, using induction, we can state

$$\left\| u_n^M - u_n \right\| \leq \tau \leq c_2 z_{MN} \tag{13.2.26}$$

for all $u_0^M \in U(u_*, r_1)$, $r_1 = \frac{3}{4} r_2$, where τ is given by

$$\tau = \frac{2 \left(\frac{1}{2} + \|u_0 - u_*\| \right) z_{MN}}{b^2 + \sqrt{b^2 - 6L\delta \left(\frac{1}{2} + \|u_0 - u_*\| \right) z_{MN}}}$$

$$\leq \frac{\left(\frac{1}{2} + \|u_0 - u_*\| \right) z_{MN}}{b^2} =: c_2 z_{MN}$$

with $b = 1 + \frac{1}{2} z_{MN} - 2\delta L \left\| u_0^M - u_* \right\|$. The constant τ is well defined, since the inequalities

$$6L\delta \left(\frac{1}{2} + \|u_0 - u_*\| \right) z_{MN} \leq \frac{2L\delta + 1}{4\left((2L_0 + 3L)\delta + 1 \right)} < \frac{1}{4} \text{ and}$$

$$b \geq 1 - 2\delta L \rho_1 \geq \frac{1}{2}$$

imply $b^2 \geq \frac{1}{4} \geq 6L\delta \left(\frac{1}{2} + \|u_0 - u_*\| \right) z_{MN}$.

While the assertion (13.2.26) is fulfilled by assumption for $n = 0$, the induction step is based on the simple decomposition

$$u_{i+1}^M - u_{i+1} = F_N'' \left(u_i^M \right)^{-1} \left\{ F_N'' \left(u_i^M \right) \left(u_i^M - u_i \right) - F_N' \left(u_i^M \right) + F_N' (u_i) \right. \tag{13.2.27}$$

$$+ \left(F_N'' \left(u_i^M \right) - F_N'' (u_i) \right) F'' (u_i)^{-1} F' (u_i)$$

$$+ F_N'' (u_i) F'' (u_i)^{-1} F' (u_i) - F' (u_i)$$

$$\left. + F' (u_i) - F_N' (u_i) \right\}.$$

Assumptions C1–C4, equation (13.2.26), and the definition of z_{MN} imply

$$\delta \left\| F_N'' \left(u_i^M \right) - F'' (u_i) \right\| \leq \delta \left\| F_N'' \left(u_i^M \right) - F'' (u_i) \right\| + \delta \left\| F_N'' (u_i) - F'' (u_i) \right\|$$

$$\leq \delta \left(L\tau + z_N^{(1)} \right)$$

$$\leq \delta L\tau + \frac{1}{2} z_{MN}$$

$$\leq \frac{\delta L z_{MN} + 2 \|u_0 - u_*\| \delta L z_{MN}}{1 - 2\delta L \|u_0 - u_*\|} + \frac{1}{2} z_{MN}$$

$$\le \frac{\frac{1}{3}\delta L + \frac{1}{4}}{(2L_0 + 3L)\delta + 1} < 1,$$

resulting in the inequality $\left\| F_N'' \left(u_i^M \right)^{-1} \right\| \le \frac{\delta}{1 - \left(L\delta\tau + \frac{1}{2} z_{MN} \right)}$. We obtain

$$\left\| F_N'' \left(u_i^M \right) \left(u_i^M - u_i \right) - F_N' \left(u_i^M \right) + F_N' (u_i) \right\| \le \frac{1}{2} L \left\| u_i^M - u_i \right\|^2 \le \frac{1}{2} L\tau^2,$$

and the convergence assertion $\| u_i - u_* \| \le \| u_0 - u_* \|$ yields

$$\left\| \left(F_N'' \left(u_i^M \right) - F_N'' (u_i) \right) F'' (u_i)^{-1} F' (u_i) \right\| \le L \left\| u_i^M - u_i \right\| \| u_i - u_{i+1} \|$$
$$\le 2L\tau \| u_0 - u_* \|.$$

The assumptions of the theorem lead to

$$\left\| F_N'' (u_i) F'' (u_i)^{-1} F' (u_i) - F' (u_i) \right\| \le$$
$$\le \left\| -F_N'' (u_i) (u_{i+1} - u_i) + F'' (u_i) (u_{i+1} - u_i) \right\|$$
$$\le \left\| F_N'' (u_i) - F'' (u_i) \right\| \| u_{i+1} - u_i \|$$
$$\le z_n^{(1)} 2 \| u_0 - u_* \|$$
$$\le \frac{1}{\delta} z_{MN} \| u_0 - u_* \|$$

and $\| F' (u_i) - F_N' (u_i) \| \le z_n^{(1)} \le \frac{1}{2\delta} z_{MN}$. Using the decomposition (13.2.27), the last inequalities complete the induction proof by

$$\left\| u_{i+1}^M - u_{i+1} \right\| \le$$
$$\le \frac{\delta}{1 - \left(L\delta\tau + \frac{1}{2} z_{MN} \right)} \left\{ \frac{1}{2} L\tau^2 + 2L \| u_0 - u_* \| \tau + \left(\frac{1}{2} + \| u_0 - u_* \| \right) \frac{z_{MN}}{\delta} \right\}$$
$$= \tau.$$

The last equality is based on the fact that τ is equal to the smallest solution of the quadratic equation $3L\delta\tau^2 - 2b\tau + 2z_{MN} \left(\frac{1}{2} + \| u_0 - u_* \| \right) = 0$.

Finally, inequality (13.2.26) is shown by

$$\left\| F_N' \left(u_n^M \right) - F' (u_n) \right\| \le \left\| F_N' \left(u_n^M \right) - F_N' (u_n) \right\| + \left\| F_N' (u_n) - F' (u_n) \right\|$$
$$\le L \left\| u_n^N - u_n \right\| + z_{MN}$$
$$\le (Lc_2 + 1) z_{MN} =: c_3 z_{MN},$$

and inequality (13.2.23) results from

$$\left\| \left(u_n^M - u_*^M \right) - (u_n - u_*) \right\| \le \left\| u_n^M - u_n \right\| + \left\| u_*^M - u_* \right\|$$
$$\le c_2 z_{MN} + z_{MN}$$
$$\le (c_2 + 1) z_{MN} =: c_4 z_{MN}. \qquad \square$$

Remark 13.2.8. *The upper bounds on z_{MN} and r_1 were defined in [6] by*

$$\frac{1}{6} \min \left\{ \frac{\hat{r}_*}{4}, \frac{1}{6L\delta + 1} \right\} \qquad (13.2.28)$$

and

$$\frac{3}{4} \min \left\{ \frac{1}{3L\delta}, \frac{\hat{r}_*}{2} \right\}, \qquad (13.2.29)$$

respectively. By comparing (13.2.21), (13.2.22) with (13.2.28) and (13.2.29), respectively, we conclude that our choices of z_{MN} and r_1 (or r_2, or r_3) are finer than those in [6], using L_1 instead of L in Theorem 13.2.7.

REFERENCES

[1] E.L. Allgower, K. Böhmer, F.A. Potra, W.C. Rheinboldt, A mesh-independence principle for operator equations and their discretizations, SIAM J. Numer. Anal. 23 (1986) 160–169.
[2] I.K. Argyros, A.A. Magreñán, Iterative Algorithms I, Nova Science Publishers Inc., New York, 2016.
[3] I.K. Argyros, A.A. Magreñán, Iterative Methods and Their Dynamics With Applications, CRC Press, New York, 2017.
[4] I.K. Argyros, On an improved convergence analysis of Newton's method, Appl. Math. Comput. 225 (2013) 372–386.
[5] L.V. Kantorovich, G.P. Akilov, Functional Analysis in Normed Spaces, Pergamon Press, Oxford, 1982.
[6] M. Laumen, Newton's mesh independence principle for a class of optimal design problems, SIAM J. Control Optim. 37 (1999) 1070–1088.

Chapter 14

Osada method

14.1 INTRODUCTION

Many problems in applied sciences and also in engineering can be written in the form

$$f(x) = 0, \qquad (14.1.1)$$

using mathematical modeling, where $f : D \subseteq \mathbb{R} \longrightarrow \mathbb{R}$ is sufficiently many times differentiable and D is a convex subset in $\mathbb{R}$. In the present study, we pay attention to the case of a solution p of multiplicity $m > 1$, namely, $f(x_*) = 0$, $f^{(i)}(x_*) = 0$ for $i = 1, 2, \ldots, m - 1$, and $f^{(m)}(x_*) \neq 0$. The determination of solutions of multiplicity m is of great interest [1–4,8–11,13–22]. In the study of electron trajectories, when the electron reaches a plate of zero speed, the distance from the electron to the plate function has a solution of multiplicity two. Multiplicity of solutions appears in connection to van der Waals equation of state and other phenomena. The convergence order of iterative methods decreases if the equation has solutions of multiplicity m. Modifications in the iterative function are made to improve the order of convergence. The modified Newton's method (MN) defined for each $n = 0, 1, 2, \ldots$ by

$$x_{n+1} = x_n - m f'(x_n)^{-1} f(x_n), \qquad (14.1.2)$$

where $x_0 \in D$ is an initial point, is an alternative to Newton's method in the case of solutions with multiplicity m that has second order of convergence. The Osada method (Osada method) [19–21] is defined for $n = 0, 1, 2, \ldots$ by

$$x_{n+1} = x_n - \frac{m(m+1)}{2} f'(x_n^{-1} f(x_n)) + \frac{(m-1)^2}{2} f''(x_n)^{-1} f(x_n). \quad (14.1.3)$$

Other iterative methods of high convergence order can be found in [5–9,12,13, 16,19] and the references therein.

Let $U(p, \lambda) := \{x \in U_1 : |x - p| < \lambda\}$ denote an open ball and $\bar{U}(p, \lambda)$ denote its closure. It is said that $U(p, \lambda) \subseteq D$ is a convergence ball for an iterative method, if the sequence generated by this iterative method converges to p, provided that the initial point $x_0 \in U(p, \lambda)$. But how close should x_0 be to x_* so that convergence can take place? Extending the ball of convergence is very important, since it shows the difficulty we confront to pick initial points. It is desirable to be able to compute the largest convergence ball. This is usually depending

A Contemporary Study of Iterative Methods. DOI: 10.1016/B978-0-12-809214-9.00014-0

197

on the iterative method and the conditions imposed on the function f and its derivatives. We can unify these conditions by expressing them as:

$$\|(f^{(m)}(x_*))^{-1}f^{(m+1)}(x)\| \le \varphi_0(\|x - x_*\|), \tag{14.1.4}$$

$$\|(f^{(m)}(x_*))^{-1}(f^{(m+1)}(x) - f^{(m+1)}(y))\| \le \varphi(\|x - y\|) \tag{14.1.5}$$

for all $x, y \in D$, where $\varphi_0, \varphi : \mathbb{R}_+ \cup \{0\} \longrightarrow \mathbb{R}_+ \cup \{0\}$ are continuous and nondecreasing functions satisfying $\varphi_0(0) = 0$. If $m \ge 1$, $\varphi_0(t) = \mu_0$, and

$$\varphi(t) = \mu t^q, \mu_0 > 0, \mu > 0, q \in [0, 1], \tag{14.1.6}$$

then, we obtain the conditions under which the preceding methods were studied [1–4,8–11,13–22]. However, there are cases where even (14.1.6) does not hold. Moreover, the smaller the functions φ_0, φ, the larger the radius of convergence. The technique, which we present next, can be used for all preceding methods, as well as in methods where $m = 1$. However, in the present study, we only use it for Osada method. The authors in [21] have used the incorrect Taylor's expansion with the integral remainder:

$$F(x) = \sum_{k=0}^{n} \frac{F^{(k)}(x_*)}{k!}(x - x_*)^k + \bar{E}_n(x),$$

where

$$\bar{E}_n(x) = \frac{1}{n!} \int_{x_*}^{x} (F^{(n+1)}(x) - F^{(n+1)}(x_*))(x - \theta)^n d\theta,$$

leading to the wrong value of the radius of convergence for Osada method under conditions (14.1.4) and (14.1.5) in the special cases when functions φ_0 and φ are given in (14.1.6). The correct Taylor's expansion is

$$F(x) = \sum_{k=0}^{n} \frac{F^{(k)}(x_*)}{k!}(x - x_*)^k + E_n(x),$$

where

$$E_n(x) = \frac{1}{n!} \int_{0}^{x} (x - \theta)^n F^{(n+1)}(\theta) d\theta,$$

or

$$E_n(x) = \frac{1}{(n-1)!} \int_{x_*}^{x} (F^{(n)}(\theta) - F^{(n)}(x_*))(x - \theta)^{n-1} d\theta.$$

In the present study we find the correct radius of convergence and in the more general setting of conditions (14.1.4) and (14.1.5).

The rest of the chapter is structured as follows: Section 14.2 contains some auxiliary results on divided differences and derivatives. The ball convergence of Osada method is given in Section 14.3. The numerical examples are in the concluding Section 14.4.

14.2 AUXILIARY RESULTS

In order to make the chapter as self-contained as possible, we restate some standard definitions and properties for divided differences [4,8,9,13–22].

Definition 14.2.1. *The divided differences $f[y_0, y_1, \ldots, y_k]$ on $k+1$ distinct points $y_0, y_1, \ldots y_k$ of a function $f(x)$ are defined by*

$$
\begin{aligned}
f[y_0] &= f(y_0), \\
f[y_0, y_1] &= \frac{f[y_0] - f[y_1]}{y_0 - y_1}, \\
&\vdots \\
f[y_0, y_1, \ldots, y_k] &= \frac{f[y_0, y_1, \ldots, y_{k-1}] - f[y_0, y_1, \ldots, y_k]}{y_0 - y_k}.
\end{aligned}
\tag{14.2.1}
$$

If the function f is sufficiently smooth, then its divided differences $f[y_0, y_1, \ldots, y_k]$ can be defined if some of the arguments y_i coincide, for instance, if $f(x)$ has k-th derivative at y_0, then it makes sense to define

$$
f[\underbrace{y_0, y_1, \ldots, y_k}_{k+1}] = \frac{f^{(k)}(y_0)}{k!}.
\tag{14.2.2}
$$

Lemma 14.2.2. *The divided differences $f[y_0, y_1, \ldots, y_k]$ are symmetric functions of their arguments, i.e., they are invariant to permutations of the $y_0, y_1, \ldots, y_k$.*

Lemma 14.2.3. *If a function f has k-th derivative, and $f^{(k)}(x)$ is continuous on the interval $I_x = [\min(y_0, y_1, \ldots, y_k), \max(y_0, y_1, \ldots, y_k)]$, then*

$$
f[y_0, y_1, \ldots, y_k] = \int_0^1 \cdots \int_0^1 \theta_1^{k-1} \theta_2^{k-1} \cdots \theta^{k-1} f^{(k)}(\theta) d\theta_1 \cdots d\theta_k,
\tag{14.2.3}
$$

where $\theta = y_0 + (y_1 - y_0)\theta_1 + (y_2 - y_1)\theta_1\theta_2 + \cdots + (y_k - y_{k-1})\theta_1 \cdots \theta_k$.

Lemma 14.2.4. *If a function f has $(k+1)$-th derivative, then for every argument x, the following formula holds*

$$
\begin{aligned}
f(x) &= f[v_0] + f[v_0, v_1](x - v_0) + \cdots \\
&+ f[v_0, v_1, \ldots, v_k](x - v_0) \cdots (x - v_k) \\
&+ f[v_0, v_1, \ldots, v_k, x]\lambda(x),
\end{aligned}
\tag{14.2.4}
$$

where

$$\lambda(x) = (x - v_0)(x - v_1)\cdots(x - v_k). \qquad (14.2.5)$$

Lemma 14.2.5. *Assume that a function f has continuous $(m+1)$-th derivative, and x_* is a zero of multiplicity m. We define functions g_0, g, and g_1 as*

$$g_0(x) = f[\underbrace{x_*, x_*, \ldots, x_*}_{m}, x, x], \; g(x) = f[\underbrace{x_*, x_*, \ldots, x_*}_{m}, x],$$

$$g_1(x) = f[\underbrace{x_*, x_*, \ldots, x_*}_{m}, x, x, x]. \qquad (14.2.6)$$

Then

$$g'(x) = g_0(x), \; g''(x) = g_1(x). \qquad (14.2.7)$$

Lemma 14.2.6. *If a function f has $(m+1)$-th derivative, and x_* is a zero of multiplicity m, then for every argument x, the following formulae hold:*

$$
\begin{aligned}
f(x) &= f[\underbrace{x_*, x_*, \ldots, x_*}_{m}, x](x - x_*)^m = g(x)(x - x_*)^m, \\
f'(x) &= f[\underbrace{x_*, x_*, \ldots, x_*}_{m}, x, x](x - x_*)^m \\
&\quad + m f[\underbrace{x_*, x_*, \ldots, x_*}_{m}, x](x - x_*)^{m-1} \\
&= g_0(x)(x - x_*)^m + m g(x)(x - x_*)^{m-1},
\end{aligned}
\qquad (14.2.8)
$$

and

$$
\begin{aligned}
f''(x) &= 2f[\underbrace{x_*, x_*, \ldots, x_*}_{m}, x, x, x](x - x_*)^m \\
&\quad + 2m f[\underbrace{x_*, x_*, \ldots, x_*}_{m}, x, x](x - x_*)^{m-1} \\
&\quad + m(m-1) f[\underbrace{x_*, x_*, \ldots, x_*}_{m}, x](x - x_*)^{m-2} \\
&= 2g_1(x)(x - x_*)^m + 2m g_0(x)(x - x_*)^{m-1} \\
&\quad + m(m-1)g(x)(x - x_*)^{m-2},
\end{aligned}
\qquad (14.2.9)
$$

where $g_0(x)$, $g(x)$, and $g_1(x)$ were defined previously.

14.3 LOCAL CONVERGENCE

It is convenient for the local convergence analysis that follows to define some real functions and parameters. Define the function ψ_0 on $\mathbb{R}_+ \cup \{0\}$ by

$$\psi_0(t) = \frac{\varphi_0(t)t}{m+1} - 1.$$

We have $\psi_0(0) = -1 < 0$ and $\psi_0(t) > 0$, if

$$\varphi_0(t)t \longrightarrow \text{a positive number or } +\infty \qquad (14.3.1)$$

as $t \to \infty$. It then follows from the intermediate value theorem that function ψ_0 has a zero in the interval $(0, +\infty)$. Denote by ρ_0 the smallest such zero. Define functions $\varphi_0^{(m)}, \varphi^{(m)}, h_0, h_1$ on the interval $[0, \rho_0)$ by

$$\varphi_0^{(m)}(t) = m! \int_0^1 \cdots \int_0^1 \theta_1^m \theta_2^{m-1} \cdots \theta_m \varphi_0(\theta_1, \ldots, \theta_m t) d\theta_1 \cdots d\theta_{m+1},$$

$$\varphi^{(m)}(t) = m! \int_0^1 \cdots \int_0^1 \theta_1^m \theta_2^{m-1} \cdots \theta_m \varphi(\theta_1, \ldots, \theta_{m-1}(1 - \theta_m)t) d\theta_1 \cdots d\theta_{m+1},$$

$$
\begin{aligned}
h_0(t) \quad &= \quad \frac{4\varphi_0^{(m)}(t)(m+1)\varphi^{(m)}(t)t^2}{m+1-\varphi_0(t)t} \\
&+ \frac{(m+1)^3(\varphi_0^{(m)}(t))^2 t^2}{m+1-\varphi_0(t)t}, \\
&+ 2m(1-m)\varphi^{(m)}(t)t,
\end{aligned}
$$

and

$$
\begin{aligned}
h_1(t) \quad = \quad & 2m^2(m-1) - 2m(m^2 + 2m - 1)\varphi_0^{(m)}(t)t \\
& -4m\varphi_0^{(m)}(t)t - \frac{4m(m+1)(\varphi_0^{(m)}(t))^2 t^2}{m+1-\varphi_0(t)t} \\
& -\frac{4(m+1)\varphi_0^{(m)}(t)\varphi^{(m)}(t)t^2}{m+1-\varphi_0(t)t}.
\end{aligned}
$$

We get that $h_1(0) = 2m^2(m-1) > 0$ and $h_1(t) \longrightarrow -\infty$ as $t \longrightarrow \rho_0^-$. Denote by r_0 the smallest zero of function h_1 in the interval $(0, \rho_0)$. Define function h on $[0, r_0)$ by

$$h(t) = \frac{h_0(t)}{h_1(t)} - 1.$$

We obtain that $h(0) = -1 < 0$ and $h(t) \longrightarrow +\infty$ as $t \longrightarrow r_0^-$. Denote by r the smallest zero of function h on the interval $(0, r_0)$. Then, we have that for each

$t \in [0, r)$,

$$0 \leq h(t) < 1. \tag{14.3.2}$$

The local convergence analysis is based on conditions $(\mathcal{A})$:

$(\mathcal{A}_1)$ Function $f : D \subseteq \mathbb{R} \longrightarrow \mathbb{R}$ is $(m + 1)$-times differentiable, and x_* is a zero of multiplicity m.

$(\mathcal{A}_2)$ Conditions (14.1.4) and (14.1.5) hold.

$(\mathcal{A}_3)$ $\bar{U}(x_*, r) \subseteq D$, where the radius of convergence r is defined previously.

$(\mathcal{A}_4)$ Condition (14.3.1) holds.

Theorem 14.3.1. *Suppose that the* $(\mathcal{A})$ *conditions hold. Then sequence* $\{x_n\}$ *generated for* $x_0 \in U(x_*, r) - \{x_*\}$ *by Osada method is well defined in* $U(x_*, r)$, *remains in* $U(x_*, r)$ *for each* $n = 0, 1, 2 \ldots$, *and converges to* x_*.

We need the following auxiliary result.

Lemma 14.3.2. *Suppose that the* $(\mathcal{A})$ *conditions hold. Then for all* $x_0 \in I :=$ $[x_* - \rho_0, x_* + \rho_0]$ *and* $\delta_0 = x_0 - x_*$, *the following assertions hold:*

(i) $|g(x_*)^{-1} g_0(x_0)| \leq \varphi_0^{(m)}(|\delta_0|)$,

(ii) $|g(x_0)^{-1} g(x_*)| \leq \dfrac{m+1}{m + 1 - \varphi_0(|\delta_0|)|\delta_0|}$,

(iii) $|g(x_*)^{-1} g_1(x_0)\delta_0| \leq \varphi^{(m)}(|\delta_0|)$,

(iv) $|g(x_0)^{-1} g_0(x_0)| \leq \dfrac{(m+1)\varphi_0^{(m)}(|\delta_0|)}{m + 1 - \varphi_0(|\delta_0|)|\delta_0|}$,

and

(v) $|g(x_0)^{-1} g_1(x_0)\delta_0| \leq \dfrac{(m+1)\varphi^{(m)}(|\delta_0|)}{m + 1 - \varphi_0(|\delta_0|)|\delta_0|}$.

Proof. (i) Using (14.2.2) and (14.2.6), we can write $g(x_*) = f[\underbrace{x_*, x_*, \ldots, x_*}_{m+1}] =$

$\dfrac{f^{(m)}(x_*)}{m!}$. Then, by (14.2.2), (14.2.6), (14.1.4), (14.1.5), and (14.2.3), we get

$|g(x_*)^{-1} g_0(x_0)|$

$= |g(x_*)^{-1} \int_0^1 \cdots \int_0^1 \theta_1^m \theta_2^m \cdots \theta_m f^{(m+1)}(x_* + \delta_0 \theta_1 \cdots \theta_m)d\theta_1 \cdots d\theta_{m+1}|$

$= \varphi_0^{(m)}(|\delta_0|)$.

(ii) We also have by (14.1.4) and the definition of ρ_0 that

$$|1 - g(x_*)^{-1} g(x_0)| = |g(x_*)^{-1}(g(x_*) - g(x_0))|$$
$$\leq |g'(x_*)^{-1} g'(y_0)\delta_0|$$
$$\leq \dfrac{\varphi_0^{(m)}(|\delta_0|)|\delta_0|}{m+1} < 1$$

(for some y_0 between x_* and x_0). It follows from the Banach lemma on invertible functions [1–3,11,19] that $g(x_0) \neq 0$ and

$$
|g(x_0)^{-1}g(x_*)| \leq \cfrac{1}{1 - \cfrac{\varphi_0^{(m)}(|\delta_0|)|\delta_0|}{m+1}}
$$

$$
= \frac{m+1}{m+1 - \varphi_0^{(m)}(|\delta_0|)|\delta_0|}.
$$

(iii) By (14.2.3) we get in turn that

$$
|g(x_*)^{-1}g_1(x_0)\delta_0|
$$

$$
= |g(x_*)^{-1} \int_0^1 \cdots \int_0^1 \theta_1^m \theta_2^m \cdots \theta_m
$$

$$
\times [f^{(m+1)}(x_* + \delta_0\theta_1 \cdots \theta_{m-1}) - f^{(m+1)}(x_* + \delta_0\theta_1 \cdots \theta_m)]d\theta_1 \cdots d\theta_{m+1}|
$$

$$
\leq \varphi^{(m)}(|\delta_0|).
$$

Items (iv) and (v) follow immediately from (i)–(iii). $\qquad\qquad\square$

Proof Theorem 14.3.1. We base the proof on mathematical induction. By Osada method, for $n = 0$ we get

$$
x_1 = x_0 - \frac{m(m+1)}{2}\frac{f(x_0)}{f'(x_0)} + \frac{(m-1)^2}{2}\frac{f'(x_0)}{f''(x_0)}. \tag{14.3.3}
$$

By Lemma 14.2.6, we can write

$$
f(x) = g(x)(x - x_*)^m, \tag{14.3.4}
$$

where $g(x)$ has been defined in Lemma 14.2.6. We get by Lemma 14.2.5:

$$
f(x_0) = g(x_0)\delta_0^m,
$$

$$
f'(x_0) = g_0(x_0)\delta_0^m + g(x_0)m\delta_0^{m-1}, \tag{14.3.5}
$$

$$
f''(x_0) = 2g_1(x_0)\delta_0^m + 2g_0(x_0)m\delta_0^{m-1} + g(x_0)m(m-1)\delta_0^{m-2}.
$$

Using (14.3.5) in (14.3.3), we get

$$
\delta_1 = \frac{4g_0(x_0)g(x_0)\delta_0^4 + g_0(x_0)^2(m+1)^2\delta_0^3 + 2g(x_0)g_1(x_0)m(1-m)\delta_0^3}{4g_0(x_0)g_1(x_0)\delta_0^3 + 4g_0(x_0)^4m\delta_0^2 + \Gamma},
$$

where $\Gamma = 2g(x_0)g_0(x_0)m(3m-1)\delta_0 + 4mg(x_0)g_1(x_0)\delta_0^2 + 2g^2(x_0)m^2(m-1)$. Set N equal to the numerator divided by δ_0 and D the denominator, dividing both by $g(x_*)$ and $g(x_0)$. Denoting the new terms by N_0 and D_0, we can write

$$\delta_1 = \frac{N}{D}\delta_0 = \frac{g(x_*)^{-1}g(x_0)^{-1}N}{g(x_*)^{-1}g(x_0)^{-1}D}\delta_0 = \frac{N_0}{D_0}\delta_0. \qquad (14.3.6)$$

Next we use (14.1.4), (14.1.5), and Lemma 14.3.2 to find an upper bound for (14.3.6). First, we bound the numerator:

$$
\begin{aligned}
|N_0| &= |g(x_*)^{-1}g(x_0)^{-1}(4g_0(x_0)g_1(x_0)\delta_0^3 + g_0^2(x_0)(m+1)^2\delta_0^2 \\
&\qquad + 2g(x_0)g_1(x_0)m(1-m)\delta_0^2| \\
&\leq 4|g(x_*)^{-1}g_0(x_0)g_0(x_0)^{-1}g_1(x_0)\delta_0|\delta_0^2 \\
&\qquad + (m+1)^2|g(x_*)^{-1}g_0(x_0)g(x_0)^{-1}g_0(x_0)|\delta_0^2 \\
&\qquad + 2m(1-m)|g(x_*)^{-1}g_1(x_0)g(x_0)^{-1}g(x_0)\delta_0|\delta_0 \\
&\leq \frac{4\varphi_0^{(m)}(|\delta_0|)(m+1)\varphi^{(m)}(|\delta_0|)|\delta_0|^2}{m+1-\varphi_0(|\delta_0|)|\delta_0|} \\
&\qquad + \frac{(m+1)^2\varphi_0^{(m)}(|\delta_0|)(m+1)\varphi_0^{(m)}(|\delta_0|)|\delta_0|^2}{m+1-\varphi_0(|\delta_0|)|\delta_0|} \\
&\qquad + 2m(1-m)\varphi^{(m)}(|\delta_0|)|\delta_0| \\
&\leq h_0(|\delta_0|).
\end{aligned}
$$

$$(14.3.7)$$

Second, we bound the denominator by:

$$
\begin{aligned}
|D_0| &= |g(x_*)^{-1}g(x_0)^{-1}(4g_0(x_0)g_1(x_0)\delta_0^3 + 4g_0(x_0)^2m\delta_0^2 \\
&\qquad + 2g(x_0)g_0(x_0)m(3m-1)\delta_0 \\
&\qquad + 4mg(x_0)g_0(x_0)\delta_0^2 + 2g(x_0)^2m^2(m-1))| \\
&= |(4g(x_*)^{-1}g_0(x_0)g(x_0)^{-1}g_1(x_0)\delta_0^3 \\
&\qquad + 4mg(x_*)^{-1}g_0(x_0)g(x_0)^{-1}g_0(x_0)\delta_0^2 \\
&\qquad + 2m(3m-1)g(x_*)^{-1}g_0(x_0)g(x_0)^{-1}g(x_*)^{-1}g(x_0)\delta_0 \\
&\qquad + 4mg(x_*)^{-1}g_1(x_0)g(x_0)^{-1}g(x_0)\delta_0^2 \\
&\qquad + 2m^2(m-1)g(x_*)^{-1}g(x_0)g(x_0)^{-1}g(x_0))| \\
&= (2m^2(m-1)g(x_*)^{-1}g(x_0) + 2m(3m-1)g(x_*)^{-1}g_0(x_0)\delta_0 \\
&\qquad + 4mg(x_*)^{-1}g_1(x_0)\delta_0^2 \\
&\qquad + 4mg(x_*)^{-1}g_0(x_0)g(x_0)^{-1}g(x_0)\delta_0^2 \qquad (14.3.8)
\end{aligned}
$$

$$+ 4g(x_*)^{-1}g_0(x_0)g(x_0)^{-1}g_1(x_0)\delta_0^3)|$$

$$\geq 2m^2(m-1) - 2m^2(m-1)|g(x_*)^{-1}(g(x_*) - g(x_0))|$$

$$- 2m(3m-1)|g(x_*)^{-1}g_0(x_0)|\delta_0 - 4m|g(x_*)^{-1}g_1(x_0)\delta_0|\delta_0$$

$$- 4m|g(x_*)^{-1}g_0(x_0)g(x_0)^{-1}g_0(x_0)|\delta_0^2$$

$$- 4|g(x_*)^{-1}g_0(x_0)g(x_0)^{-1}g_1(x_0)\delta_0|\delta_0^2$$

$$\geq 2m^2(m-1) - 2m^2(m-1)\varphi_0^{(m)}(|\delta_0|)|\delta_0|$$

$$- 2m(3m-1)\varphi_0^{(m)}(|\delta_0|)|\delta_0|$$

$$- 4m\varphi^{(m)}(|\delta_0|)|\delta_0| - \frac{4m\varphi_0^{(m)}(|\delta_0|)(m+1)\varphi_0^{(m)}(|\delta_0|)}{m+1-\varphi_0(|\delta_0|)|\delta_0|}|\delta_0|^2$$

$$- 4\frac{\varphi_0^{(m)}(|\delta_0|)(m+1)\varphi^{(m)}(|\delta_0|)|\delta_0|^2}{m+1-\varphi_0(|\delta_0|)|\delta_0|} = h_1(|\delta_0|).$$

By simply replacing x_0, x_1 with x_k, x_{k+1} in the preceding estimates, we get

$$|x_{k+1} - x_*| \leq c|x_k - x_*| < r, \tag{14.3.9}$$

where $c = \dfrac{h_0(|\delta_0|)}{h_1(|\delta_0|)} \in [0, 1)$, so $\lim_{k \to +\infty} x_k = x_*$ and $x_{k+1} \in U(x_*, r)$. $\quad\square$

Next we present a uniqueness result for the solution x_*.

Proposition 14.3.3. *Suppose that the $(\mathcal{A})$ conditions hold. Then the limit point x_* is the only solution of equation $f(x) = 0$ in $D_1 = D \cap \bar{U}(x_*, \rho_0)$.*

Proof. Let x_{**} be a solution of equation $f(x) = 0$ in D_1. We can write by (14.2.8) that

$$f(x_{**}) = g(x_{**})(x_{**} - x_*)^m. \tag{14.3.10}$$

Using (14.1.4) and the properties of divided differences, we get in turn that

$$\begin{aligned}
|1 - g'(x_*)^{-1}g(x_{**})| &= |g(x_*)^{-1}(g(x_*) - g(x_{**}))| \\
&= |g(x_*)^{-1}\frac{f^{(m+1)}(z_0)}{(m+1)!}(x_{**} - x_*)| \\
&\leq \frac{\varphi_0(|x_{**} - x_*|)|x_{**} - x_*)}{m+1} \\
&< 1
\end{aligned} \tag{14.3.11}$$

for some point between x_{**} and x_*. It follows from (14.3.10) and (14.3.11) that $x_{**} = x_*$. $\quad\square$

14.4   NUMERICAL EXAMPLE

Example 14.4.1. *Let*

$$D = [0, 1],$$
$$m = 2,$$
$$p = 0,$$

and define function f on D by

$$f(x) = \frac{4}{35}x^{\frac{7}{2}} + \frac{1}{6}x^3 + \frac{1}{2}x^2.$$

We have

$$f'(x) = \frac{2}{5}x^{\frac{5}{2}} + \frac{x^2}{2} + x,$$
$$f''(x) = x^{\frac{3}{2}} + x + 1,$$

and

$$f''(0) = 1.$$

Function f'' cannot satisfy (14.1.5) with ψ given by (14.1.6). Hence the results in [4,8,16,17,20,21] cannot apply. However, the new results apply for

$$\varphi(t) = \frac{3}{2}t^{\frac{1}{2}}$$

and

$$\varphi_0(t) = \frac{5}{2}.$$

Moreover, the convergence radius is

$$r = 0.2564.$$

Example 14.4.2. *Let*

$$D = [-1, 1],$$
$$m = 2,$$
$$p = 0,$$

and define function f on D by

$$f(x) = e^x - x - 1.$$

We get

$$\varphi_0(t) = et,$$
$$\varphi(t) = et,$$

and the convergence radius is

$$r = 0.9487.$$

REFERENCES

[1] S. Amat, M.A. Hernández, N. Romero, Semilocal convergence of a sixth order iterative method for quadratic equations, Appl. Numer. Math. 62 (2012) 833–841.

[2] I.K. Argyros, Computational theory of iterative methods, in: C.K. Chui, L. Wuytack (Eds.), Stud. Comput. Math., vol. 15, Elsevier Publ. Co., New York, USA, 2007.

[3] I.K. Argyros, A.A. Magreñán, Iterative Methods and Their Dynamics With Applications, CRC Press, New York, 2017.

[4] W. Bi, H.M. Ren, Q. Wu, Convergence of the modified Halley's method for multiple zeros under Hölder continuous derivatives, Numer. Algor. 58 (4) (2011) 497–512.

[5] A. Cordero, J.M. Gutiérrez, Á.A. Magreñán, J.R. Torregrosa, Stability analysis of a parametric family of iterative methods for solving nonlinear models, Appl. Math. Comput. 285 (2016) 26–40.

[6] A. Cordero, Á.A. Magreñán, C. Quemada, J.R. Torregrosa, Stability study of eighth-order iterative methods for solving nonlinear equations, J. Comput. Appl. Math. 291 (2016) 348–357.

[7] A. Cordero, L. Feng, Á.A. Magreñán, J.R. Torregrosa, A new fourth-order family for solving nonlinear problems and its dynamics, J. Math. Chem. 53 (3) (2014) 893–910.

[8] C. Chun, B. Neta, A third order modification of Newton's method for multiple roots, Appl. Math. Comput. 211 (2009) 474–479.

[9] E. Hansen, M. Patrick, A family of root finding methods, Numer. Math. 27 (1977) 257–269.

[10] A.A. Magreñán, Different anomalies in a Jarratt family of iterative root finding methods, Appl. Math. Comput. 233 (2014) 29–38.

[11] A.A. Magreñán, A new tool to study real dynamics: the convergence plane, Appl. Math. Comput. 248 (2014) 29–38.

[12] Á.A. Magreñán, I.K. Argyros, On the local convergence and the dynamics of Chebyshev–Halley methods with six and eight order of convergence, J. Comput. Appl. Math. 298 (2016) 236–251.

[13] B. Neta, New third order nonlinear solvers for multiple roots, Appl. Math. Comput. 202 (2008) 162–170.

[14] N. Obreshkov, On the numerical solution of equations, Annuaire Univ. Sofia. Fac. Sci. Phy. Math. 56 (1963) 73–83 (in Bulgarian).

[15] N. Osada, An optimal multiple root-finding method of order three, J. Comput. Appl. Math. 52 (1994) 131–133.

[16] M.S. Petkovic, B. Neta, L. Petkovic, J. Džunič, Multipoint Methods for Solving Nonlinear Equations, Elsevier, 2013.

[17] H.M. Ren, I.K. Argyros, Convergence radius of the modified Newton method for multiple zeros under Hölder continuity derivative, Appl. Math. Comput. 217 (2010) 612–621.

[18] E. Schröder, Über unendlich viele Algorithmen zur Auflösung der Gleichungen, Math. Ann. 2 (1870) 317–365.

[19] J.F. Traub, Iterative Methods for the Solution of Equations, AMS Chelsea Publishing, 1982.

[20] X. Zhou, Y. Song, Convergence radius of Osada's method for multiple roots under Hölder and center-Hölder continuous conditions, in: ICNAAM 2011 (Greece), in: AIP Conf. Proc., vol. 1389, 2011, pp. 1836–1839.

[21] X. Zhou, Y. Song, Convergence radius of Osada's method under center-Hölder continuous condition, Appl. Math. Comput. 243 (2014) 809–816.

[22] X. Zhou, X. Chen, Y. Song, On the convergence radius of the modified Newton method for multiple roots under the center-Hölder condition, Numer. Algor. 65 (2) (2014) 221–232.

Chapter 15

Newton's method to solve equations with solutions of multiplicity greater than one

15.1 INTRODUCTION

Many problems in applied sciences and also in engineering can be written in the form

$$F(x) = 0, \qquad (15.1.1)$$

using mathematical modeling, where $F : \Omega \subseteq \mathcal{X} \longrightarrow \mathcal{Y}$ is sufficiently many times differentiable and $\Omega, \mathcal{X}, \mathcal{Y}$ are convex subsets in $\mathbb{R}$. In the present study, we pay attention to the case of a solution p of multiplicity $m > 1$, namely, $F(p) = 0, F^{(i)}(p) = 0$ for $i = 1, 2, \ldots, m - 1$, and $F^{(m)}(p) \neq 0$. The determination of solutions of multiplicity m is of great interest. In the study of electron trajectories, when the electron reaches a plate of zero speed, the distance from the electron to the plate function has a solution of multiplicity two. Multiplicity of solution appears in connection to van der Waals equation of state and other phenomena. The convergence order of iterative methods decreases if the equation has solutions of multiplicity m. Modifications in the iterative function are made to improve the order of convergence. The modified Newton's method (MN) defined for each $n = 0, 1, 2, \ldots$ by

$$x_{n+1} = x_n - m F'(x_n)^{-1} F(x_n), \qquad (15.1.2)$$

where $x_0 \in \Omega$ is an initial point is an alternative to Newton's method in the case of solutions with multiplicity m that has second order of convergence. A method having third order convergence is defined by

$$x_{n+1} = x_n - \left(\frac{m+1}{2m} F'(x_n) - \frac{F''(x_n) F(x_n)}{2 F'(x_n)} \right)^{-1} F(x_n). \qquad (15.1.3)$$

Method (15.1.3) is an extension of the classical Halley's method of the third order. Another cubical convergence method was given by Traub [19]:

$$x_{n+1} = x_n - \frac{m(3-m)}{2} \frac{F(x_n)}{F'(x_n)} - \frac{m^2}{2} \frac{F(x_n)^2 F''(x_n)}{F'(x_n)^3}. \qquad (15.1.4)$$

A Contemporary Study of Iterative Methods. DOI: 10.1016/B978-0-12-809214-9.00015-2

Method (15.1.4) is an extension of the Chebyshev's method of the third order. Other iterative methods of high convergence order can be found in [6–8,5,9,12, 13,16,19] and the references therein. Methods of high convergence order can be found in [5,9–11,13–16,18,19] and the references therein.

Let $B(p, \lambda) := \{x \in B_1 : |x - p| < \lambda\}$ denote an open ball and $\bar{B}(p, \lambda)$ denote its closure. It is said that $B(p, \lambda) \subseteq \Omega$ is a convergence ball for an iterative method, if the sequence generated by this iterative method converges to p, provided that the initial point $x_0 \in B(p, \lambda)$. But how close should x_0 be to p so that convergence can take place? Extending the ball of convergence is very important, since it shows the difficulty we confront to pick initial points. It is desirable to be able to compute the largest convergence ball. This usually depends on the iterative method and the conditions imposed on the function F and its derivatives. We can unify these conditions by expressing them as

$$\|(F^{(m)}(p))^{-1}(F^{(m)}(x) - F^{(m)}(y)\| \leq \psi(\|x - y\|) \qquad (15.1.5)$$

for all $x, y \in \Omega$, where $\psi : \mathbb{R}_+ \cup \{0\} \longrightarrow \mathbb{R}_+ \cup \{0\}$ is a continuous and nondecreasing function satisfying $\psi(0) = 0$. If we specialize function ψ, by taking $m \geq 1$ and letting

$$\psi(t) = \mu t^2, \mu > 0, q \in [0, 1], \qquad (15.1.6)$$

then we obtain the conditions under which the preceding methods were studied [4,5,16,17,20,21]. However, there are cases where even (15.1.6) does not hold. Moreover, the smaller the function ψ, the larger the radius of convergence. The technique, which we present next, can be used for all preceding methods, as well as in methods where $m = 1$. However, in the present study, we only use it for MN. This way, in particular, we extend the results in [4,5,16,17,20,21]. In view of (15.1.5), there always exists a continuous, nondecreasing function $\varphi_0 : \mathbb{R}_+ \cup \{0\} \longrightarrow \mathbb{R}_+ \cup \{0\}$ satisfying

$$\|(F^{(m)}(p))^{-1}(F^{(m)}(x) - F^{(m)}(p)\| \leq \varphi_0(\|x - y\|) \qquad (15.1.7)$$

for all $x \in \Omega$. We can always choose $\varphi_0(t) = \psi(t)$ for all $t \geq 0$. However, in general,

$$\varphi_0(t) \leq \psi(t), t \geq 0 \qquad (15.1.8)$$

holds, and $\dfrac{\psi}{\varphi_0}$ can be arbitrarily large [2]. Denote by r_0 the smallest positive solution of the equation $\varphi_0(t) = 1$. Set $\Omega_0 := \Omega \cup B(p, r_0)$. We have again by (15.1.5) that there exists a continuous, nondecreasing function $\varphi : [0, r_0) \longrightarrow \mathbb{R}_+ \cup \{0\}$ satisfying $\varphi(0) = 0$ such that for each $x, y \in \Omega_0$

$$\|(F^{(m)}(p))^{-1}(F^{(m)}(x) - F^{(m)}(y)\| \leq \varphi(\|x - y\|). \qquad (15.1.9)$$

Clearly, we have

$$\varphi(t) \leq \psi(t), \text{ for all } t \in [0, r_0), \qquad (15.1.10)$$

since $\Omega_0 \subseteq \Omega$. It turns out that the more precise (15.1.7) (see (15.1.8)) than (15.1.5) can be used to estimate upper bounds on the inverses of the functions involved (see (15.3.10) or (15.3.17)). Moreover, for the upper bounds on the numerators (see (15.3.11) or (15.3.18)), we can use (15.1.9) which is tighter than (15.1.5) (see (15.1.10)). This way we obtain (15.3.12) and (15.3.19), which are tighter than the corresponding ones using only ψ (or its special case (15.1.6)). This way we obtain a larger radius of convergence, leading to a wider choice of initial guesses and at least as tight error bounds on the distances $|x_n - p|$, resulting in the computation of at least as few iterates to obtain a desired error tolerance (see also the numerical examples). It is worth noticing that the preceding advantages are obtained under the same computational cost as in earlier studies, since in practice the computation of function ψ (or (15.1.6)) requires the computation of functions φ_0 and ψ as special cases.

The rest of the chapter is structured as follows: Section 15.2 contains some auxiliary results on divided differences and derivatives. The ball of convergence of MN is given in Section 15.3. The numerical examples are in the concluding Section 15.4.

15.2 AUXILIARY RESULTS

We need the definition of divided differences and their properties, which can be found in [4,17,20,21].

Definition 15.2.1. *The divided differences $F[y_0, y_1, \ldots, y_k]$ on $k + 1$ distinct points $y_0, y_1, \ldots, y_k$ of a function $F(x)$ are defined by*

$$F[y_0] = F(y_0),$$

$$F[y_0, y_1] = \frac{F[y_0] - F[y_1]}{y_0 - y_1},$$

$$\vdots$$

$$F[y_0, y_1, \ldots, y_k] = \frac{F[y_0, y_1, \ldots, y_{k-1}] - F[y_0, y_1, \ldots, y_k]}{y_0 - y_k}.$$

If the function F is sufficiently smooth, then its divided differences $F[y_0, y_1, \ldots, y_k]$ can be defined if some of the arguments y_i coincide. For instance, if $F(x)$ has k-th derivative at y_0, then it makes sense to define

$$F[\underbrace{y_0, y_1, \ldots, y_k}_{k+1}] = \frac{F^{(k)}(y_0)}{k!}. \tag{15.2.1}$$

Lemma 15.2.2. *The divided differences $F[y_0, y_1, \ldots, y_k]$ are symmetric functions of their arguments, i.e., they are invariant to permutations of the $y_0, y_1, \ldots, y_k$.*

Lemma 15.2.3. *If the function F has $(k+1)$-th derivative, and p is a zero of multiplicity m, then for every argument x, the following formula holds:*

$$F(x) = F[y_0] + \sum_{i=1}^{k} F[y_0, y_1, \ldots, y_k] \prod_{j=0}^{i-1} (x - y_j)$$

$$+ F[y_0, y_1, \ldots, y_k, x] \prod_{i=0}^{k} (x - y_i).$$

(15.2.2)

Lemma 15.2.4. *If a function F has $(m+1)$-th derivative, and p is a zero of multiplicity m, then for every argument x, the following formulae hold:*

$$F(x) = F[\underbrace{p, p, \ldots, p}_{m}, x](x - p)^m,$$

(15.2.3)

$$F'(x) = F[\underbrace{p, p, \ldots, p}_{m}, x, x](x - p)^m + m F[\underbrace{p, p, \ldots, p}_{m}, x](x - p)^{m-1}.$$

(15.2.4)

We need the following lemma on Genocchi's integral expression formula for divided differences:

Lemma 15.2.5. *If a function F has continuous k-th derivative, then the following formula holds for any points $y_0, y_1, \ldots, y_k$:*

$$F[y_0, y_1, \ldots, y_k] = \int_0^1 \cdots \int_0^1 F^{(k)}(y_0 + \sum_{i=1}^{k}(y_i - y_{i-1}) \prod_{j=1}^{i} \theta_j) \prod_{i=1}^{k}(\theta_i^{k-i} d\theta_i).$$

(15.2.5)

We shall also use the following Taylor expansion with the integral form remainder.

Lemma 15.2.6. *Suppose that $F(x)$ is n-times differentiable in the ball $B(x_0, r), r > 0$, and $F^{(n)}(x)$ is integrable from a to $x \in B(a, r)$. Then*

$$F(x) = F(a) + F'(a)(x - a) + \frac{1}{2} F''(a)(x - a)^2 + \cdots + \frac{1}{n!} F^{(n)}(a)(x - a)^n$$

$$+ \frac{1}{(n-1)!} \int_0^1 [F^{(n)}(a + t(x - a)) - F^{(n)}(a)](x - a)^n (1 - t)^{n-1} dt$$

and

$$F'(x) = F'(a) + F''(a)(x - a) + \frac{1}{2}F'''(a)(x - a)^2$$

$$+ \cdots + \frac{1}{(n-1)!}F^{(n)}(a)(x - a)^{n-1}$$

$$+ \frac{1}{(n-2)!}\int_0^1 [F^{(n)}(a + t(x - a)) - F^{(n)}(a)](x - a)^{n-1}(1 - t)^{n-2}dt.$$

15.3 BALL OF CONVERGENCE

The ball of convergence uses some auxiliary real functions and parameters. Let $\varphi_0 : \mathbb{R}_+ \cup \{0\} \longrightarrow \mathbb{R}_+ \cup \{0\}$ be a continuous and nondecreasing function satisfying $\varphi_0(0) = 0$. Define function $\beta : \mathbb{R}_+ \cup \{0\} \longrightarrow \mathbb{R}_+ \cup \{0\}$ by

$$\beta(t) = (m-1)! [\int_0^1 \cdots \int_0^1 \varphi_0(t \prod_{i=1}^{m-1} \theta_i)$$

$$+ (m-1) \int_0^1 \cdots \int_0^1 \varphi_0(t \prod_{i=1}^{m} \theta_i)] \prod_{i=1}^{m} \theta_i^{m-i} d\theta_i.$$

Notice that $\beta(0) = 0$ and function β is continuous and nondecreasing on $\mathbb{R}_+ \cup \{0\}$. Suppose that

$$\beta(t) \longrightarrow 1 \text{ as } t \longrightarrow \text{ a positive number or } + \infty. \tag{15.3.1}$$

It follows from the intermediate value theorem that equation $\beta(t) = 1$ has a solution in $(0, +\infty)$. Denote by r_0 the smallest positive solution of equation $\beta(t) = 1$. Let $\varphi : [0, r_0) \longrightarrow \mathbb{R}_+ \cup \{0\}$ be a continuous and nondecreasing function satisfying $\varphi(0) = 0$. Moreover, define an α function on $[0, r_0)$ by

$$\alpha(t) = (m-1)! \int_0^1 \cdots \int_0^1 \varphi(t \prod_{i=1}^{m-1} \theta_i(1 - \theta_m)) \prod_{i=1}^{m} \theta_i^{m-i} d\theta_i d\theta_m.$$

Furthermore, define functions δ and γ on $[0, r_0)$ by

$$\gamma(t) = \frac{\alpha(t)}{1 - \beta(t)}$$

and

$$\delta(t) = \gamma(t) - 1.$$

We get that $\delta(0) = -1 < 0$ and $\delta(t) \longrightarrow +\infty$ as $t \longrightarrow r_0^-$. Denote by r the smallest solution of equation $\delta(t) = 0$ in $(0, r_0)$. Then we have that for each

$t \in [0, r)$

$$0 \le \beta(t) < 1 \qquad (15.3.2)$$

and

$$0 \le \delta(t) < 1. \qquad (15.3.3)$$

First, we provide the ball of convergence of the modified Newton's method under conditions $(\mathcal{A})$:

$(\mathcal{A}_1)$ $F : \Omega \subseteq \mathcal{X} \longrightarrow \mathcal{Y}$ is continuously m-times Fréchet-differentiable.

$(\mathcal{A}_2)$ Function F has a zero p of multiplicity m, $m = 1, 2, \ldots$

$(\mathcal{A}_3)$ There exists a continuous and nondecreasing function $\varphi_0 : \mathbb{R}_+ \cup \{0\} \longrightarrow \mathbb{R}_+ \cup \{0\}$ satisfying $\varphi_0(0) = 0$ such that for each $x \in \Omega$

$$\| F^{(m)}(p)^{-1}(F^{(m)}(x) - F^{(m)}(p)) \| \le \varphi_0(\|x - p\|).$$

Let $\Omega_0 = \Omega \cup B(p, r_0)$, where r_0 is defined previously.

$(\mathcal{A}_4)$ There exists a continuous and nondecreasing $\varphi : [0, r) \longrightarrow \mathbb{R}_+ \cup \{0\}$ satisfying $\varphi(0) = 0$ such that for each $x, y \in \Omega_0$,

$$\| F^{(m)}(p)^{-1}(F^{(m)}(x) - F^{(m)}(y)) \| \le \varphi(\|x - p\|).$$

$(\mathcal{A}_5)$ Condition (15.3.1) holds.

$(\mathcal{A}_6)$ $\bar{B}(p, r) \subseteq \Omega$.

Theorem 15.3.1. *Suppose that Conditions $(\mathcal{A})$ hold. Then, for starting point $x_0 \in B(p, r) - \{p\}$, the sequence $\{x_n\}$ generated by MN is well defined in $B(p, r)$, remains in $B(p, r)$ for all $n = 0, 1, 2, \ldots$, and converges to p.*

Proof. We shall use mathematical induction. It is convenient to define functions $g(x)$ and $g_0(x)$ as follows:

$$g(x) = F[\underbrace{p, p, \ldots, p}_{m}, x], \quad g_0(x) = F[\underbrace{p, p, \ldots, p}_{m}, x, x]. \qquad (15.3.4)$$

Let $e_n = x_n - p$. Using Lemma 15.2.4, we can write:

$$F(x_0) = g(x_0)e_0^m, \qquad (15.3.5)$$

$$F'(x_0) = [g_0(x_0)e_0 + mg(x_0)]e_0^{m-1}. \qquad (15.3.6)$$

By NM, (15.3.5), and (15.3.6), we have

$$
\begin{aligned}
e_1 &= e_0 - \frac{mg_0(x_0)e_0^m}{[g_0(x_0)e_0 + mg(x_0)]e_0^{m-1}} \\
&= e_0 - \frac{mg(x_0)e_0}{g_0(x_0)e_0 + mg(x_0)} \\
&= \frac{(mg(p))^{-1}g_0(x_0)e_0}{(mg(p))^{-1}[g_0(x_0)e_0 + mg_0(x_0)]}e_0.
\end{aligned}
\tag{15.3.7}
$$

We suppose that $g_0(x_0)e_0 + mg(x_0) \neq 0$ (which will be shown later). In view of the definition of divided differences, we have

$$
g_0((x_0)e_0 = F[\underbrace{p, p, \ldots, p}_{m-1}, x_0, x_0] - g(x_0).
\tag{15.3.8}
$$

Then, we obtain from (15.2.1) and (15.3.8) that

$$
\begin{aligned}
&|1 - (mg(p))^{-1}[h_0(x_0)e_0 + mg(x_0)]| \\
&= |(mg(p))^{-1}[g_0(x_0)e_0 + mg(x_0) - mg(p)]| \\
&= (m-1)!F^{(m)}(p)^{-1}(F[\underbrace{p, p, \ldots, p}_{m-1}, x_0, x_0] \\
&\quad -g(p) + (m-1)[g(x_0) - g(p)])|.
\end{aligned}
$$

By Lemma 15.2.5, we get

$$
\begin{aligned}
&F[\underbrace{p, p, \ldots, p}_{m-1}, x_0, x_0] \\
&= \int_0^1 \cdots \int_0^1 F^{(m)}(p + e_0 \prod_{i=1}^{m-1}\theta_i)\prod_{i=1}^m(\theta_i^{m-1}d\theta_i), \\
g(x_0) &= \int_0^1 \cdots \int_0^1 F^{(m)}(p + e_0 \prod_{i=1}^m\theta_i)\prod_{i=1}^m(\theta_i^{m-1}d\theta_i), \\
g(p) &= \int_0^1 \cdots \int_0^1 F^{(m)}(p)\prod_{i=1}^m(\theta_i^{m-1}d\theta_i).
\end{aligned}
$$

Next, using Condition $(\mathcal{A}_3)$, $x_0 \in B(p, r)$, and the definition of r, we get

$$
\begin{aligned}
&|1 - (mg(p))^{-1}]g_0(x_0)e_0 + mg(x_0)]| \\
&= (m-1)!|\int_0^1 \cdots \int_0^1 F^{(m)}(p)^{-1}(F^{(m)}(p + e_0 \prod_{i=1}^{m-1}\theta_i) \\
&\quad -F^{(m)}(p))\prod_{i=1}^m(\theta_i^{m-i}d\theta_i) \\
&\quad +(m-1)F^{(m)}(p)^{-1}(F^{(m)}(p) + e_0 \prod_{i=1}^{m-1}\theta_i) \\
&\quad -F^{(m)}(p))\prod_{i=1}^m(\theta_i^{m-i}d\theta_i)|
\end{aligned}
$$

$$\leq (m-1)!(\int_0^1 \cdots \int_0^1 |F^{(m)}(p)^{-1}(F^{(m)}(p+e_0 \prod_{i=1}^{m-1}\theta_i)$$
$$-F^{(m)}(p))| \prod_{i=1}^m (\theta_i^{m-i}d\theta_i) \quad (15.3.9)$$
$$+(m-1)\int_0^1 \cdots \int_0^1 |F^{(m)}(p)^{-1}(F^{(m)}(p)+e_0 \prod_{i=1}^{m-1}\theta_i)$$
$$-F^{(m)}(p))| \prod_{i=1}^m (\theta_i^{m-i}d\theta_i)|$$
$$\leq (m-1)![\int_0^1 \cdots \int_0^1 \varphi_0(|e_0| \prod_{i=1}^{m-1}\theta_i) \prod_{i=1}^m \theta_i^{m-i}d\theta_i$$
$$+(m-1)\int_0^1 \cdots \int_0^1 \varphi_0(|e_0| \prod_{i=1}^m \theta_i) \prod_{i=1}^m \theta_i^{m-i}d\theta_i]$$
$$\leq \beta(|e_0|) < \beta(r) < 1.$$

It follows from the Banach perturbation lemma [1,3] and (15.3.9) that $g_0(x_0)e_0 + mg(x_0) \neq 0$ and

$$|(mg(p))^{-1}]g_0(x_0)e_0 + mg(x_0))^{-1}| \leq \frac{1}{1-\beta(|e_0|)} < \frac{1}{1-\beta(r)}. \quad (15.3.10)$$

Moreover, using (15.3.8) and Condition ($\mathcal{A}_4$), we have in turn that

$$|(mg(p))^{-1}]g_0(x_0)e_0|$$
$$= (m-1)!|\int_0^1 \cdots \int_0^1 F^{(m)}(p)^{-1}(F^{(m)}(p+e_0 \prod_{i=1}^{m-1}\theta_i)$$
$$-F^{(m)}(p+e_0 \prod_{i=1}^m \theta_i)) \prod_{i=1}^m (\theta_i^{m-i}d\theta_i)|$$
$$= (m-1)!|\int_0^1 \cdots \int_0^1 |F^{(m)}(p)^{-1}(F^{(m)}(p+e_0 \prod_{i=1}^{m-1}\theta_i)$$
$$-F^{(m)}(p+e_0 \prod_{i=1}^m \theta_i))| \prod_{i=1}^m (\theta_i^{m-i}d\theta_i)|$$
$$\leq (m-1)!|\int_0^1 \cdots \int_0^1 \varphi_0(|e_0| \prod_{i=1}^{m-1}\theta_i(1-\theta_m)) \prod_{i=1}^m \theta_i^{m-i}d\theta_i d\theta_m$$
$$= \alpha(|e_0|) < \alpha(r) < 1.$$
$$(15.3.11)$$

Furthermore, by (15.3.7), (15.3.10), (15.3.11), and the definition of r, we get that

$$|e_1| \leq |e_0|\frac{\alpha(|e_0|)}{1-\beta(|e_0|)}$$
$$\leq |e_0|\frac{\alpha(r)}{1-\beta(r)} < |e_0| < r. \quad (15.3.12)$$

Hence we deduce that $x_1 \in B(p,r)$ and $|e_1| \leq c|e_0|$ where $c = \frac{\alpha(|e_0|)}{1-\beta(|e_0|)} \in$ [0, 1). By simply replacing x_0, x_1 with x_k, x_{k+1}, we arrive at

$$|x_{k+1} - p| \leq c|x_k - p| < r, \quad (15.3.13)$$

which shows that $\lim_{k\to+\infty} x_k = p$ and $x_{k+1} \in B(p,r)$. □

Concerning the uniqueness of the solution p, we have:

Proposition 15.3.2. *Suppose that Conditions ($\mathcal{A}$) and the inequality*

$$\frac{m}{(s_2 - s_1)^m} \int_{s_1}^{s_2} \varphi_0(|t - s_1|)|s_2 - t|^{m-1}dt < 1, \qquad (15.3.14)$$

for all s_1, t, s_2 with $0 \le s_1 \le t \le s_2 \le \bar{r}$ for some $\bar{r} \ge r$, hold. Then the solution p of equation $F(x) = 0$ is unique in $\Omega_0 = \Omega \cup \bar{B}(p, \bar{r})$.

Proof. Suppose that $p^* \in \Omega_0$ is a solution of equation $F(x) = 0$ with $p \neq p^*$. Without loss of generality, suppose $p < p^*$. We can write

$$F(p^*) - F(p) = \frac{1}{(m-1)!} \int_p^{p^*} F^{(m)}(t)(p^* - t)^{m-1}dt. \qquad (15.3.15)$$

Using Condition ($\mathcal{A}_3$) and (15.3.14), we obtain in turn that

$$\left| 1 - \left(\frac{p^* - p)^m}{m} F^{(m)}(p) \right)^{-1} \int_p^{p^*} F^{(m)}(t)(p^* - t)^{m-1}dt \right|$$

$$= \left| \left(\frac{(p^* - p)^m}{m} F^{(m)}(p) \right)^{-1} \int_p^{p^*} [F^{(m)}(t) - F^{(m)}(p)](p^* - t)^{m-1}dt \right|$$

$$\le \frac{m}{(p^* - p)^m} \int_p^{p^*} \varphi_0(|t - p|)|p^* - t|^{m-1}dt < 1,$$

so $\left(\frac{(p^* - p)^m}{m} F^{(m)}(p) \right)^{-1} \int_p^{p^*} F^{(m)}(t)(p^* - t)^{m-1}dt$ is invertible, i.e., $\int_p^{p^*} F^{(m)}(t)(p^* - t)^{m-1}dt$ is invertible. $\qquad \square$

Next, in an analogous way, we shall show a ball convergence result for NM by dropping Condition ($\mathcal{A}_4$) from Conditions ($\mathcal{A}$). Consider again functions $\bar{\alpha}, \bar{\beta}, \bar{\gamma}$, and $\bar{\delta}$ defined by

$$\bar{\alpha}(t) = \int_0^1 \varphi_0(t\theta)|1 - m\theta|d\theta,$$

$$\bar{\beta}(t) = (m - 1) \int_0^1 \varphi_0(t\theta)(1 - \theta)^{m-2}d\theta,$$

$$\bar{\gamma}(t) = \frac{\bar{\alpha}(t)}{1 - \bar{\beta}(t)},$$

and

$$\bar{\delta}(t) = \bar{\gamma}(t) - 1$$

with corresponding radii, ρ_0 and ρ. Replace r_0, r by ρ_0 and ρ, respectively, and drop Condition $(\mathcal{A}_4)$ from Conditions $(\mathcal{A})$. Denote the resulting conditions by $(\mathcal{A}')$. Then Theorem 15.3.1 and Proposition 15.3.2 can be reproduced in this weaker setting.

Theorem 15.3.3. *Suppose that Conditions $(\mathcal{A}')$ hold. Then the conclusions of Theorem 15.3.1 remain true.*

Proof. Using Lemma 15.2.6 and MN, we get instead of (15.3.7), (15.3.9), (15.3.10), (15.3.11), (15.3.12), respectively,

$$e_1 = \frac{\int_0^1 [F^{(m)}(p + \theta e_0) - F^{(m)}(p)][(m-1)(1-t)^{m-2} - m(1-t)^{m-1}]dt e_0}{F^{(m)}(p) + (m-1)\int_0^1 [F^{(m)}(p + te_0) - F^{(m)}(p)](1-t)^{m-2}dt},$$
(15.3.16)

$$\begin{aligned} & \left| 1 - (F^{(m)}(p))^{-1} \right. \\ & \left. \times [F^{(m)}(p) + (m-1)\int_0^1 [F^{(m)}(p + \theta e_0) - F^{(m)}(p)](1-t)^{m-2}dt] \right| \\ = & \ (m-1)\left| \int_0^1 (F^{(m)}(p))^{-1}[F^{(m)}(p + te_0) - F^{(m)}(p)](1-t)^{m-2}dt \right| \\ \leq & \ (m-1)\int_0^1 \varphi_0(t|e_0|)(1-t)^{m-2}dt = \bar{\beta}(|e_0|) < \bar{\beta}(\rho) < 1, \end{aligned}$$

so $F^{(m)}(p) + (m-1)\int_0^1 [F^{(m)}(p + te_0) - F^{(m)}(p)](1-t)^{m-2}dt \neq 0$,

$$\begin{aligned} & |(F^{(m)}(p) + (m-1)\int_0^1 [F^{(m)}(p + te_0) - F^{(m)}(p)] \\ & (1-t)^{m-2}dt)^{-1}F^{(m)}(p)| \\ \leq & \ \frac{1}{1 - \bar{\beta}(|e_0|)} \\ < & \ \frac{1}{1 - \bar{\beta}(\rho)}, \end{aligned}$$
(15.3.17)

$$\begin{aligned} & |\int_0^1 F^{(m)}(p)^{-1}[F^{(m)}(p + te_0) - F^{(m)}(p)] \\ & [(m-1)(1-t)^{m-2} - m(1-t)^{m-1}]dt| \\ \leq & \ \int_0^1 \varphi_0(t|e_0|)|1 - mt|dt = \bar{\alpha}(e_0) < \bar{\alpha}(\rho), \end{aligned}$$
(15.3.18)

$$|e_1| \leq \frac{\bar{\alpha}(|e_0|)}{1 - \bar{\beta}(|e_0|)}|e_0| \leq \bar{c}\gamma(\rho) < |e_0|,$$
(15.3.19)

and

$$|x_{k+1} - p| \leq \bar{c}|x_k - p| < \rho,$$
(15.3.20)

where $\bar{c} = \bar{\gamma}(|x_0 - p|) \in [0, 1)$. $\qquad\square$

Proposition 15.3.4. *Suppose that Conditions* $(\mathcal{A}')$ *and the inequality*

$$\frac{m}{(s_2 - s_1)^m} \int_{s_1}^{s_2} \varphi_0(|t - s_1|)|s_2 - t|^{m-1} dt < 1, \qquad (15.3.21)$$

for all s_1, t, s_2 *with* $0 \le s_1 \le t \le s_2 \le \bar{\rho}$ *for some* $\bar{\rho} \ge \rho$, *hold. Then the solution* p *of equation* $F(x) = 0$ *is unique in* $\Omega_1 = \cup \bar{B}(p, \bar{\rho})$.

Proof. Simply replace Ω_0, (15.3.14), $r, \bar{r}$ by Ω_1, (15.3.21), $\rho, \bar{\rho}$, respectively, in the proof of Proposition 15.3.2. $\qquad\qquad\qquad\qquad\square$

15.4 NUMERICAL EXAMPLES

Example 15.4.1. *Let*

$$\mathcal{X} = \mathcal{Y} = \mathbb{R},$$
$$\Omega = [0, 1],$$
$$m = 2,$$
$$p = 0,$$

and define function F *on* Ω *by*

$$F(x) = \frac{4}{35} x^{\frac{7}{2}} + \frac{1}{6} x^3 + \frac{1}{2} x^2.$$

We have

$$F'(x) = \frac{2}{5} x^{\frac{5}{2}} + \frac{x^2}{2} + x,$$
$$F''(x) = x^{\frac{3}{2}} + x + 1,$$

and

$$F''(0) = 1.$$

Function F'' *cannot satisfy (15.1.5) with* ψ *given by (15.1.6). Hence the results in [4,5,16,17,20,21] cannot apply. However, the new results apply for*

$$\varphi_0(t) = \varphi(t) = t^{\frac{3}{2}} + t.$$

Moreover, the convergence radius is

$$r = 1.5341,$$

so r can be chosen to be 1 and

$$\rho = 1.6861.$$

Example 15.4.2. *Let*

$$\mathcal{X} = \mathcal{Y} = \mathbb{R},$$
$$\Omega = [-1, 1],$$
$$m = 2,$$
$$p = 0,$$

and define function F on Ω by

$$F(x) = e^x - x - 1.$$

We get

$$r_0 = 2.0951,$$
$$\varphi(t) = \psi(t) = et,$$

and

$$\varphi_0(t) = (e - 1)t.$$

Notice that

$$\varphi_0(t) < \varphi(t) = \psi(t) \text{ for all } t \geq 0.$$

Then the old results give

$$r^* = 0.5518191617571,$$

while the new results give

$$r = 1.745930120607978$$

and

$$\rho = 0.762085987032,$$

so we choose

$$r = 1.$$

Notice that

$$r^* < \rho < r$$

and

$$\delta^*(t) \leq \bar{\delta}(t) < \delta(t).$$

Hence we obtain a larger radius of convergence and a smaller ratio of convergence than before [4,17,20,21].

REFERENCES

[1] S. Amat, M.A. Hernández, N. Romero, Semilocal convergence of a sixth order iterative method for quadratic equations, Appl. Numer. Math. 62 (2012) 833–841.

[2] I.K. Argyros, Computational theory of iterative methods, in: C.K. Chui, L. Wuytack (Eds.), Stud. Comput. Math., vol. 15, Elsevier Publ. Co., New York, USA, 2007.

[3] I.K. Argyros, On the convergence and application of Newtons method under Hölder continuity assumptions, Int. J. Comput. Math. 80 (2003) 767–780.

[4] W. Bi, H.M. Ren, Q.B. Wu, Convergence of the modified Halley's method for multiple zeros under Höder continuous derivatives, Numer. Algor. 58 (2011) 497–512.

[5] C. Chun, B. Neta, A third order modification of Newton's method for multiple roots, Appl. Math. Comput. 211 (2009) 474–479.

[6] A. Cordero, J.M. Gutiérrez, Á.A. Magreñán, J.R. Torregrosa, Stability analysis of a parametric family of iterative methods for solving nonlinear models, Appl. Math. Comput. 285 (2016) 26–40.

[7] A. Cordero, Á.A. Magreñán, C. Quemada, J.R. Torregrosa, Stability study of eighth-order iterative methods for solving nonlinear equations, J. Comput. Appl. Math. 291 (2016) 348–357.

[8] A. Cordero, L. Feng, Á.A. Magreñán, J.R. Torregrosa, A new fourth-order family for solving nonlinear problems and its dynamics, J. Math. Chem. 53 (3) (2014) 893–910.

[9] E. Hansen, M. Patrick, A family of root finding methods, Numer. Math. 27 (1977) 257–269.

[10] A.A. Magreñán, Different anomalies in a Jarratt family of iterative root finding methods, Appl. Math. Comput. 233 (2014) 29–38.

[11] A.A. Magreñán, A new tool to study real dynamics: the convergence plane, Appl. Math. Comput. 248 (2014) 29–38.

[12] Á.A. Magreñán, I.K. Argyros, On the local convergence and the dynamics of Chebyshev–Halley methods with six and eight order of convergence, J. Comput. Appl. Math. 298 (2016) 236–251.

[13] B. Neta, New third order nonlinear solvers for multiple roots, Appl. Math. Comput. 202 (2008) 162–170.

[14] N. Obreshkov, On the numerical solution of equations, Annuaire Univ. Sofia. Fac. Sci. Phy. Math. 56 (1963) 73–83 (in Bulgarian).

[15] N. Osada, An optimal multiple root-finding method of order three, J. Comput. Appl. Math. 52 (1994) 131–133.

[16] M.S. Petkovic, B. Neta, L. Petkovic, J. Džunič, Multipoint Methods for Solving Nonlinear Equations, Elsevier, 2013.

[17] H.M. Ren, I.K. Argyros, Convergence radius of the modified Newton method for multiple zeros under Hölder continuity derivative, Appl. Math. Comput. 217 (2010) 612–621.

[18] E. Schröder, Über unendlich viele Algorithmen zur Auflösung der Gleichungen, Math. Ann. 2 (1870) 317–365.

[19] J.F. Traub, Iterative Methods for the Solution of Equations, AMS Chelsea Publishing, 1982.

[20] X. Zhou, Y. Song, Convergence radius of Osada's method for multiple roots under Hölder and center-Hölder continuous conditions, in: ICNAAM 2011 (Greece), in: AIP Conf. Proc., vol. 1389, 2011, pp. 1836–1839.

[21] X. Zhou, X. Chen, Y. Song, On the convergence radius of the modified Newton method for multiple roots under the center-Hölder condition, Numer. Algor. 65 (2) (2014) 221–232.

Chapter 16

Laguerre-like method for multiple zeros

16.1 INTRODUCTION

Many problems in applied sciences can be written in the form

$$f(x) = 0, \tag{16.1.1}$$

using mathematical modeling, where $f : D \subseteq \mathbb{R} \longrightarrow \mathbb{R}$ is sufficiently many times differentiable and D is a convex subset in $\mathbb{R}$. In the present study, we pay attention to the case of a solution p of multiplicity $m > 1$, namely, $f(x_*) = 0$, $f^{(i)}(x_*) = 0$ for $i = 1, 2, \ldots m - 1$, and $f^{(m)}(x_*) \neq 0$. The determination of solutions of multiplicity m is of great interest. In the study of electron trajectories, when the electron reaches a plate of zero speed, the distance from the electron to the plate function has a solution of multiplicity two. Multiplicity of solution appears in connection to van der Waals equation of state and other phenomena. The convergence order of iterative methods decreases if the equation has solutions of multiplicity m. Modifications in the iterative function are made to improve the order of convergence. The modified Newton's method (MN) defined for each $n = 0, 1, 2, \ldots$ by

$$x_{n+1} = x_n - m f'(x_n)^{-1} f(x_n), \tag{16.1.2}$$

where $x_0 \in D$ is an initial point is an alternative to Newton's method in the case of solutions with multiplicity m that has second order of convergence. We consider the Laguerre-like method for multiple zeros (LLMM) [7] defined for each $n = 0, 1, 2, \ldots$ by

$$x_{n+1} = x_n - \cfrac{\lambda f(x_n)}{f'(x_n) \pm \sqrt{\cfrac{\lambda - m}{m} [(\lambda - 1) f'(x_n)^2 - \lambda f^2(x_n) f''(x_n)]}}, \tag{16.1.3}$$

where $x_0 \in D$ is an initial point and $\lambda \in \mathbb{R}$ is a parameter. Some special cases of the parameter λ in (16.1.3) reduce to the well-known third order methods:

- **Euler–Chebyshev-like method** ($\lambda = 2$) **[13,16]**;
- **Halley-like method** ($\lambda = 0$) **[4]**;
- **Ostrowski-like method** ($\lambda \longrightarrow \infty$) **[13,16]**;

A Contemporary Study of Iterative Methods. DOI: 10.1016/B978-0-12-809214-9.00016-4

- **Hansen and Patrick method** $(\lambda = \dfrac{1}{\mu} + 1)$ **[6].**

Let $U(p, \lambda) := \{x \in U_1 : |x - p| < \lambda\}$ denote an open ball and $\bar{U}(p, \lambda)$ denote its closure. It is said that $U(p, \lambda) \subseteq D$ is a convergence ball for an iterative method, if the sequence generated by this iterative method converges to p, provided that the initial point $x_0 \in U(p, \lambda)$. But how close should x_0 be to x_* so that convergence can take place? Extending the ball of convergence is very important, since it shows the difficulty we confront to pick initial points. It is desirable to be able to compute the largest convergence ball. This usually depends on the iterative method and the conditions imposed on the function f and its derivatives. We can unify these conditions by expressing them as:

$$\|(f^{(m)}(x_*))^{-1} f^{(m+1)}(x)\| \leq \varphi_0(\|x - x_*\|), \tag{16.1.4}$$

$$\|(f^{(m)}(x_*))^{-1} (f^{(m+1)}(x) - f^{(m+1)}(y))\| \leq \varphi(\|x - y\|) \tag{16.1.5}$$

for all $x, y \in D$, where $\varphi_0, \varphi : \mathbb{R}_+ \cup \{0\} \longrightarrow \mathbb{R}_+ \cup \{0\}$ are continuous and nondecreasing functions satisfying $\varphi(0) = 0$. If $m \geq 1$, $\varphi_0(t) = \mu_0$, and

$$\varphi(t) = \mu t^q, \mu_0 > 0, \mu > 0, q \in [0, 1], \tag{16.1.6}$$

then we obtain the conditions under which the preceding methods were studied [1–16]. However, there are cases where even (16.1.6) does not hold. Moreover, the smaller the functions φ_0, φ, the larger the radius of convergence. The technique, which we present next, can be used for all preceding methods, as well as in methods where $m = 1$.

The rest of the chapter is structured as follows. Section 16.2 contains some auxiliary results on divided differences and derivatives. The ball convergence of LLMM is given in Section 16.3. The numerical examples are given in the concluding Section 16.4.

16.2 AUXILIARY RESULTS

In order to make the chapter as self-contained as possible, we restate some standard definitions and properties for divided differences [4,6,7,13,15].

Definition 16.2.1. *The divided differences* $f[y_0, y_1, \ldots, y_k]$ *on* $k + 1$ *distinct points* $y_0, y_1, \ldots y_k$ *of a function* $f(x)$ *are defined by*

$$
\begin{aligned}
f[y_0] &= f(y_0), \\
f[y_0, y_1] &= \frac{f[y_0] - f[y_1]}{y_0 - y_1}, \\
&\vdots \\
f[y_0, y_1, \ldots, y_k] &= \frac{f[y_0, y_1, \ldots, y_{k-1}] - f[y_0, y_1, \ldots, y_k]}{y_0 - y_k}.
\end{aligned}
\tag{16.2.1}
$$

If the function f is sufficiently smooth, then its divided differences $f[y_0, y_1, \ldots, y_k]$ can be defined if some of the arguments y_i coincide. For instance, if $f(x)$ has k-th derivative at y_0, then it makes sense to define

$$\underbrace{f[y_0, y_1, \ldots, y_k]}_{k+1} = \frac{f^{(k)}(y_0)}{k!}. \qquad (16.2.2)$$

Lemma 16.2.2. *The divided differences $f[y_0, y_1, \ldots, y_k]$ are symmetric functions of their arguments, i.e., they are invariant to permutations of the $y_0, y_1, \ldots, y_k$.*

Lemma 16.2.3. *If a function f has k-th derivative, and $f^{(k)}(x)$ is continuous on the interval $I_x = [\min(y_0, y_1, \ldots, y_k), \max(y_0, y_1, \ldots, y_k)]$, then*

$$f[y_0, y_1, \ldots, y_k] = \int_0^1 \cdots \int_0^1 \theta_1^{k-1} \theta_2^{k-1} \cdots \theta^{k-1} f^{(k)}(\theta) d\theta_1 \cdots d\theta_k, \quad (16.2.3)$$

where $\theta = y_0 + (y_1 - y_0)\theta_1 + (y_2 - y_1)\theta_1\theta_2 + \cdots + (y_k - y_{k-1})\theta_1 \cdots \theta_k$.

Lemma 16.2.4. *If a function f has $(k+1)$-th derivative, then for every argument x, the following formula holds:*

$$\begin{aligned} f(x) &= f[v_0] + f[v_0, v_1](x - v_0) + \cdots \\ &\quad + f[v_0, v_1, \ldots, v_k](x - v_0) \cdots (x - v_k) \qquad (16.2.4) \\ &\quad + f[v_0, v_1, \ldots, v_k, x]\lambda(x), \end{aligned}$$

where

$$\lambda(x) = (x - v_0)(x - v_1) \cdots (x - v_k). \qquad (16.2.5)$$

Lemma 16.2.5. *Assume that a function f has continuous $(m+1)$-th derivative, and x_* is a zero of multiplicity m. We define functions $g_0, g,$ and g_1 as*

$$g_0(x) = f[\underbrace{x_*, x_*, \ldots, x_*}_{m}, x, x], \quad g(x) = f[\underbrace{x_*, x_*, \ldots, x_*}_{m}, x],$$

$$g_1(x) = f[\underbrace{x_*, x_*, \ldots, x_*}_{m}, x, x, x]. \qquad (16.2.6)$$

Then

$$g'(x) = g_0(x), \quad g''(x) = g_1(x). \qquad (16.2.7)$$

Lemma 16.2.6. *If a function f has $(m+1)$-th derivative, and x_* is a zero of multiplicity m, then for every argument x, the following formulae hold:*

$$f(x) = \underbrace{f[x_*, x_*, \ldots, x_*}_{m}, x](x - x_*)^m = g(x)(x - x_*)^m,$$

$$f'(x) = \underbrace{f[x_*, x_*, \ldots, x_*}_{m}, x, x](x - x_*)^m$$

$$+ m\underbrace{f[x_*, x_*, \ldots, x_*}_{m}, x](x - x_*)^{m-1} \qquad (16.2.8)$$

$$= g_0(x)(x - x_*)^m + mg(x)(x - x_*)^{m-1},$$

and

$$f''(x) = 2\underbrace{f[x_*, x_*, \ldots, x_*}_{m}, x, x, x](x - x_*)^m$$

$$+ 2m\underbrace{f[x_*, x_*, \ldots, x_*}_{m}, x, x](x - x_*)^{m-1}$$

$$+ m(m - 1)\underbrace{f[x_*, x_*, \ldots, x_*}_{m}, x](x - x_*)^{m-2} \qquad (16.2.9)$$

$$= 2g_1(x)(x - x_*)^m + 2mg_0(x)(x - x_*)^{m-1}$$

$$+ m(m - 1)g(x)(x - x_*)^{m-2},$$

where $g_0(x)$, $g(x)$ and $g_1(x)$ are defined previously.

16.3 LOCAL CONVERGENCE

It is convenient for the local convergence analysis that follows to define some real functions and parameters. Define the function ψ_0 on $\mathbb{R}_+ \cup \{0\}$ by

$$\psi_0(t) = \frac{\varphi_0(t)t}{m+1} - 1.$$

We have $\psi_0(0) = -1 < 0$ and $\psi_0(t) > 0$, if

$$\varphi_0(t)t \longrightarrow \text{a positive number or } +\infty \qquad (16.3.1)$$

as $t \to \infty$. It then follows from the intermediate value theorem that function ψ_0 has a zero in the interval $(0, +\infty)$. Denote by ρ_0 the smallest such zero. Define functions $\varphi_0^{(m)}, \varphi^{(m)}, h_0, h_1$ on the interval $[0, \rho_0)$ by

$$\varphi_0^{(m)}(t) = m! \int_0^1 \cdots \int_0^1 \theta_1^m \theta_2^{m-1} \cdots \theta_m \varphi_0(\theta_1, \ldots, \theta_m t) d\theta_1 \cdots d\theta_{m+1},$$

$$\varphi^{(m)}(t) = m! \int_0^1 \theta_1^m \theta_2^{m-1} \cdots \theta_m \varphi(\theta_1, \ldots, \theta_{m-1}(1 - \theta_m)t) d\theta_1 \cdots d\theta_{m+1},$$

$$h_0(t) = \varphi_0^{(m)}(t)t + |m - \lambda| \frac{\varphi_0^{(m)}(t)t}{m+1}$$

$$+\sqrt{|\lambda-1|}\varphi_0^{(m)}(t)t+\sqrt{m|m-\lambda|}\frac{\varphi_0^{(m)}(t)t}{m+1}$$

$$+\frac{2m\varphi_0^{(m)}(t)t}{\sqrt{m+1-\varphi_0(t)t}}+2m|\lambda|\sqrt{\frac{\varphi_0^{(m)}(t)\varphi^{(m)}(t)t}{m+1-\varphi_0(t)t}},$$

and

$$\begin{aligned} h_1(t) &= 1-\frac{\varphi_0(t)t}{m+1}-\sqrt{|\lambda-1|}\varphi_0^{(m)}(t)t\\ &\quad -\sqrt{m|m-\lambda|}\frac{\varphi_0^{(m)}(t)t}{m+1}-\frac{2m\varphi_0^{(m)}(t)t}{\sqrt{m+1-\varphi_0(t)t}}\\ &\quad -2m|\lambda|\sqrt{\frac{\varphi_0^{(m)}(t)\varphi^{(m)}(t)t}{m+1-\varphi_0(t)t}}. \end{aligned}$$

We get that $h_1(0) = 1 > 0$ and $h_1(t) \longrightarrow -\infty$ as $t \longrightarrow \rho_0^-$. Denote by r_0 the smallest zero of function h_1 in the interval $(0, \rho_0)$. Define function h on $[0, r_0)$ by

$$h(t) = \frac{h_0(t)}{h_1(t)} - 1.$$

We obtain that $h(0) = -1 < 0$ and $h(t) \longrightarrow +\infty$ as $t \longrightarrow r_0^-$. Denote by r the smallest zero of function h on the interval $(0, r_0)$. Then, we have that for each $t \in [0, r)$

$$0 \leq h(t) < 1. \tag{16.3.2}$$

The local convergence analysis is based on Conditions $(\mathcal{A})$:

$(\mathcal{A}_1)$ Function $f : D \subseteq \mathbb{R} \longrightarrow \mathbb{R}$ is $(m+1)$-times differentiable, and x_* is a zero of multiplicity m.

$(\mathcal{A}_2)$ Conditions (16.1.4) and (16.1.5) hold.

$(\mathcal{A}_3)$ $\bar{U}(x_*, r) \subseteq D$, where the radius of convergence r is defined previously.

$(\mathcal{A}_4)$ Condition (16.3.1) holds.

Theorem 16.3.1. *Suppose that Conditions $(\mathcal{A})$ hold. Then sequence $\{x_n\}$ generated for $x_0 \in U(x_*, r) - \{x_*\}$ by LLMM is well defined in $U(x_*, r)$, remains in $U(x_*, r)$ for each $n = 0, 1, 2 \ldots$, and converges to x_*.*

We need an auxiliary result:

Lemma 16.3.2. *Suppose that Conditions $(\mathcal{A})$ hold. Then, for all $x_0 \in I :=$ $[x_* - \rho_0, x_* + \rho_0]$ and $\delta_0 = x_0 - x_*$, the following assertions hold:*

(i) $|g(x_*)^{-1}g_0(x_0)| \leq \varphi_0^{(m)}(|\delta_0|),$

(ii) $|g(x_0)^{-1}g(x_*)| \leq \dfrac{m+1}{m+1-\varphi_0(|\delta_0|)|\delta_0|},$

(iii) $|g(x_*)^{-1}g_1(x_0)\delta_0| \le \varphi^{(m)}(|\delta_0|),$

(iv) $|g(x_0)^{-1}g_0(x_0)| \le \dfrac{(m+1)\varphi_0^{(m)}(|\delta_0|)}{m+1-\varphi_0(|\delta_0|)|\delta_0|},$

   and

(v) $|g(x_0)^{-1}g_1(x_0)\delta_0| \le \dfrac{(m+1)\varphi^{(m)}(|\delta_0|)}{m+1-\varphi_0(|\delta_0|)|\delta_0|}.$

Proof. (i) Using (16.2.2) and (16.2.6), we can write $g(x_*) = f[\underbrace{x_*, x_*, \ldots, x_*}_{m+1}] =$

$\dfrac{f^{(m)}(x_*)}{m!}$. Then, by (16.2.2), (16.2.6), (16.1.4), (16.1.5), and (16.2.3), we get

$$|g(x_*)^{-1}g_0(x_0)|$$
$$= |g(x_*)^{-1} \int_0^1 \cdots \int_0^1 \theta_1^m \theta_2^m \cdots \theta_m f^{(m+1)}(x_* + \delta_0\theta_1 \cdots \theta_m)d\theta_1 \cdots d\theta_{m+1}|$$
$$= \varphi_0^{(m)}(|\delta_0|).$$

(ii) We also have by (16.1.4) and the definition of ρ_0 that

$$
\begin{aligned}
|1 - g(x_*)^{-1}g(x_0)| &= |g(x_*)^{-1}(g(x_*) - g(x_0))| \\
&\le |g'(x_*)^{-1}g'(y_0)\delta_0| \\
&\le \frac{\varphi_0^{(m)}(|\delta_0|)|\delta_0|}{m+1} < 1
\end{aligned}
$$

(for some y_0 between x_* and x_0). It follows from the Banach lemma on invertible functions [1–3] that $g(x_0) \ne 0$ and

$$
\begin{aligned}
|g(x_0)^{-1}g(x_*)| &\le \frac{1}{1 - \dfrac{\varphi_0^{(m)}(|\delta_0|)|\delta_0|}{m+1}} \\
&= \frac{m+1}{m+1-\varphi_0^{(m)}(|\delta_0|)|\delta_0|}.
\end{aligned}
$$

(iii) By (16.2.3) we get in turn that

$$|g(x_*)^{-1}g_1(x_0)\delta_0|$$
$$= |g(x_*)^{-1} \int_0^1 \cdots \int_0^1 \theta_1^m \theta_2^m \cdots \theta_m$$
$$\times [f^{(m+1)}(x_* + \delta_0\theta_1 \cdots \theta_{m-1}) - f^{(m+1)}(x_* + \delta_0\theta_1 \cdots \theta_m)]d\theta_1 \cdots d\theta_{m+1}|$$
$$\le \varphi^{(m)}(|\delta_0|).$$

Items (iv) and (v) follow immediately from (i)–(iii). $\qquad\square$

Proof Theorem 16.3.1. We shall use mathematical induction, Lemma 16.3.2 and the estimates

$$
\begin{aligned}
f(x_0) &= g(x_0)\delta_0^m, \\
f'(x_0) &= g_0(x_0)\delta_0^m + g(x_0)m\delta_0^{m-1},
\end{aligned}
\tag{16.3.3}
$$

and

$$
f''(x_0) = 2g_1(x_0)\delta_0^m + 2g_0(x_0)m\delta_0^{m-1} + m(m-1)g(x_0)\delta_0^{m-2}.
$$

Using (16.3.3) in (16.1.3) and $\delta_0 = x_0 - x_*$, we have in turn that

$$
|\delta_1| = \left| \frac{N_1\delta_0^m}{D_0\delta_0^{m-1}} \right| = \frac{g(x_*)^{-1}|N|}{g(x_*)^{-1}|D|}|\delta_0|,
\tag{16.3.4}
$$

where

$$
N_0 = \sqrt{(\lambda-1)g_0(x_0)^2\delta_0^2 + m(\lambda-m)g_0(x_0)^2 - 2mg(x_0)g_0(x_0)\delta_0 - \Lambda},
\tag{16.3.5}
$$

with $\Lambda = 2\lambda m g(x_0)g_1(x_0))\delta_0$,

$$
N_1 = g_0(x_0)\delta_0 + (m-\lambda)g(x_0) + N_0,
\tag{16.3.6}
$$

and

$$
D_0 = g_0(x_0)\delta_0 + N_0.
\tag{16.3.7}
$$

Next we shall show that $h_0(|\delta_0|)$ is an upper bound on $|g(x_*)^{-3}N_1|$ and $h_1(|\delta_0|)$ is a lower bound on $|g(x_*)^{-3}D_0|$. Using Lemma 16.3.2, we get in turn

$$
\begin{aligned}
|g'(x_*)^{-1}g_0(x_0)| &\leq \varphi_0^{(m)}(|\delta_0|), \\
|g(x_*)^{-1}g(x_0)| &= |g'(x_*)^{-1}(g(x_*) - g(x_0))| \\
&\leq |g(x_*)^{-1}g'(y_0)\delta_0| \\
&\leq \frac{\varphi_0^{(m)}(|\delta_0|)|\delta_0|}{m+1}|g(x_*)^{-1}g(x_0)g(x_*)^{-1}g_0(x_0) \\
&\leq \frac{\varphi_0^{(m)}(|\delta_0|)|\delta_0|}{m+1}\frac{(m+1)\varphi_0^{(m)}(|\delta_0|)}{m+1-\varphi_0(|\delta_0|)|\delta_0|},
\end{aligned}
\tag{16.3.8}
$$

and

$$
|g(x_*)^{-1}g(x_0)g(x_*)^{-1}g_1(x_0)\delta_0| \leq \frac{\varphi_0^{(m)}(|\delta_0|)|\delta_0|}{m+1}\frac{(m+1)\varphi^{(m)}(|\delta_0|)}{m+1-\varphi_0(|\delta_0|)|\delta_0|}.
\tag{16.3.9}
$$

Using (16.3.5), (16.3.6), (16.3.8)–(16.3.9), the definition of function h_0, and the triangle inequality, we obtain in turn

$$
\begin{aligned}
|g(x_*)^{-1}N_1| \;\leq\; & |g(x_*)^{-1}g_0(x_0)||\delta_0| + |m - \lambda||g(x_*)^{-1}g(x_0)| \\
& + \sqrt{|\lambda - 1|}\varphi_0^{(m)}(|\delta_0|)|\delta_0| + \sqrt{m}|\lambda - m|\frac{\varphi_0^{(m)}(|\delta_0|)|\delta_0|}{m+1} \\
& + 2m\sqrt{\frac{(\varphi_0^{(m)}(|\delta_0|))^2|\delta_0|^2}{m+1 - \varphi_0(|\delta_0|)|\delta_0|}} \\
& + 2m|\lambda|\sqrt{\frac{\varphi_0^{(m)}(|\delta_0|)|\delta_0|\varphi^{(m)}(|\delta_0|)}{m+1 - \varphi_0(|\delta_0|)|\delta_0|}} = h_0(|\delta_0|).
\end{aligned}
$$

$$(16.3.10)$$

Moreover, we have from (16.3.7), (16.3.8)–(16.3.9), the definition of function h_1 and the inequality $|u + v| \geq |u| - |v|$, $u, v \in \mathbb{R}$:

$$
\begin{aligned}
|g(x_*)^{-1}D_0| \;\geq\; & 1 - |g(x_*)^{-1}(g(x_*) - g(x_0))| - |g(x_*)^{-1}N_0| \\
\geq\; & 1 - \frac{\varphi_0(|\delta_0|)|\delta_0|}{m+1} - |g(x_*)^{-1}N_0| \qquad\qquad (16.3.11) \\
\geq\; & h_1(|\delta_0|).
\end{aligned}
$$

In view of (16.3.2), (16.3.4), and (16.3.9), we get that

$$|\delta_1| \leq h(|\delta_0|)|\delta_0| \leq c|\delta_0|, \qquad\qquad (16.3.12)$$

where $c = h(|\delta_0|) \in [0, 1)$, so $x_1 \in U(x_*, r)$. By simply replacing x_0, x_1 by x_k, x_{k+1} in the preceding estimates, we get

$$|x_{k+1} - x_*| \leq c|x_k - x_*| < r, \qquad\qquad (16.3.13)$$

so $\lim_{k \to +\infty} x_k = x_*$ and $x_{k+1} \in U(x_*, r)$. $\square$

Next we present a uniqueness result for the solution x_*.

Proposition 16.3.3. *Suppose that Conditions (A) hold. Then the limit point x_* is the only solution of equation $f(x) = 0$ in $D_1 = D \cap \bar{U}(x_*, \rho_0)$.*

Proof. Let x_{**} be a solution of equation $f(x) = 0$ in D_1. We can write by (16.2.8) that

$$f(x_{**}) = g(x_{**})(x_{**} - x_*)^m. \qquad\qquad (16.3.14)$$

Using (16.1.4) and the properties of divided differences, we get in turn that

$$
\begin{aligned}
|1 - g'(x_*)^{-1}g(x_{**})| &= |g(x_*)^{-1}(g(x_*) - g(x_{**}))| \\
&= |g(x_*)^{-1}\frac{f^{(m+1)}(z_0)}{(m+1)!}(x_{**} - x_*)| \qquad (16.3.15) \\
&\leq \frac{\varphi_0(|x_{**} - x_*|)|x_{**} - x_*)}{m+1} < 1
\end{aligned}
$$

for some point between x_{**} and x_*. It follows from (16.3.14) and (16.3.15) that $x_{**} = x_*$.   □

16.4   NUMERICAL EXAMPLES

Example 16.4.1. *Let*

$$
D = [0, 1],
$$

$$
m = 2,
$$

$$
p = 0,
$$

and define function f on D by

$$
f(x) = \frac{4}{35}x^{\frac{7}{2}} + \frac{1}{6}x^3 + \frac{1}{2}x^2.
$$

We have

$$
f'(x) = \frac{2}{5}x^{\frac{5}{2}} + \frac{x^2}{2} + x,
$$

$$
f''(x) = x^{\frac{3}{2}} + x + 1,
$$

and

$$
f''(0) = 1.
$$

So function f'' cannot satisfy (16.1.5) with ψ given by (16.1.6). Hence the results in [4–16] cannot apply. However, the new results apply for

$$
\varphi(t) = \frac{3}{2}t^{\frac{1}{2}}
$$

and

$$
\varphi_0(t) = \frac{5}{2}.
$$

Moreover, the convergence radius is

$$
r = 4.2849e - 04.
$$

Example 16.4.2. *Let*

$$D = [-1, 1],$$
$$m = 2,$$
$$p = 0,$$

and define function f on D by

$$f(x) = e^x - x - 1.$$

We get

$$\varphi_0(t) = \varphi(t) = et$$

and the convergence radius is

$$r = 0.03309.$$

REFERENCES

[1] S. Amat, M.A. Hernández, N. Romero, Semilocal convergence of a sixth order iterative method for quadratic equations, Appl. Numer. Math. 62 (2012) 833–841.

[2] I.K. Argyros, Computational theory of iterative methods, in: C.K. Chui, L. Wuytack (Eds.), Stud. Comput. Math., vol. 15, Elsevier Publ. Co., New York, USA, 2007.

[3] I.K. Argyros, On the convergence and application of Newtons method under Hölder continuity assumptions, Int. J. Comput. Math. 80 (2003) 767–780.

[4] W. Bi, H.M. Ren, Q. Wu, Convergence of the modified Halley's method for multiple zeros under Hölder continuous derivatives, Numer. Algor. 58 (2011) 497–512.

[5] C. Chun, B. Neta, A third order modification of Newton's method for multiple roots, Appl. Math. Comput. 211 (2009) 474–479.

[6] E. Hansen, M. Patrick, A family of root finding methods, Numer. Math. 27 (1977) 257–267.

[7] E. Laguerre, Sur la résolution des équations numériques, Nouv. Ann. de Math., Sér. 2 17 (1878) 97–101.

[8] A.A. Magreñán, Different anomalies in a Jarratt family of iterative root finding methods, Appl. Math. Comput. 233 (2014) 29–38.

[9] A.A. Magreñán, A new tool to study real dynamics: the convergence plane, Appl. Math. Comput. 248 (2014) 29–38.

[10] B. Neta, New third order nonlinear solvers for multiple roots, Appl. Math. Comput. 202 (2008) 162–170.

[11] N. Obreshkov, On the numerical solution of equations, Annuaire Univ. Sofia. Fac. Sci. Phy. Math. 56 (1963) 73–83 (in Bulgarian).

[12] N. Osada, An optimal multiple root-finding method of order three, J. Comput. Appl. Math. 52 (1994) 131–133.

[13] M.S. Petkovic, B. Neta, L. Petkovic, J. Džunič, Multipoint Methods for Solving Nonlinear Equations, Elsevier, 2013.

[14] H.M. Ren, I.K. Argyros, Convergence radius of the modified Newton method for multiple zeros under Hölder continuity derivative, Appl. Math. Comput. 217 (2010) 612–621.

[15] E. Schröder, Über unendlich viele Algorithmen zur Auflösung der Gleichungen, Math. Ann. 2 (1870) 317–365.

[16] J.F. Traub, Iterative Methods for the Solution of Equations, AMS Chelsea Publishing, 1982.

Chapter 17

Traub's method for multiple roots

17.1 INTRODUCTION

Many problems in applied mathematics can be brought to the form

$$f(x) = 0, \tag{17.1.1}$$

using mathematical modeling, where $f : D \subseteq \mathbb{R} \longrightarrow \mathbb{R}$ is sufficiently many times differentiable and D is a convex subset in $\mathbb{R}$. In the present study, we pay attention to the case of a solution p of multiplicity $m > 1$, namely, $f(x_*) = 0$, $f^{(i)}(x_*) = 0$ for $i = 1, 2, \ldots m - 1$, and $f^{(m)}(x_*) \neq 0$. The determination of solutions of multiplicity m is of great interest. In the study of electron trajectories, when the electron reaches a plate of zero speed, the distance from the electron to the plate function has a solution of multiplicity two. Multiplicity of solution appears in connection to van der Waals equation of state and other phenomena. The convergence order of iterative methods decreases if the equation has solutions of multiplicity m. Modifications in the iterative function are made to improve the order of convergence, since in general methods loose their order of convergence when the solution has multiplicity greater than 1.

The modification of the Newton's method for roots with multiplicity m is called modified Newton's method (MN) and it is defined as

$$x_{n+1} = x_n - m f'(x_n)^{-1} f(x_n), \tag{17.1.2}$$

for each $n = 0, 1, 2, \ldots$, where $x_0 \in D$ is an initial guess. This method has quadratic order of convergence.

Traub's method (TM) [15] is defined for $n = 0, 1, 2, \ldots$ by

$$x_{n+1} = x_n - \frac{m(3-m)}{2} f'(x_n)^{-1} f(x_n) + \frac{m^2}{2} f''(x_n)^{-3} f^2(x_n) f''(x_n), \tag{17.1.3}$$

and it is an extension of Chebyshev's method of third order. Different high order iterative methods can be found in the works of several recognized authors in [5, 6,9,12,15] and the references therein.

Throughout all this chapter we will denote by $U(p, \lambda) := \{x \in U_1 : |x - p| < \lambda\}$ the open ball and by $\bar{U}(p, \lambda)$ we will denote its closure. It is said that

A Contemporary Study of Iterative Methods. DOI: 10.1016/B978-0-12-809214-9.00017-6

$U(p, \lambda) \subseteq D$ is a convergence ball for an iterative method, if the sequence generated by this iterative method converges to p, provided that the initial point $x_0 \in U(p, \lambda)$. But how close should x_0 be to x_* so that convergence can take place? Extending the ball of convergence is very important, since it shows the difficulty we confront to pick initial points. It is desirable to be able to compute the largest convergence ball. This usually depends on the iterative method and the conditions imposed on the function f and its derivatives. We can unify these conditions by expressing them as:

$$\|(f^{(m)}(x_*))^{-1} f^{(m+1)}(x)\| \leq \varphi_0(\|x - x_*\|), \qquad (17.1.4)$$

$$\|(f^{(m)}(x_*))^{-1}(f^{(m+1)}(x) - f^{(m+1)}(y))\| \leq \varphi(\|x - y\|) \qquad (17.1.5)$$

for all $x, y \in D$, where $\varphi_0, \varphi : \mathbb{R}_+ \cup \{0\} \longrightarrow \mathbb{R}_+ \cup \{0\}$ are continuous and nondecreasing functions satisfying $\varphi(0) = 0$. If $m \geq 1$, $\varphi_0(t) = \mu_0$, and

$$\varphi(t) = \mu t^q, \mu_0 > 0, \mu > 0, q \in [0, 1], \qquad (17.1.6)$$

then we obtain the conditions under which the preceding methods were studied [1–15]. However, there are cases where even (17.1.6) does not hold. Moreover, the smaller the functions φ_0, φ, the larger the radius of convergence. The technique presented in this chapter can be used for all preceding methods, as well as in methods where $m = 1$.

The rest of the chapter is structured as follows. Section 17.2 contains some auxiliary results on divided differences and derivatives. The convergence ball associated to Traub's method is given in Section 17.3. The numerical examples are in the concluding Section 17.4.

17.2 AUXILIARY RESULTS

In order to make the chapter as self-contained as possible, we restate some standard definitions and properties for divided differences [4,13,15].

Definition 17.2.1. *The divided differences $f[y_0, y_1, \ldots, y_k]$ on $k + 1$ distinct points $y_0, y_1, \ldots y_k$ of a function $f(x)$ are defined by*

$$
\begin{aligned}
f[y_0] &= f(y_0), \\
f[y_0, y_1] &= \frac{f[y_0] - f[y_1]}{y_0 - y_1}, \\
&\vdots \\
f[y_0, y_1, \ldots, y_k] &= \frac{f[y_0, y_1, \ldots, y_{k-1}] - f[y_0, y_1, \ldots, y_k]}{y_0 - y_k}.
\end{aligned}
\qquad (17.2.1)
$$

If the function f is sufficiently smooth, then its divided differences $f[y_0, y_1, \ldots, y_k]$ can be defined if some of the arguments y_i coincide. For instance, if $f(x)$

has k-th derivative at y_0, then it makes sense to define

$$f[\underbrace{y_0, y_1, \ldots, y_k}_{k+1}] = \frac{f^{(k)}(y_0)}{k!}. \tag{17.2.2}$$

Lemma 17.2.2. *The divided differences $f[y_0, y_1, \ldots, y_k]$ are symmetric functions of their arguments, i.e., they are invariant to permutations of the $y_0, y_1, \ldots, y_k$.*

Lemma 17.2.3. *If a function f has k-th derivative, and $f^{(k)}(x)$ is continuous on the interval $I_x = [\min(y_0, y_1, \ldots, y_k), \max(y_0, y_1, \ldots, y_k)]$, then*

$$f[y_0, y_1, \ldots, y_k] = \int_0^1 \cdots \int_0^1 \theta_1^{k-1} \theta_2^{k-1} \cdots \theta^{k-1} f^{(k)}(\theta) d\theta_1 \cdots d\theta_k, \tag{17.2.3}$$

where $\theta = y_0 + (y_1 - y_0)\theta_1 + (y_2 - y_1)\theta_1\theta_2 + \cdots + (y_k - y_{k-1})\theta_1 \cdots \theta_k$.

Lemma 17.2.4. *If a function f has $(k + 1)$-th derivative, then for every argument x, the following formula holds*

$$\begin{aligned} f(x) &= f[v_0] + f[v_0, v_1](x - v_0) + \cdots \\ &\quad + f[v_0, v_1, \ldots, v_k](x - v_0) \cdots (x - v_k) \tag{17.2.4} \\ &\quad + f[v_0, v_1, \ldots, v_k, x]\lambda(x), \end{aligned}$$

where

$$\lambda(x) = (x - v_0)(x - v_1) \cdots (x - v_k). \tag{17.2.5}$$

Lemma 17.2.5. *Assume that a function f has continuous $(m + 1)$-th derivative, and x_* is a zero of multiplicity m. We define functions g_0, g, and g_1 as*

$$g_0(x) = f[\underbrace{x_*, x_*, \ldots, x_*}_{m}, x, x], \quad g(x) = f[\underbrace{x_*, x_*, \ldots, x_*}_{m}, x], \tag{17.2.6}$$

$$g_1(x) = f[\underbrace{x_*, x_*, \ldots, x_*}_{m}, x, x, x].$$

Then

$$g'(x) = g_0(x), \quad g''(x) = g_1(x). \tag{17.2.7}$$

Lemma 17.2.6. *If a function f has an $(m + 1)$-th derivative, and x_* is a zero of multiplicity m, then for every argument x, the following formulae hold:*

$$f(x) = f[\underbrace{x_*, x_*, \ldots, x_*}_{m}, x](x - x_*)^m = g(x)(x - x_*)^m,$$

$$f'(x) = f[\underbrace{x_*, x_*, \ldots, x_*}_{m}, x, x](x - x_*)^m$$

$$+ mf[\underbrace{x_*, x_*, \ldots, x_*}_{m}, x](x - x_*)^{m-1}$$

$$(17.2.8)$$

$$= g_0(x)(x - x_*)^m + mg(x)(x - x_*)^{m-1},$$

and

$$f''(x) = 2f[\underbrace{x_*, x_*, \ldots, x_*}_{m}, x, x, x](x - x_*)^m$$

$$+ 2mf[\underbrace{x_*, x_*, \ldots, x_*}_{m}, x, x](x - x_*)^{m-1}$$

$$+ m(m-1)f[\underbrace{x_*, x_*, \ldots, x_*}_{m}, x](x - x_*)^{m-2}$$

$$(17.2.9)$$

$$= 2g_1(x)(x - x_*)^m + 2mg_0(x)(x - x_*)^{m-1}$$

$$+ m(m-1)g(x)(x - x_*)^{m-2},$$

where $g_0(x)$, $g(x)$, and $g_1(x)$ are defined previously.

17.3 LOCAL CONVERGENCE

It is convenient for the local convergence analysis that follows to define some real functions and parameters. Define the function ψ_0 on $\mathbb{R}_+ \cup \{0\}$ by

$$\psi_0 = \frac{\varphi_0(t)t}{m+1} - 1.$$

We have $\psi_0(0) = -1 < 0$ and $\psi_0(t) > 0$, if

$$\varphi_0(t)t \longrightarrow \text{a positive number or } +\infty \qquad (17.3.1)$$

as $t \to \infty$. It then follows from the intermediate value theorem that function ψ_0 has a zero in the interval $(0, +\infty)$. Denote by ρ_0 the smallest such zero. Define functions $\varphi_0^{(m)}, \varphi^{(m)}, h_0, h_1$ on the interval $[0, \rho_0)$ by

$$\varphi_0^{(m)}(t) = m! \int_0^1 \cdots \int_0^1 \theta_1^m \theta_2^{m-1} \cdots \theta_m \varphi_0(\theta_1, \ldots, \theta_m t) d\theta_1 \cdots d\theta_{m+1},$$

$$\varphi^{(m)}(t) = m! \int_0^1 \theta_1^m \theta_2^{m-1} \cdots \theta_m \varphi(\theta_1, \ldots, \theta_{m-1}(1 - \theta_m)t) d\theta_1 \cdots d\theta_{m+1},$$

$$h_0(t) = \left(\frac{\varphi_0^{(m)}(t)}{m+1}\right)^2 [(m+1)(m^2 + 5m + 2)\varphi_0^{(m)}(t) + m\varphi^{(m)}(t)]t^2,$$

and

$$h_1(t) = 2(m - \frac{2m+1}{m+1}\varphi_0^{(m)}(t)t)^3.$$

We get that $h_1(0) = 2m^3 > 0$ and $h_1(t) \longrightarrow -\infty$ as $t \longrightarrow \rho_0^-$. Denote by r_0 the smallest zero of function h_1 in the interval $(0, \rho_0)$. Define function h on $[0, r_0)$ by

$$h(t) = \frac{h_0(t)}{h_1(t)} - 1.$$

We obtain that $h(0) = -1 < 0$ and $h(t) \longrightarrow +\infty$ as $t \longrightarrow r_0^-$. Denote by r the smallest zero of function h on the interval $(0, r_0)$. Then we have that for each $t \in [0, r)$

$$0 \le h(t) < 1. \tag{17.3.2}$$

The local convergence analysis is based on Conditions $(\mathcal{A})$:

$(\mathcal{A}_1)$ Function $f : D \subseteq \mathbb{R} \longrightarrow \mathbb{R}$ is $(m+1)$-times differentiable, and x_* is a zero of multiplicity m.

$(\mathcal{A}_2)$ Conditions (17.1.4) and (17.1.5) hold.

$(\mathcal{A}_3)$ $\bar{U}(x_*, r) \subseteq D$, where the radius of convergence r is defined previously.

$(\mathcal{A}_4)$ Condition (17.3.1) holds.

Theorem 17.3.1. *Suppose that Conditions $(\mathcal{A})$ hold. Then sequence $\{x_n\}$ generated for $x_0 \in U(x_*, r) - \{x_*\}$ by TM is well defined in $U(x_*, r)$, remains in $U(x_*, r)$ for each $n = 0, 1, 2 \ldots$, and converges to x_*.*

We need an auxiliary result:

Lemma 17.3.2. *Suppose that Conditions $(\mathcal{A})$ hold. Then for all $x_0 \in I := [x_* - \rho_0, x_* + \rho_0]$ and $\delta_0 = x_0 - x_*$ the following assertions hold:*

(i) $|g(x_*)^{-1}g_0(x_0)| \le \varphi_0^{(m)}(|\delta_0|),$

(ii) $|g(x_0)^{-1}g(x_*)| \le \dfrac{m+1}{m+1-\varphi_0(|\delta_0|)|\delta_0|},$

(iii) $|g(x_*)^{-1}g_1(x_0)\delta_0| \le \varphi^{(m)}(|\delta_0|),$

(iv) $|g(x_0)^{-1}g_0(x_0)| \le \dfrac{(m+1)\varphi_0^{(m)}(|\delta_0|)}{m+1-\varphi_0(|0|)},$

and

(v) $|g(x_0)^{-1}g_1(x_0)\delta_0| \le \dfrac{(m+1)\varphi^{(m)}(|\delta_0|)}{m+1-\varphi_0(|\delta_0|)|\delta_0|}.$

Proof. (i) Using (17.2.2) and (17.2.6), we can write $g(x_*) = f[\underbrace{x_*, x_*, \ldots, x_*}_{m+1}] = \dfrac{f^{(m)}(x_*)}{m!}$. Then, by (17.2.2), (17.2.6), (17.1.4), (17.1.5), and (17.2.3), we get

$$|g(x_*)^{-1}g_0(x_0)|$$
$$= |g(x_*)^{-1} \int_0^1 \cdots \int_0^1 \theta_1^m \theta_2^m \cdots \theta_m f^{(m+1)}(x_* + \delta_0\theta_1 \cdots \theta_m)d\theta_1 \cdots d\theta_{m+1}|$$
$$= \varphi_0^{(m)}(|\delta_0|).$$

(ii) We also have by (17.1.4) and the definition of ρ_0 that

$$
\begin{aligned}
|1 - g(x_*)^{-1}g(x_0)| &= |g(x_*)^{-1}(g(x_*) - g(x_0))| \\
&\leq |g'(x_*)^{-1}g'(y_0)\delta_0| \\
&\leq \dfrac{\varphi_0^{(m)}(|\delta_0|)|\delta_0|}{m+1} < 1
\end{aligned}
$$

(for some y_0 between x_* and x_0). It follows from the Banach lemma on invertible functions [1–3] that $g(x_0) \neq 0$ and

$$
\begin{aligned}
|g(x_0)^{-1}g(x_*)| &\leq \dfrac{1}{1 - \dfrac{\varphi_0^{(m)}(|\delta_0|)|\delta_0|}{m+1}} \\
&= \dfrac{m+1}{m+1 - \varphi_0^{(m)}(|\delta_0|)|\delta_0|}.
\end{aligned}
$$

(iii) By (17.2.3) we get in turn that

$$|g(x_*)^{-1}g_1(x_0)\delta_0|$$
$$= |g(x_*)^{-1} \int_0^1 \cdots \int_0^1 \theta_1^m \theta_2^m \cdots \theta_m$$
$$\times [f^{(m+1)}(x_* + \delta_0\theta_1 \cdots \theta_{m-1}) - f^{(m+1)}(x_* + \delta_0\theta_1 \cdots \theta_m)]d\theta_1 \cdots d\theta_{m+1}|$$
$$\leq \varphi^{(m)}(|\delta_0|).$$

Items (iv) and (v) follow immediately from (i)–(iii). □

Proof Theorem 17.3.1. The proof is based on mathematical induction, Lemma 17.3.2 and the estimates

$$
\begin{aligned}
f(x_0) &= g(x_0)\delta_0^m, \\
f'(x_0) &= g_0(x_0)\delta_0^m + g(x_0)m\delta_0^{m-1},
\end{aligned}
\tag{17.3.3}
$$

and

$$f''(x_0) = 2g_1(x_0)\delta_0^m + 2g_0(x_0)m\delta_0^{m-1} + g(x_0)m(m-1)\delta_0^{m-2}.$$

Using (17.3.3) in (17.1.3) and $\delta_0 = x_0 - x_*$, we have that

$$
\begin{aligned}
\delta_1 &= \delta_0 + \frac{m(m-3)}{2}\frac{g(x_0)\delta_0^m}{(g_0(x_0)\delta_0 + mg(x_0))\delta_0^{m-1}} \\
&\quad - \frac{m^2}{2}\frac{\delta_0^{2m}(2g_0(x_0)\delta_0^m + 2g_0(x_0)m\delta_0^{m-1} + g(x_0)m(m-1)\delta_0^{m-2})}{(g_0(x_0)\delta_0^m + g(x_0)m\delta_0^{m-1})^3}
\end{aligned}
$$

so

$$|\delta_1| = \frac{|N|}{|D|}|\delta_0|, \tag{17.3.4}$$

where

$$N = 2g_0(x_0)^3\delta_0^3 + mg(x_0)((m+3)g_0(x_0)^2 - 2mg(x_0)g_1(x_0))\delta_0^2 \tag{17.3.5}$$

and

$$D = 2(g_0(x_0)\delta_0 + g(x_0)m)^3. \tag{17.3.6}$$

Next we shall find an upper bound $h_0(|\delta_0|)$ on $|g'(x_*)^{-3}N|$ and a lower bound on $|g'(x_*)^{-3}D|$ given by $h_1(|\delta_0|)$. We have by Lemma 17.3.2 in turn that

$$
\begin{aligned}
|g'(x_*)^{-3}N| &\le 2|g'(x_*)^{-1}g_0(x_0)|^3|\delta_0|^3 \\
&\quad + m|g'(x_*)^{-1}g(x_0)|[(m+3)|g'(x_*)^{-1}g_0(x_0)|^2|\delta_0|^2 \\
&\quad + 2m|g'(x_*)^{-1}g(x_0)||g'(x_*)^{-1}g_1(x_0)||\delta_0|^2] \\
&\le 2\varphi_0^{(m)}(|\delta_0|)^3|\delta_0|^3 + \frac{m\varphi_0(|\delta_0|)|\delta_0|}{m+1} \\
&\quad \times [(m+3)\varphi_0^{(m)}(|\delta_0|)^3|\delta_0|^2 + \frac{2m}{m+1}\varphi_0(|\delta_0|)\varphi^{(m)}(|\delta_0|)|\delta_0|^2] \\
&= \frac{\varphi_0^{(m)}(|\delta_0|)^2}{(m+1)^2}[(m+1)(m^2+5m+2)\varphi_0^{(m)}(|\delta_0|) \\
&\quad + m\varphi^{(m)}(|\delta_0|)]|\delta_0|^2 \\
&= h_0(|\delta_0|)
\end{aligned}
$$

$$\tag{17.3.7}$$

and

$$|mg'(x_*)^{-1}g(x_0) + g'(x_*)^{-1}g_0(x_0)\delta_0|$$
$$\geq m - m|g'(x_*)^{-1}(g(x_*) - g(x_0))|$$
$$-|g'(x_*)^{-1}g_0(x_0)||\delta_0|$$
$$\geq m - \frac{m}{m+1}\varphi^{(m)}(|\delta_0|)|\delta_0| - \varphi_0^{(m)}(|\delta_0|)|\delta_0|$$
$$= m - \frac{2m+1}{m+1}\varphi_0^{(m)}(|\delta_0|)|\delta_0|,$$

(17.3.8)

so by (17.3.8)

$$|g'(x_*)^{-3}D| \geq h_1(|\delta_0|). \tag{17.3.9}$$

It then follows from (17.3.2), (17.3.4), (17.3.11), and (17.3.9) that

$$|\delta_1| \leq \frac{h_0(|\delta_0|)}{h_1(|\delta_0|)}|\delta_0| \leq c|\delta_0|, \tag{17.3.10}$$

where $c = h(|\delta_0|) \in [0, 1)$, so $x_1 \in U(x_*, r)$. By simply replacing x_0, x_1 with x_k, x_{k+1} in the preceding estimates, we get

$$|x_{k+1} - x_*| \leq c|x_k - x_*| < r, \tag{17.3.11}$$

so $\lim_{k \longrightarrow +\infty} x_k = x_*$ and $x_{k+1} \in U(x_*, r)$. □

Related to the uniqueness result for the solution x_* we present the following result.

Proposition 17.3.3. *Suppose that Conditions $(\mathcal{A})$ hold. Then the limit point x_* is the only solution of equation $f(x) = 0$ in $D_1 = D \cap \bar{U}(x_*, \rho_0)$.*

Proof. Let x_{**} be a solution of equation $f(x) = 0$ in D_1. We can write by (17.2.8) that

$$f(x_{**}) = g(x_{**})(x_{**} - x_*)^m. \tag{17.3.12}$$

Using (17.1.4) and the properties of divided differences, we get in turn that

$$|1 - g'(x_*)^{-1}g(x_{**})| = |g(x_*)^{-1}(g(x_*) - g(x_{**}))|$$
$$= |g(x_*)^{-1}\frac{f^{(m+1)}(z_0)}{(m+1)!}(x_{**} - x_*)| \tag{17.3.13}$$
$$\leq \frac{\varphi_0(|x_{**} - x_*|)|x_{**} - x_*)}{m+1} < 1$$

for some point between x_{**} and x_*. It follows from (17.3.12) and (17.3.13) that $x_{**} = x_*$. □

17.4 NUMERICAL EXAMPLES

Example 17.4.1. *Let*

$$D = [0, 1],$$
$$m = 2,$$
$$p = 0,$$

and define function f on D by

$$f(x) = \frac{4}{35}x^{\frac{7}{2}} + \frac{1}{6}x^3 + \frac{1}{2}x^2.$$

We have

$$f'(x) = \frac{2}{5}x^{\frac{5}{2}} + \frac{x^2}{2} + x,$$
$$f''(x) = x^{\frac{3}{2}} + x + 1,$$

and

$$f''(0) = 1$$

The function f'' cannot satisfy (17.1.5) with ψ given by (17.1.6). As a consequence, the results in [4,5,12,13,15] cannot apply.

However, our new results apply for

$$\varphi(t) = \frac{3}{2}t^{\frac{1}{2}}$$

and

$$\varphi_0(t) = \frac{5}{2}.$$

Moreover, the convergence radius is

$$r = 0.9467.$$

Example 17.4.2. *Let*

$$D = [-1, 1],$$
$$m = 2,$$
$$p = 0.$$

Define function f on D by

$$f(x) = e^x - x - 1.$$

We get $\varphi_0(t) = \varphi(t) = et$. Then the convergence radius is

$$r = 0.9628.$$

REFERENCES

[1] S. Amat, M.A. Hernández, N. Romero, Semilocal convergence of a sixth order iterative method for quadratic equations, Appl. Numer. Math. 62 (2012) 833–841.

[2] I.K. Argyros, Computational theory of iterative methods, in: C.K. Chui, L. Wuytack (Eds.), Stud. Comput. Math., vol. 15, Elsevier Publ. Co., New York, USA, 2007.

[3] I.K. Argyros, On the convergence and application of Newton's method under Hölder continuity assumptions, Int. J. Comput. Math. 80 (2003) 767–780.

[4] W. Bi, H.M. Ren, Q. Wu, Convergence of the modified Halley's method for multiple zeros under Hölder continuous derivatives, Numer. Algor. 58 (2011) 497–512.

[5] C. Chun, B. Neta, A third order modification of Newton's method for multiple roots, Appl. Math. Comput. 211 (2009) 474–479.

[6] E. Hansen, M. Patrick, A family of root finding methods, Numer. Math. 27 (1977) 257–269.

[7] A.A. Magreñán, Different anomalies in a Jarratt family of iterative root finding methods, Appl. Math. Comput. 233 (2014) 29–38.

[8] A.A. Magreñán, A new tool to study real dynamics: the convergence plane, Appl. Math. Comput. 248 (2014) 29–38.

[9] B. Neta, New third order nonlinear solvers for multiple roots, Appl. Math. Comput. 202 (2008) 162–170.

[10] N. Obreshkov, On the numerical solution of equations, Annuaire Univ. Sofia. Fac. Sci. Phy. Math. 56 (1963) 73–83 (in Bulgarian).

[11] N. Osada, An optimal multiple root-finding method of order three, J. Comput. Appl. Math. 52 (1994) 131–133.

[12] M.S. Petkovic, B. Neta, L. Petkovic, J. Džunič, Multipoint Methods for Solving Nonlinear Equations, Elsevier, 2013.

[13] H.M. Ren, I.K. Argyros, Convergence radius of the modified Newton method for multiple zeros under Hölder continuity derivative, Appl. Math. Comput. 217 (2010) 612–621.

[14] E. Schröder, Über unendlich viele Algorithmen zur Auflösung der Gleichungen, Math. Ann. 2 (1870) 317–365.

[15] J.F. Traub, Iterative Methods for the Solution of Equations, AMS Chelsea Publishing, 1982.

Chapter 18

Shadowing lemma for operators with chaotic behavior

18.1 INTRODUCTION

It is well known that complicated behavior of dynamical systems can easily be detected via numerical experiments. However, it is very difficult to prove in a mathematical way in general that a given system behaves chaotically.

Several authors have worked on various aspects of this problem, see, for example, the works [4–6], and the references therein. In particular, the shadowing lemma [4, p. 1684] proved via the celebrated Newton–Kantorovich theorem [3] was used in [4] to present a computer-assisted method that allows us to prove that a discrete dynamical system admits the shift operator as a subsystem. Motivated by this work and using a weaker version of the Newton–Kantorovich theorem reported by us in [1], [2], [7] (see Theorem 18.2.1 that follows) we show that it is possible to weaken the assumptions of the shadowing lemma on which the work in [4] is based. In particular, we show that, under weaker hypotheses and the same computational cost, a larger upper bound on the crucial norm of operator M^{-1} (see (18.2.2)) can be found and the information on location of the shadowing orbit is more precise. Other advantages have already been reported in [1]. Clearly, this approach widens the applicability of the shadowing lemma.

18.2 THE SHADOWING LEMMA

We need the following definitions: Let $D \subseteq \mathbb{R}^k$ be an open subset of $\mathbb{R}^k$ (with k being a natural number), and let $f : D \rightarrow D$ be an injective operator. Then the pair (D, f) is a discrete dynamical system. Denote by $S = l^{\infty}\left(\mathbb{Z}, \mathbb{R}^k\right)$ the space of $\mathbb{R}^k$ valued bounded sequences $x = \{x_n\}$ with norm $\|x\| = \sup_{n \in \mathbb{Z}} |x_n|_2$. Here we use the Euclidean norm in $\mathbb{R}^k$ and denote it by $|\cdot|$, omitting the index 2. A δ_0-pseudo-orbit is a sequence $y = \{y_n\} \in D^{\mathbb{Z}}$ with $|y_{n+1} - f(y_n)| \leq \delta_0$ ($n \in \mathbb{Z}$). An r-shadowing orbit $x = \{x_n\}$ of a δ_0-pseudo-orbit y is an orbit of (D, f) with $|y_n - x_n| \leq 2$ ($n \in \mathbb{Z}$).

We need the following semilocal convergence theorem for Newton's method [4, page 132, Case 3 for $\delta = \delta_0$].

Theorem 18.2.1. *Let $F : D \subseteq X \rightarrow Y$ be a Fréchet differentiable operator. Assume that there exist $y_0 \in D$ and positive constants η, β, L_0, L, and γ such*

A Contemporary Study of Iterative Methods. DOI: 10.1016/B978-0-12-809214-9.00018-8

that:

$$F'(y_0)^{-1} \in L(Y, X),$$

$$\left\| F'(y_0)^{-1} \right\| \leq \beta \left\| F'(y_0)^{-1} F(y_0) \right\| \leq \eta,$$

$$\left\| F'(x) - F'(y_0) \right\| \leq L_0 \|x - y_0\|, \text{ for all } x \in D,$$

$$\left\| F'(x) - F'(y) \right\| \leq L \|x - y\|, \text{ for all } x, y \in D_0$$

$$D_0 = D \cap U(x_1, \frac{1}{\beta L_0} - \|F'(y_0)^{-1} F(y_0)\|),$$

$$|F'(y_1) - F'(y_0)\| \leq \gamma \|y_1 - y_0\|,$$

$$\|F'(y_0 + \theta(y_1 - y_0)) - F'(y_0)\| \leq \delta\theta \|y_1 - y_0\| \text{ for all } \theta \in [0, 1],$$

$$h_A = \beta L_1 \eta \leq 1,$$

and

$$\bar{U}(y_0, s^*) = \left\{ x \in X : \|x - y_0\| \leq s^* \right\} \subseteq D,$$

where $y_1 = y_0 - F'(y_0)^{-1} F(y_0),$

$$s^* = \lim_{n \to \infty} s_n,$$

$$s_0 = 0, \, s_1 = \eta, \, s_2 = s_1 + \frac{\beta \delta (s_1 - s_0)^2}{2(1 - \beta \gamma s_1)}$$

$$s_{n+2} = s_{n+1} + \frac{\beta L (s_{n+1} - s_n)^2}{2(1 - \beta L_0 s_{n+1})}, \quad n = 1, 2, \dots,$$

$$L_1^{-1} = \begin{cases} \dfrac{1}{L_0 + \gamma}, \\ \qquad \text{if } b = L\delta + \alpha L_0(\delta - 2\gamma) = 0, \\[2mm] 2\dfrac{-\alpha(L_0 + \gamma) + \sqrt{\alpha^2 (L_0 + \gamma)^2 + \alpha b}}{b}, \quad \text{if } b > 0, \\[2mm] -2\dfrac{\alpha(L_0 + \gamma) + \sqrt{\alpha^2 (L_0 + \gamma)^2 + \alpha b}}{b}, \quad \text{if } b < 0, \end{cases}$$

where $\alpha = \dfrac{2L}{L + \sqrt{L^2 + 8 L_0 L}}$. *Then sequence* $\{y_n\}$ $(n \geq 0)$ *generated by Newton's method*

$$y_{n+1} = y_n - F'(y_n)^{-1} F(y_n) \quad (n \geq 0)$$

is well defined, remains in $\bar{U}(y_0, s^*)$ *for all* $n = 1, 2, \dots,$ *and converges to a unique solution* $y^* \in \bar{U}(y x_0, s^*)$, *so that estimates*

$$\|y_{n+1} - y_n\| \leq s_{n+1} - s_n$$

and

$$\|y_n - y^*\| \le s^* - s_n$$

hold for all $n \ge 0$. Moreover, y^* is the unique solution of equation $F(y) = 0$ in $U(x_0, R)$ provided that

$$L_0 (s^* + R) \le 2$$

and

$$U(x_0, R) \subseteq D.$$

The advantages of Theorem 18.2.1 over the Newton–Kantorovich theorem [3] have been explained in detail in [1] and [2].

From now on we set $X = Y = \mathbb{R}^k$.

Sufficient conditions for a λ_0-pseudo-orbit y to admit a unique r-shadowing orbit are given in the following main result.

Theorem 18.2.2. *Let $D \subseteq \mathbb{R}^k$ be open, $f \in C^{1,Lip}(D, D)$ be injective, $y = \{y_n\} \in D^{\mathbb{Z}}$ be a given sequence, $\{A_n\}$ be a bounded sequence of $k \times k$ matrices, and let $\gamma_0, \lambda_0, \lambda, \ell_0, \ell$ be positive constants. Assume that the operator*

$$M : S \to S \text{ with } \{M z\}_n = z_{n+1} - A z_n \tag{18.2.1}$$

is invertible and

$$\|M^{-1}\| \le a = \frac{1}{\lambda + \sqrt{\ell_1 \lambda_0}}, \tag{18.2.2}$$

where ℓ_1 is L_1, $\ell_0 = L$, and $\ell = L$. Then the numbers t^* and R given by

$$t^* = \lim_{n \to \infty} t_n \text{ and } R = \frac{2}{\beta \ell_0} - t^*$$

satisfy $0 < t^* \le R$, where sequence $\{t_n\}$ is given by

$$t_0 = 0, t_1 = \eta, t_2 = t_1 + \frac{\beta \delta (t_1 - t_0)^2}{2(1 - \beta \gamma t_1)},$$

$$t_{n+2} = t_{n+1} + \frac{\beta \ell (t_{n+1} - t_n)^2}{2(1 - \beta \ell_0 t_{n+1})}, \quad n = 1, 2, \dots,$$

$$\eta = \frac{\delta_0}{\dfrac{1}{\|M^{-1}\|} - \delta}, \tag{18.2.3}$$

and

$$\beta = \left(\frac{1}{\|M^{-1}\|} - \lambda\right)^{-1}. \tag{18.2.4}$$

Let $r \in [t^*, R]$. Moreover, assume that

$$\overline{\bigcup_{n \in \mathcal{Z}} U(y_n, r)} \subseteq D \qquad (18.2.5)$$

and for every $n \in \mathbb{Z}$

$$|y_{n+1} - f(y_n)| \le \delta_0, \qquad (18.2.6)$$

$$|A_n - Df(y_n)| \le \delta, \qquad (18.2.7)$$

$$\left| F'(u) - F'(0) \right| \le \ell_0 |u|, \qquad (18.2.8)$$

and

$$\left| F'(u) - F'(v) \right| \le \ell |u - v|, \qquad (18.2.9)$$

for all $u, v \in U(y_n, r)$. Then there is a unique t^*-shadowing orbit $x^* = \{x_n\}$ of y. Moreover, there is no orbit $\bar{x}$ other than x^* such that

$$\|\bar{x} - y\| \le r. \qquad (18.2.10)$$

Proof. We shall solve the difference equation

$$x_{n+1} = f(x_n) \quad (n \ge 0) \qquad (18.2.11)$$

provided that x_n is close to y_n. Setting

$$x_n = y_n + z_n \qquad (18.2.12)$$

and

$$g_n(z_n) = f(z_n + y_n) - A_n z_n - y_{n+1}, \qquad (18.2.13)$$

we have

$$z_{n+1} = A_n z_n + g_n(z_n). \qquad (18.2.14)$$

Define $D_0 = \{z = \{z_n\} : \|z\| \le 2\}$ and nonlinear operator $G : D_0 \to S$ by

$$(G(z))_n = g_n(z_n). \qquad (18.2.15)$$

Operator G can naturally be extended to a neighborhood of D_0. Equation (18.2.14) can be rewritten as

$$F(x) = M x - G(x) = 0, \qquad (18.2.16)$$

where F is an operator from D_0 into S.

We will show the existence and uniqueness of a solution $x^* = \{x_n\}$ $(n \geq 0)$ of equation (18.2.16) with $\|x^*\| \leq r$ using Theorem 18.2.1. Clearly, we need to express $\eta, L_0, L,$ and β in terms of $\|M^{-1}\|, \delta_0, \delta, \ell_0,$ and ℓ.

(i) $\|F'(0)^{-1} F(0)\| \leq \eta$.

Using (18.2.6), (18.2.7), and (18.2.13), we get $\|F(0)\| \leq \delta_0$ and $\|G'(0)\| \leq \delta$, since $[G'(0)(w)]_n = (F'(y_n) - A_n) w_n$.

By (18.2.2) and the Banach lemma on invertible operators [3], we get that $F'(0)^{-1}$ exists and

$$\|F'(0)^{-1}\| \leq \left(\frac{1}{\|M^{-1}\|} - \delta\right)^{-1}. \tag{18.2.17}$$

That is, η can be given by (18.2.3).

(ii) $\|F'(0)^{-1}\| \leq \beta$.

By (18.2.17), we can set $\beta = \left(\frac{1}{\|M^{-1}\|} - \delta\right)^{-1}$.

(iii) $\|F'(u) - F'(v)\| \leq L \|u - v\|$.

We have using (18.2.9)

$$\left|(F'(u) - F'(v))(w)_n\right| = \left|(F'(y_n + u_n) - F'(y_n + v_n)) w_n\right|$$
$$\leq \ell |u_n - v_n| |w_n|.$$

Hence, we can set $L = \ell$.

(iv) $\|F'(u) - F'(0)\| \leq L_0 \|u\|$.

By (18.2.10), we get

$$\left|(F'(u) - F'(0))(w)_n\right| = \left|(F'(y_n + u_n) - F'(y_n + 0)) w_n\right|$$

$$\leq \ell_0 |u_n| |w_n|. \tag{18.2.18}$$

That is, we can take $L_0 = \ell_0$.

The crucial condition is satisfied due to (18.2.2) and with the above choices of $\eta, \beta, L,$ and L_0. Therefore the claims of Theorem 18.2.2 follow immediately from the conclusions of Theorem 18.2.1. □

Remark 18.2.3. *In general,*

$$\gamma \leq \delta \leq L_0 \text{ and } L \leq \bar{L}, \tag{18.2.19}$$

where $\bar{L}$ is the Lipschitz constant on D. If $\gamma = \delta = L_0 = L = \bar{L}$, Theorem 18.2.2 reduces to Theorem 1 in [4, p. 1684], otherwise our Theorem 18.2.2 improves Theorem 1 in [4]. Indeed, the upper bound in [6, p. 1684] is given by

$$\|M^{-1}\| \leq p = \frac{1}{\delta + \sqrt{2\ell\delta_0}}. \tag{18.2.20}$$

By comparing (18.2.2) with (18.2.20), we deduce

$$p < a$$

(if $L_0 < L$).
That is, we have justified the claims made in the introduction.

REFERENCES

[1] E.L. Allgower, K. Böhmer, F.A. Potra, W.C. Rheinboldt, A mesh-independence principle for operator equations and their discretizations, SIAM J. Numer. Anal. 23 (1986) 160–169.

[2] I.K. Argyros, A.A. Magreñán, Iterative Algorithms I, Nova Science Publishers Inc., New York, 2016.

[3] I.K. Argyros, A.A. Magreñán, Iterative Methods and Their Dynamics With Applications, CRC Press, New York, 2017.

[4] I.K. Argyros, On an improved convergence analysis of Newton's method, Appl. Math. Comput. 225 (2013) 372–386.

[5] L.V. Kantorovich, G.P. Akilov, Functional Analysis in Normed Spaces, Pergamon Press, Oxford, 1982.

[6] M. Laumen, Newton's mesh independence principle for a class of optimal design problems, SIAM J. Control Optim. 37 (1999) 1070–1088.

[7] W.C. Rheinboldt, An adaptive continuation process for solving systems of nonlinear equations, in: Mathematical models and Numerical Methods, in: Banach Center Publ., vol. 3, PWN, Warsaw, 1978, pp. 129–142.

Chapter 19

Inexact two-point Newton-like methods

19.1 INTRODUCTION

In this chapter we are concerned with the problem of approximating a locally unique solution $x^\star$ of the nonlinear equation

$$F(x) + G(x) = 0, \qquad (19.1.1)$$

where F is a Fréchet-differentiable operator defined on a convex subset $\mathcal{D}$ of a Banach space $\mathcal{X}$ with values in a Banach space $\mathcal{Y}$ and $G : \mathcal{D} \to \mathcal{Y}$ is a continuous operator.

Many problems in applied sciences reduce to solving an equation in the form (19.1.1) using mathematical modeling [1–13,16,23,29–33,37]. These solutions can be rarely found in closed form. That is why the most solution methods for these equations are iterative. The convergence analysis of iterative methods is usually divided into two categories: semilocal and local convergence analysis. In the semilocal convergence analysis one derives convergence criteria from the information around an initial point whereas in the local analysis one finds estimates of the radii of convergence balls, as well as the convergence order, from the information around a solution.

In this chapter we will study the inexact Newton-like method which is defined as

$$x_{n+1} = x_n - A_n^{-1}(F(x_n) + G(x_n) + r_n), \qquad (19.1.2)$$

for each $n = 0, 1, 2, \ldots$, where $x_{-1}, x_0 \in \mathcal{D}$ are initial guesses, $A_n = A(x_n, x_{n-1}) \in L(\mathcal{X}, \mathcal{Y})$ and $\{r_n\} \in \mathcal{Y}$ is a sequence converging to zero to generate a sequence $\{x_n\}$ approximating x^*.

Inexact Newton's method has been used in many areas, and now it is widely considered that various forms of inexact Newton's method are among the most effective tools for solving systems of nonlinear equations [10,13,17]. If $A(x, y) = A(x, x)$ for each $x, y \in \mathcal{D}$ and $r_n = 0$ for each $n = 0, 1, 2, \ldots$, we obtain a Krasnosel'skii–Zincenko-type iteration [6,10,15]:

$$x_{n+1} = x_n - A_n^{-1}(F(x_n) + G(x_n)), \quad \text{for each } n = 0, 1, 2, \ldots \qquad (19.1.3)$$

A Contemporary Study of Iterative Methods. DOI: 10.1016/B978-0-12-809214-9.00019-X

Moreover, if $A_n = F'(x_{n-1})$ for each $x \in \mathcal{D}$, we obtain Zabrejko–Nguen and Zabrejko–Zlepko [6,10,15] iterations, respectively, defined by

$$x_{n+1} = x_n - F'(x_n)^{-1}(F(x_n) + G(x_n)), \quad \text{for each } n = 0, 1, 2, \ldots, \quad (19.1.4)$$

$$x_{n+1} = x_n - F'(x_0)^{-1}(F(x_n) + G(x_n)), \quad \text{for each } n = 0, 1, 2, \ldots \quad (19.1.5)$$

If $G = 0$ in $\mathcal{D}$, methods (19.1.4) and (19.1.5) reduce to Newton's method and modified Newton's method [10,12]

$$x_{n+1} = x_n - F'(x_n)^{-1}F(x_n), \quad \text{for each } n = 0, 1, 2, \ldots, \quad (19.1.6)$$

$$x_{n+1} = x_n - F'(x_0)^{-1}F(x_n), \quad \text{for each } n = 0, 1, 2, \ldots, \quad (19.1.7)$$

respectively. If $G = 0$ on $\mathcal{D}$, $r_n = 0$ for each $n = 0, 1, 2, \ldots$ and $A(x, y) = [x, y; F]$ we obtain from (19.1.2) the secant method or the modified secant method [6,11,13,14,17,19–22,24–26,32],

$$x_{n+1} = x_n - [x_n, x_{n-1}; F]^{-1}F(x_n), \quad \text{for each } n = 0, 1, 2, \ldots, \quad (19.1.8)$$

$$x_{n+1} = x_n - [x_0, x_{-1}; F]^{-1}F(x_n), \quad \text{for each } n = 0, 1, 2, \ldots, \quad (19.1.9)$$

respectively, where $[x, y; F] \in L(\mathcal{X}, \mathcal{Y})$ is a divided difference of order 1 for operator F at the points $(x, y) \in \mathcal{D} \times \mathcal{D}$ [10,12,29,32]. If $r_n = 0$ for each $n = 0, 1, 2, \ldots$ and $A(y, x) = F'(y) + [y, x; F]$, we obtain the method

$$x_{n+1} = x_n - (F'(x_n) + [x_n, x_{n-1}; F])^{-1}(F(x_n) + G(x_n)), \quad (19.1.10)$$

studied by Cătinaş in [14]. There exist lots of results related to the semilocal convergence for the preceding special cases of methods under Lipschitz-type conditions [1–3,6–15,17,19–22,24–26,18,27–29,32–36].

In the present chapter we will focus our attention in the semilocal, as well the local, convergence of method (19.1.2) under more general conditions than in earlier studies. In special cases our sufficient convergence conditions reduce to the earlier ones.

The rest of the chapter is organized as follows. Section 19.2 contains the convergence analysis of method (19.1.2). The numerical examples are presented in the concluding Section 19.3.

19.2 CONVERGENCE ANALYSIS FOR METHOD (19.1.2)

The semilocal convergence analysis will be shown for $\Gamma_n = (F, G, r_n, x_0, x_{-1})$ belonging to the class $A_n = A_n(L_i, M, N_1, N, K, K_0, b, c)$ for $i = 1, 2, \ldots, 7$ defined as follows:

Definition 19.2.1. Let $L_i \geq 0$, for $i = 1, 2, \ldots, 7$, $M > 0$, $N_1 > 0$, $N \geq 0$, $K_0 \geq 0$, $K \geq 0$, $b \geq 0$, and $c \geq 0$ be given parameters satisfying

$$L_7 + K + K_0 + N < 1, \quad (19.2.1)$$

$$0 < L_7 + N + K + K_0 + (L_6 - L_2 - L_4)b < 1 \qquad (19.2.2)$$

$$2 \max\{L_3, N_1, L_1 + L_4 + L_5, L_2 + L_3 - L_1\} \le M, \qquad (19.2.3)$$

and

$$2M(b + c) \le (1 - (L_7 + (L_6 - L_2 - L_4)b + N + K + K_0))^2. \qquad (19.2.4)$$

We say that Γ_n belongs to the class A_n if:

(H_1) F, G are nonlinear operators defined on a convex subset $\mathcal{D}$ of a Banach space $\mathcal{X}$ with values in a Banach space $\mathcal{Y}$.

(H_2) The points x_0 and x_{-1} belong to the interior $\overset{\circ}{\mathcal{D}}$ of $\mathcal{D}$ and satisfy

$$\|x_0 - x_{-1}\| \le b. \qquad (19.2.5)$$

(H_3) F is Fréchet-differentiable on $\overset{\circ}{\mathcal{D}}$; G is continuous on $\overset{\circ}{\mathcal{D}}$;

There exists a mapping $A(y, x) : \overset{\circ}{\mathcal{D}} \times \overset{\circ}{\mathcal{D}} \to L(\mathcal{X}, \mathcal{Y})$ such that the following hold: $A_0 = A(x_0, x_{-1})$, $A(x_0, x_{-1})^{-1} \in L(\mathcal{Y}, \mathcal{X})$ and for each $x, y \in \overset{\circ}{\mathcal{D}}$

$$\|A_0^{-1}(F(x_0) + G(x_0))\| \le c, \qquad (19.2.6)$$

$$\|A_0^{-1}(F'(x) - F'(y))\| \le M\|x - y\|, \qquad (19.2.7)$$

$$\|A_0^{-1}(A_0 - A(y, x))\| \le L_1\|y - x\| + L_2\|x - x_0\| + L_3\|x - x_{-1}\|$$
$$+ L_4\|y - x_0\| + L_5\|y - x_{-1}\| + L_6\|x_0 - x_{-1}\| + L_7, \qquad (19.2.8)$$

$$\|A_0^{-1}(F'(y) - A(y, x))\| \le N_1\|y - x\| + N, \qquad (19.2.9)$$

and

$$\|A_0^{-1}(G(x) - G(y))\| \le K\|x - y\|; \qquad (19.2.10)$$

(H_4) Sequence $\{r_n\}$ converges to zero and

$$\|A_0^{-1}(r_{n+1} - r_n)\| \le K_0\|x_{n+1} - x_n\| \quad \text{for each } n = 0, 1, 2, \dots; \qquad (19.2.11)$$

(H_5) The set $\mathcal{D}_c = \{x \in \mathcal{D}; F, G \text{ are continuous at } x\}$ contains the closed ball $\bar{U}(x_0, t^* - b)$, where

$$t^* = (1 - L_7 + (L_6 - L_2 - L_4)b - N - K - K_0$$
$$- \sqrt{(1 - (L_7 + (L_6 - L_2 - L_4)b + N + K + K_0))^2 - 2M(b + c))}$$
$$\times M^{-1}.$$

$$\qquad (19.2.12)$$

It is convenient for the study of the semilocal convergence of method (19.1.2) to associate with the class A_n the scalar sequence $\{t_n\}$ defined by

$$t_{-1} = 0, \quad t_0 = b, \quad t_1 = b + c,$$

$$t_{n+1} = t_n - \frac{(t_n - t_{n-1})h(t_n)}{h(t_n) - h(t_{n-1})} \quad \text{for each } n = 1, 2, \ldots \tag{19.2.13}$$

where

$$h(t) = \frac{M}{2}t^2 - (1 - (L_7 + (L_6 - L_2 - L_4)b + N + K + K_0))t + b + c. \tag{19.2.14}$$

The main semilocal convergence result is presented as:

Theorem 19.2.2. *If Γ_n belongs to the class A_n, then the sequence $\{x_n\}$ generated by method (19.1.2) is well defined, remains in $\bar{U}(x_0, t^* - b)$ for each $n = 0, 1, 2, \ldots$, and converges to a solution $x^* \in \bar{U}(x_0, t^* - b)$ of equation $F(x) + G(x) = 0$. Moreover, the following estimates hold:*

$$\|x_n - x^*\| \le t^* - t_n. \tag{19.2.15}$$

Proof. It follows from (19.2.2) and (19.2.4) that function $h(t)$ has two positive roots t^* and t^{**}, $t^* \le t^{**}$, $t_m \le t_{m+1}$, so that sequence $\{t_m\}$ increasingly converges to t^* given by (19.2.11) and

$$t^* = (1 - L_7 + (L_6 - L_2 - L_4)b + N + K + K_0$$

$$+ \sqrt{(1 - (L_7 + (L_6 - L_2 - L_4)b + N + K + K_0))^2 - 2M(b + c)}$$

$$\times M^{-1}.$$

Following an inductive process on the subindex n we shall show that

$$\|x_{n+1} - x_n\| \le t_{n+1} - t_n. \tag{19.2.16}$$

Estimate (19.2.16) holds for $n = -1, 0$ since by (19.1.2) (for $n = 0$), (19.2.5), (19.2.6) and (19.2.12), $\|x_{-1} - x_0\| \le b = t_0 - t_{-1}$ and $\|x_1 - x_0\| \le c = t_1 - t_0$. Suppose that (19.2.16) holds for all $m \le n$. Then we have $\|x_{m+1} - x_0\| \le t_{m+1} - b$. Using (19.2.3), (19.2.8), (19.2.14), and the induction hypotheses, we have in turn that

$$\|A_0^{-1}(A_0 - A(x_{m+1}, x_m))\|$$

$$\le L_1\|x_{m+1} - x_m\| + L_2\|x_m - x_0\|$$

$$+ L_3\|x_{m+1} - x_{-1}\| + L_4\|x_{m+1} - x_0\| + L_5\|x_{m+1} - x_{-1}\|$$

$$+ L_6\|x_0 - x_{-1}\| + L_7$$

$$\le L_1(t_{m+1} - t_m) + L_2(t_m - t_0) + L_3(t_m - t_0 + t_0 - t_{-1}) +$$

$$L_4(t_{m+1} - t_0) + L_5(t_{m+1} - t_0 + t_0 - t_{-1}) + L_6 b + L_7$$
$$= (L_1 + L_4 + L_5)t_{m+1} + (L_2 + L_3 - L_1)t_m$$
$$+ (L_6 - L_2 - L_4)b + L_7$$
$$\leq \frac{M}{2}(t_{m+1} + t_m)$$
$$+ (L_6 - L_2 - L_4)b + L_7 + N$$
$$+ K + K_0 < 1.$$

$$(19.2.17)$$

It follows from (19.2.17) and the Banach lemma on invertible operators [10,11] that $A_{m+1}^{-1} \in L(\mathcal{Y}, \mathcal{X})$ and

$$\|A_{m+1}^{-1} A_0\| \leq \frac{1}{1 - \dfrac{M}{2}(t_{m+1} + t_m) + (L_6 - L_2 - L_4)b + L_7 + N + K + K_0}$$
$$\leq -\frac{t_{m+1} - t_m}{h(t_{m+1}) - h(t_m)}.$$

$$(19.2.18)$$

Using the induction hypotheses (19.1.2), (19.2.1)–(19.2.3), (19.2.7)–(19.2.10), (19.2.12)–(19.2.14), we obtain in turn that

$$\|A_0^{-1}(F(x_{m+1}) + G(x_{m+1}) + r_{m+1})\|$$
$$= \|A_0^{-1}(F(x_{m+1}) + G(x_{m+1}) + r_{m+1} - F(x_m) - G(x_m) - r_m -$$
$$A_m(x_{m+1} - x_m) + F'(x_m)(x_{m+1} - x_m) - F'(x_m)(x_{m+1} - x_m))\|$$
$$\leq \| \int_0^1 A_0^{-1}(F'(x_m + \theta(x_{m+1} - x_m)) - F'(x_m))(x_{m+1} - x_m)d\theta \|$$
$$+ \|A_0^{-1}(F'(x_m) - A_m)(x_{m+1} - x_m)\| + \|A_0^{-1}(G(x_{m+1}) - G(x_m))\|$$
$$+ \|A_0^{-1}(x_{m+1} - x_m)\|$$
$$\leq (\frac{M}{2}\|x_{m+1} - x_m\| + N_1\|x_m - x_{m-1}\| + N + K + K_0)\|x_{m+1} - x_m\|$$
$$\leq (\frac{M}{2}(t_{m+1} - t_m) + N_1(t_m - t_{m-1}) + N + K + K_0)(t_{m+1} - t_m)$$
$$+ h(t_m) + (t_{m+1} - t_m)(\frac{M}{2}(t_m + t_{m-1})$$
$$- (1 - (L_7 + (L_6 - L_2 - L_4)b + N + K + K_0)))$$
$$= h(t_{m+1}) - (N + K + K_0)t_m + \frac{1}{2}(M - 2N_1)(t_{m+1} - t_m)(t_{m-1} - t_m)$$
$$\leq h(t_{m+1})$$

$$(19.2.19)$$

since $N \geq 0$, $K \geq 0$, $K_0 \geq 0$, $M \geq 2N_1$, and $t_{m-1} \leq t_m \leq t_{m+1}$. Then it follows from (19.1.2), (19.2.18), and (19.2.19) that

$$\|x_{m+2} - x_{m+1}\| \le \|A_{m+1}^{-1} A_0\| \|A_0^{-1}(F(x_{m+1}) + G(x_{m+1}) + r_{m+1})\|$$

$$\le -\frac{(t_{m+1} - t_m)h(t_{m+1})}{h(t_{m+1}) - h(t_m)} = t_{m+2} - t_{m+1},$$

which completes the induction for (19.2.16). Moreover, we obtain

$$
\begin{aligned}
\|x_{m+2} - x_0\| &\le \|x_{m+2} - x_{m+1}\| + \|x_{m+1} - x_m\| + \cdots + \|x_1 - x_0\| \\
&\le (t_{m+2} - t_{m+1}) + (t_{m+1} - t_m) + \cdots + (t_1 - t_0) \\
&= t_{m+2} - t_0,
\end{aligned}
$$

so $x_{m+2} \in \bar{U}(x_0, t_{m+2} - t_0) \subset \bar{U}(x_0, t^* - b)$. It follows from (19.2.16) that $\{x_n\}$ is a Cauchy sequence (since $\{t_n\}$ is Cauchy as a convergent sequence) in a Banach space $\mathcal{X}$ and as such it converges to some $x^* \in \bar{U}(x_0, t^* - b)$ (since $\bar{U}(x_0, t^* - b)$ is a closed set). By letting $m \to \infty$ in (19.2.19), we get $F(x^*) + G(x^*) = 0$. Moreover, estimate (19.2.15) follows from (19.2.16) by using standard majorization techniques [10,11].   □

Related to the uniqueness of the solution x^* of equation (19.1.1), we present the following result.

Proposition 19.2.3. *Suppose that the hypotheses of Theorem 19.2.2 are satisfied. Moreover, suppose that there exist $T \ge t^* - b$, $a \ge 0$, $y^* \in \bar{U}(x_0, T)$ such that $F(y^*) + G(y^*) = 0$,*

$$\|A_0^{-1} r_n\| \le a\|x_n - y^*\| \quad \text{for each } n = 0, 1, 2, \ldots, \tag{19.2.20}$$

and

$$
\begin{aligned}
&\frac{M}{2}T + K + a + (L_2 + L_3 + L_4 + L_5 + N_1)t^* \\
&+ (L_6 - L_2 - L_4 - N_1)b + L_7 + N < 1.
\end{aligned}
\tag{19.2.21}
$$

Then the limit point x^* is the only solution of equation $F(x) + G(x) = 0$ in $\bar{U}(x_0, T) \cap D$.

Proof. The existence of the solution x^* has been established in Theorem 19.2.2. Let $y^* \in \bar{U}(x_0, T)$ be a solution of equation $F(x) + G(x) = 0$. Using (19.2.17) and (19.2.20)–(19.2.21), we obtain in turn that

$$
\begin{aligned}
&\|x_{m+1} - y^*\| \\
&= \|y^* - x_m + A_m^{-1}(F(x_m) + G(x_m) + r_m) - A_m^{-1}(F(y^*) + G(y^*))\| \\
&\le \|A_m^{-1} A_0\| \Big[\|\int_0^1 A_0^{-1}(F'(y^* + \theta(x_m - y^*)) - F'(x_m))(x_m - y^*)d\theta\| \\
&\quad + \|A_0^{-1}(G(x_m) - G(y^*))\| + \|A_0^{-1}(F'(x_m) - A_m)(x_m - y^*)\| + \|A_0^{-1} r_m\|\Big]
\end{aligned}
$$

$$\leq \frac{\left[\frac{M}{2}\|x_m - y^*\| + N_1\|x_m - x_{m-1}\| + N + K + a\right]\|x_m - y^*\|}{1 - ((L_1 + L_4 + L_5)t_{m-1} + (L_2 + L_3 - L_1)t_{m-2} + (L_6 - L_2 - L_4)b + L_7)}$$

$$\leq \frac{\frac{M}{2}T + N_1(t^* - b) + K + a}{1 - ((L_2 + L_3 + L_4 + L_5)t^* + (L_6 - L_2 - L_4)b + L_7)}\|x_m - y^*\|$$

$$< \|x_m - y^*\|,$$

(19.2.22)

which implies $\lim_{m\to\infty} x_m = y^*$. Then $x^* = y^*$ and therefore the limit point x^* is the only solution of equation $F(x) + G(x) = 0$ in $\bar{U}(x_0, T) \cap D$. $\qquad\square$

The local convergence analysis will be performed for $\bar{\Gamma}_n = (F, G, r_n, x^*)$ belonging to the class $\bar{A}_n = \bar{A}_n(l_i, \alpha, \xi, \lambda, \mu)$, $i = 1, 2, 3, 4$, defined as follows

Definition 19.2.4. *Let $l_i \geq 0$ for $i = 1, 2, 3, 4$, $\alpha \geq 0$, $\xi \geq 0$, $\lambda \geq 0$, and $\mu \geq 0$ be given parameters satisfying*

$$\xi + \lambda + \alpha + l_4 < 1.$$

(19.2.23)

We say that $\bar{\Gamma}_n$ belongs to the class $\bar{A}_n$ if

($\bar{H}_1$) Condition (H_1) holds;

($\bar{H}_2$) There exists a solution $x^* \in D$ of equation (19.1.1);

($\bar{H}_3$) F is Fréchet-differentiable on $\overset{\circ}{D}$, G is continuous on $\overset{\circ}{D}$, and there exists a mapping such that $A(y, x) : \overset{\circ}{D} \times \overset{\circ}{D} \to L(\mathcal{X}, \mathcal{Y})$ such that for $A_* = A(x^*, x^*)$, $A_*^{-1} \in L(\mathcal{Y}, \mathcal{X})$ and each $x, y \in \overset{\circ}{D}$

$$\|A_*^{-1}(F'(x) - F'(y))\| \leq \mu\|x - y\|,$$

(19.2.24)

$$\|A_*^{-1}(A_* - A(y, x))\| \leq l_1\|y - x^*\| + l_2\|x - x^*\| + l_3\|y - x\| + l_4,$$

(19.2.25)

$$\|A_*^{-1}(F'(y) - A(y, x))\| \leq \xi_1\|y - x\| + \xi,$$

(19.2.26)

and

$$\|A_*^{-1}(G(x) - G(y))\| \leq \lambda\|x - y\|;$$

(19.2.27)

($\bar{H}_4$) Sequence $\{r_n\}$ is null and

$$\|A_*^{-1}r_n\| \leq \alpha\|x_n - x^*\| \quad \text{for each } n = 1, 2, \ldots;$$

(19.2.28)

and

($\bar{H}_5$)

$$\bar{U}(x^*, R) \subseteq D,$$

(19.2.29)

where

$$R = \frac{1 - (\xi + \lambda + \alpha + l_4)}{\frac{\mu}{2} + l_1 + l_2 + 2l_3 + 2\xi_1}. \qquad (19.2.30)$$

Using the above notation, we can show the main local convergence result for method (19.1.2).

Theorem 19.2.5. *If $\bar{\Gamma}_n$ belongs to the class $\bar{A}_n$, then the sequence $\{x_n\}$ generated by method (19.1.2) for $x_{-1}, x_0 \in U(x^*, R)$ is well defined, remains in $U(x_0, t^* - b)$ for each $n = 0, 1, 2, \ldots$, and converges to x^*. Moreover, the following estimates hold:*

$$\|x_{n+1} - x^*\| \le \delta_n \|x_n - x^*\| \le \|x_n - x^*\| < R, \qquad (19.2.31)$$

where

$$\delta_n = \frac{\frac{\mu}{2}\|x_n - x^*\| + \xi_1\|x_n - x_{n-1}\| + \xi + \lambda + \alpha}{1 - (l_1\|x_n - x^*\| + l_2\|x_{n-1} - x^*\| + l_3\|x_n - x_{n-1}\| + l_4)}. \qquad (19.2.32)$$

Proof. From the imposed hypotheses we have that $x_{-1}, x_0 \in U(x^*, R)$. Moreover, suppose that x_m are defined for all $m \le n$. Then, using (19.2.23), (19.2.25), (19.2.29), and (19.2.30), we get that

$$
\begin{aligned}
\|A_*^{-1}(A_* - A_m)\| &\le\ l_1\|x_m - x^*\| + l_2\|x_{m-1} - x^*\| + l_3\|x_m - x_{m-1}\| + l_4 \\
&\le\ l_1 R + l_2 R + l_3(\|x_m - x^*\| + \|x^* - x_{m-1}\|) + l_4 \\
&\le\ (l_1 + l_2 + 2l_3)R + l_4 < 1.
\end{aligned}
$$
$$(19.2.33)$$

It follows from (19.2.33) and the Banach lemma on invertible operators that $A_m^{-1} \in L(\mathcal{Y}, \mathcal{X})$ and

$$
\begin{aligned}
&\|A_m^{-1} A_*\| \\
&\le \frac{1}{1 - (l_1\|x_m - x^*\| + l_2\|x_{m-1} - x^*\| + l_3(\|x_m - x^*\| + \|x^* - x_{m-1}\|) + l_4)}.
\end{aligned}
$$
$$(19.2.34)$$

Then x_{m+1} is well defined by method (19.1.2). Using the identity

$$x^* - x_{m+1} = x^* - x_m + A_m^{-1}(F(x_m) + G(x_m) + r_m) - A_m^{-1}(F(x^*) + G(x^*)) \qquad (19.2.35)$$

(19.2.24)–(19.2.28), (19.2.34), and (19.2.35), we obtain in turn that

$$\|x_{m+1} - x^*\|$$

$$\leq \|A_m^{-1} A_*\| \|A_*^{-1}(F(x_m) + G(x_m) + r_m)\|$$

$$\leq \|A_m^{-1} A_*\| \| \int_0^1 A_*^{-1}(F'(x^* + \theta(x_m - x^*)) - F'(x_m))(x_m - x^*)d\theta\|$$

$$+ \|A_*^{-1}(F'(x_m) - A_m)\| + \|A_*^{-1}(G(x_m) - G(x^*))\| + \|A_*^{-1} r_m\|$$

$$\leq \frac{(\frac{\mu}{2}\|x_m - x^*\| + \lambda + \alpha + \xi_1\|x_m - x_{m-1}\| + \xi)\|x_m - x^*\|}{1 - (l_1\|x_m - x^*\| + l_2\|x_{m-1} - x^*\| + l_3\|x_m - x_{m-1}\| + l_4)}$$

$$\leq \delta_m \|x_m - x^*\|$$

$$(19.2.36)$$

$$< \frac{(\frac{\mu}{2} R + \xi_1(R + R) + \lambda + \alpha + \xi)\|x_m - x^*\|}{1 - (l_1 R + l_2 R + l_3(R + R) + l_4)} = \|x_m - x^*\| < R, \quad (19.2.37)$$

by the definition of R, which shows (19.2.31). It follows from (19.2.36) that

$$\|x_{m+1} - x^*\| < \|x_m - x^*\| < R.$$

Hence we conclude that $x_{m+1} \in U(x^*, R)$ and $\lim_{m \to \infty} x_m = x^*$. $\qquad\square$

Finally, we can state the following remark.

Remark 19.2.6. *Special choices of A, G, r_n and the parameters lead to well known sufficient convergence criteria:*

- *Semilocal case*

  (i) *Newton's method. Let $G = 0$, $r_n = 0$, $A(y, x) = A(x, x) = F'(x)$, and $x_0 = x_{-1}$ to obtain Newton's method. Then we can choose $b = 0$, $L_1 = L_3 = L_4 = L_5 = K_0 = K = N_1 = N = 0$ and $L_2 = M$. In this case conditions (19.2.1)–(19.2.4) reduce to the famous for its simplicity and clarity Newton–Kantorovich hypothesis*

  $$2Mc \leq 1. \qquad (19.2.38)$$

  (ii) *Secant method. Let $G = 0$, $r_n = 0$ and $A(y, x) = [y, x; F]$ to obtain the secant method. Then we can choose $L_1 = L_2 = L_5 = L_6 = L_7 = K_0 = K = N = 0$, $L_3 = L_4 = N_1 = M/2$, and conditions (19.2.1)–(19.2.4) reduce to popular hypothesis for the convergence of the secant method [32]*

  $$\sqrt{2Mc} + \frac{M}{2} b \leq 1. \qquad (19.2.39)$$

- *Local case*

(iii) *Newton's method. We can choose* $\lambda = \alpha = \xi = l_1 = l_3 = l_4 = \xi_1 = 0$. *Then we obtain*

$$R = \frac{2}{\mu + 2l_2}. \tag{19.2.40}$$

This value of R was obtained by us in [6]. It is worth noticing that

$$l_2 \le \mu \tag{19.2.41}$$

holds in general, and $\dfrac{\mu}{l_2}$ *can be arbitrarily large [6,10–12]. If* $l_2 = \mu$, *then (19.2.40) reduces to*

$$R_{TR} = \frac{2}{3\mu}. \tag{19.2.42}$$

The value of r was given independently by W. Rheinboldt [35] and J. Traub. Notice that

$$r \le R,$$

$$r < R \quad \text{if} \quad l_2 < \mu,$$

and

$$\frac{r}{R} \to \frac{1}{3} \quad \text{as} \quad \frac{l_2}{\mu} \to 0.$$

(iv) *Condition (19.2.12) can hold in many cases. For example, if* $r_n = 0$, *then we can choose* $K_0 = 0$. *Another choice for* K_0 *is given by*

$$r_n = L^{-1} A_0^{-1} (F(x_n) + G(x_n)),$$

where $L \in L(\mathcal{X}, \mathcal{Y})$ *is a given invertible operator such that*

$$\|A_0^{-1} L^{-1} A_0\| \le p \quad \text{for some } p > 0.$$

Moreover, suppose that

$$\|A_0^{-1} F'(x)\| \le q \quad \text{for some } q > 0 \text{ and each } x \in \mathcal{D}.$$

Then we have under the hypotheses of Theorem 19.2.2 that

$$
\begin{aligned}
\|A_0^{-1}(r_{n+1} - r_n)\| &\le \quad \|A_0^{-1} L^{-1} A_0\| \|A_0^{-1}((F(x_{n+1}) - F(x_n)) \\
&\quad + (G(x_{n+1}) - G(x_n)))\| \\
&\le \quad p\Big[\|\textstyle\int_0^1 A_0^{-1} F'(x_n + \theta(x_{n+1} - x_n))d\theta\| \\
&\quad \times \|x_{n+1} - x_n\| \\
&\quad + \|A_0^{-1}(G(x_{n+1}) - G(x_n))\|\Big]
\end{aligned}
$$

$$\leq \quad p(q\|x_{n+1} - x_n\| + K\|x_{n+1} - x_n\|)$$
$$= \quad p(q + K)\|x_{n+1} - x_n\|.$$

Consequently, condition (19.2.12) is satisfied for $K_0 = p(q + K)$. Furthermore, as

$$
\begin{aligned}
\|A_0^{-1} r_n\| &\leq \quad \|A_0^{-1} L^{-1} A_0\| \|A_0^{-1}((F(x_n) - F(x^*)) \\
&\quad + (G(x_n) - G(x^*)))\| \\
&\leq \quad p\left[\| \int_0^1 A_0^{-1} F'(x^* + \theta(x_n - x^*))(x_n - x^*)d\theta\| \right. \\
&\quad \left. + \|A_0^{-1}(G(x_n) - G(x^*))\|\right] \\
&\leq \quad p(q\|x_n - x^*\| + K\|x_n - x^*\|) \\
&= \quad p(q + K)\|x_n - x^*\|,
\end{aligned}
$$

we can choose $a = p(q + K)$. In a similar way, related to the local convergence case, one can choose

$$r_n = L^{-1} A_*^{-1}(F(x_n) + G(x_n))$$

and suppose

$$\|A_*^{-1} L^{-1} A_*\| \leq p_* \quad \text{for some } p_* > 0,$$

$$\|A_*^{-1} F'(x)\| \leq q_* \quad \text{for some } q_* > 0 \quad \text{and } x \in \mathcal{D}.$$

Next, choosing

$$\alpha = p_*(q_* + \lambda),$$

condition (19.2.28) is satisfied.

Other choices for sequence $\{r_n\}$ and operator A can be found in [6,10,11, 17].

19.3 NUMERICAL EXAMPLES

Example 19.3.1. *Let $\mathcal{X} = \mathcal{Y} = \mathbb{R}$, $x_{-1} = 1.001$, $x_0 = 1$, $\gamma \in [0, 1)$, $\mathcal{D} = [\gamma, 2 - \gamma]$ and define functions F, G on $\mathcal{D}$ by*

$$F(x) = x^3 - \gamma = 0 \text{ and } G(x) = \delta|x - 1| \text{ for some real number } \delta \neq 3.$$

Then function F is differentiable on $\mathcal{D}$ and function G is continuous on $\mathcal{D}$ but not differentiable at $x_0 = 1$. Moreover, define function $A(y, x) : \overset{\circ}{\mathcal{D}} \times \overset{\circ}{\mathcal{D}} \to L(\mathcal{Y}, \mathcal{X})$ by $A(y, x) = F'(y) + [y, x; G]$, where $[y, x; G]$ is a divided difference *of order 1 for operator G satisfying*

$$[y, x; G] = \frac{G(y) - G(x)}{y - x} \quad \textit{for each } x, y \in \overset{\circ}{\mathcal{D}} \textit{ with } x \neq y.$$

Then we have by the definition of function A that $A_0 = 3 - \delta \neq 0$. Suppose we choose a sequence converging to zero $\{r_n\}$ such that $\|r_{n+1} - r_n\| \leq \|\beta(x_{n+1} - x_n)\|$ (see also Remark 2.6(iv)) for some real number β.

Choose $\gamma = 0.9$, $\beta = 0.0298$, and $\delta = 0.02$. Then we have that hypotheses (H_1)–(H_5) are satisfied if

$$b = 0.001, \quad c = 0.033046893, \quad M = 2.1810949, \quad K_0 = 0.090174889,$$

$$N = 0.00660937877\ldots,$$

$$L_1 = L_2 = L_3 = L_4 = L_6 = L_7 = N_1 = K = 0, \quad \textit{and} \quad L_5 = 0.991406816\ldots$$

Conditions (19.2.1)–(19.2.4) respectively become

$$0.0967842 < 1,$$

$$0 < 0.0967842 < 1,$$

$$1.98281363 < 2.1810949,$$

and

$$0.1485190 < 0.8157986,$$

which are satisfied. Hence sequence $\{x_n\}$ generated for $x_{-1} = 1.001$, $x_0 = 1$ by method (19.1.2) is converging to $x^* = \sqrt[3]{0.9} = 0.965489385$. We get $t^* = 0.03958739\ldots$

Example 19.3.2. *Let $X = Y = C[0, 1]$, the space of continuous functions defined on $[0, 1]$, be equipped with the max-norm. Let $D = \{x \in C[0, 1]; \|x\| \leq R\}$ with $R = 5$ and F be defined on D by*

$$F(x)(s) = x(s) - f(s) - \frac{1}{8} \int_0^1 G(s, t)x(t)^3 \, dt, \quad x \in C[0, 1], \ s \in [0, 1],$$

where $f \in C[0, 1]$ is a given function and the kernel G is the Green's function

$$G(s, t) = \begin{cases} (1 - s)t, & t \leq s, \\ s(1 - t), & s \leq t. \end{cases}$$

In this case, for each $x \in D$, $F'(x)$ is a linear operator defined on D by the following expression:

$$[F'(x)(v)](s) = v(s) - \frac{3}{8} \int_0^1 G(s, t)x(t)^2 v(t) \, dt, \quad v \in C[0, 1], \ s \in [0, 1].$$

If we choose $x_0(s) = f(s) = s$, we obtain

$$\|F(x_0)\| \le \frac{1}{64}.$$

Define the divided difference defined by

$$[x, y; F] = \int_0^1 F'(\tau x + (1 - \tau)y)d\tau.$$

Taking into account that

$$
\begin{aligned}
\|[x, y; F] - F'(z)\| &\le \int_0^1 \|F'(\tau x + (1 - \tau)y) - F'(z)\| d\tau \\
&\le \frac{3}{64} \int_0^1 \left(3\tau^2(\|x^2 - z^2\| + \|y^2 - z^2\|)\right. \\
&\quad \left. + 6\tau(1 - \tau)\|xy - z^2\|\right)\tau \\
&\le \frac{1}{64} \left(\|x^2 - z^2\| + \left(\|y^2 - z^2\|\right) + \left(\|xy - z^2\|\right)\right),
\end{aligned}
$$

and considering $G = 0$, $A(y, x) = F'(y)$, it is easy to see that

$$b = N_1 = N = L_1 = L_3 = L_4 = L_5 = L_6 = L_7 = 0,$$

$$L_2 = 0.229508, \quad M = 0.5625, \text{ and } K_0 = \frac{64\beta}{61}.$$

Suppose we choose a converging to zero sequence $\{r_n\}$ such that $\|r_{n+1} - r_n\| \le \|0.1(x_{n+1} - x_n)\|$. Then conditions (19.2.1)–(19.2.4) respectively become

$$0.104918 < 1,$$

$$0 < 0.104918 < 1,$$

$$0.459016 < 0.5625,$$

and

$$0.0184426 < 0.801172,$$

which are satisfied. Hence sequence $\{x_n\}$ generated for $x_0 = 1$ by method (19.1.2) is converging to a solution of the equation $F(x) = 0$. We get $t^* = 0.0373591\ldots$

REFERENCES

[1] S. Amat, I.K. Argyros, S. Busquier, R. Castro, S. Hilout, S. Plaza, Newton-type methods on Riemannian manifolds under Kantorovich-type conditions, Appl. Math. Comput. 227 (2014) 762–787.

[2] S. Amat, S. Busquier, J.M. Gutiérrez, Geometric constructions of iterative functions to solve nonlinear equations, J. Comput. Appl. Math. 157 (1) (2003) 197–205.

[3] S. Amat, S. Busquier, R. Castro, S. Plaza, Third-order methods on Riemannian manifolds under Kantorovich conditions, J. Comput. Appl. Math. 255 (2014) 106–121.

[4] S. Amat, S. Busquier, Á.A. Magreñán, Reducing chaos and bifurcations in Newton-type methods, Abst. Appl. Anal. 2013 (2013) 726701, https://doi.org/10.1155/2013/726701.

[5] S. Amat, Á.A. Magreñán, N. Romero, On a family of two step Newton-type methods, Appl. Math. Comput. 219 (4) (2013) 11341–11347.

[6] I.K. Argyros, A unifying local–semilocal convergence analysis and applications for two-point Newton-like methods in Banach space, J. Math. Anal. Appl. 298 (2004) 374–397.

[7] I.K. Argyros, A semilocal convergence analysis for directional Newton methods, Math. Comp. 80 (2011) 327–343.

[8] I.K. Argyros, S. Hilout, Expanding the applicability of Newton's method using the Smale alpha theory, J. Comput. Appl. Math. 261 (2014) 183–200.

[9] I.K. Argyros, S. Hilout, Expanding the applicability of Inexact Newton's method using the Smale alpha theory, Appl. Math. Comput. 224 (2014) 224–237.

[10] I.K. Argyros, Computational theory of iterative methods, in: C.K. Chui, L. Wuytack (Eds.), Stud. Comput. Math., vol. 15, Elsevier Publ. Co., New York, USA, 2007.

[11] I.K. Argyros, S. Hilout, Numerical Methods in Nonlinear Analysis, World Scientific Publ. Comp., New Jersey, 2013.

[12] I.K. Argyros, S. Hilout, Weaker conditions for the convergence of Newton's method, J. Complexity 28 (2012) 364–387.

[13] E. Cătinaş, The inexact, inexact perturbed, and quasi-Newton methods are equivalent models, Math. Comp. 74 (249) (2005) 291–301.

[14] E. Cătinaş, On some iterative methods for solving nonlinear equations, Rev. Anal. Numer. Theor. Approx. 23 (1) (1994) 47–53.

[15] J. Chen, T. Yamamoto, Convergence domains of certain iterative methods for solving nonlinear equations, Numer. Funct. Anal. Optimiz. 10 (1989) 37–48.

[16] R.S. Dembo, S.C. Eisenstat, T. Steihaug, Inexact Newton methods, SIAM J. Numer. Anal. 19 (2) (1982) 400–408.

[17] J.E. Dennis, On Newton-like methods, Numer. Math. 11 (1968) 324–330.

[18] M. Frontini, E. Sormani, Some variant of Newton's method with third-order convergence, Appl. Math. Comput. 140 (2003) 419–426.

[19] J.M. Gutiérrez, M.Á. Hernández, Third-order iterative methods for operators with bounded second derivative, J. Comput. Appl. Math. 82 (1997) 171–183.

[20] J.M. Gutiérrez, M.Á. Hernández, Recurrent relations for the super-Halley method, Comput. Math. Appl. 36 (1998) 1–8.

[21] J.M. Gutiérrez, M.Á. Hernández, N. Romero, Dynamics of a new family of iterative processes for quadratic polynomials, J. Comput. Appl. Math. 233 (2010) 2688–2695.

[22] J.M. Gutiérrez, M.Á. Hernández, M. Á Salanova, Quadratic equations in Banach spaces, Numer. Funct. Anal. Optimiz. 7 (1–2) (1996) 113–121.

[23] J.M. Gutiérrez, Á.A. Magreñán, N. Romero, On the semilocal convergence of Newton–Kantorovich method under center-Lipschitz conditions, Appl. Math. Comput. 221 (2013) 79–88.

[24] M.Á. Hernández, Chebyshev's approximation algorithms and applications, Comput. Math. Appl. 41 (2001) 433–455.

[25] M.Á. Hernández, M.A. Salanova, Modification of the Kantorovich assumptions for semilocal convergence of the Chebyshev method, J. Comput. Appl. Math. 126 (2000) 131–143.

[26] M.Á. Hernández, N. Romero, On a characterization of some Newton-like methods of R-order at least three, J. Comput. Appl. Math. 183 (2005) 53–66.

[27] E. Kahya, A new unidimensional search method for optimization: linear interpolation method, Appl. Math. Comput. 171 (2) (2005) 912–926.

[28] J. Kou, Y. Li, X. Wang, On modified Newton methods with cubic convergence, Appl. Math. Comput. 176 (1) (2006) 123–127.
[29] D.G. Luenberger, Linear and Nonlinear Programming, second ed., Addison-Wesley Publishing Company, Inc., 1984.
[30] Á.A. Magreñán, Different anomalies in a Jarratt family of iterative root-finding methods, Appl. Math. Comput. 233 (2014) 29–38.
[31] Á.A. Magreñán, A new tool to study real dynamics: the convergence plane, Appl. Math. Comput. 248 (2014) 215–224.
[32] F.A. Potra, V. Pták, Nondiscrete Induction and Iterative Processes, Pitman, New York, 1984.
[33] R.L. Rardin, Optimization in Operations Research, Prentice-Hall, Inc., NJ, 1998.
[34] H. Ren, Q. Wu, Convergence balls of a modified Secant method for finding zero of derivatives, Appl. Math. Comput. 174 (2009) 24–33.
[35] W.C. Rheinboldt, An adaptive continuation process for solving systems of nonlinear equations, in: Banach Ctr. Publ., vol. 3, Polish Academy of Science, 1977, pp. 129–142.
[36] S. Weerakoon, T.G.I. Fernando, A variant of Newton's method with accelerated third-order convergence, Appl. Math. Lett. 13 (2000) 87–93.
[37] P.P. Zabrejko, D.F. Nguen, The majorant method in the theory of Newton–Kantorovich approximations and the Pták error estimates, Numer. Funct. Anal. Optimiz. 9 (5–6) (1987) 671–684.

Chapter 20

Two-step Newton methods

20.1 INTRODUCTION

In this chapter we are concerned with the problem of approximating a locally unique solution $x^\star$ of the nonlinear equation

$$F(x) = 0, \qquad (20.1.1)$$

where F is a Fréchet-differentiable operator defined on a convex subset $\mathcal{D}$ of a Banach space $\mathcal{X}$ with values in a Banach space $\mathcal{Y}$. Many problems in applied sciences reduce to solving an equation in the form (20.1.1). The methods used to solve this kind of equation are usually iterative. The convergence analysis of iterative methods is usually divided into two categories: semilocal and local convergence analysis depending on the criteria used to ensure the convergence.

The most well-known and used method is Newton's method, which is defined as

$$x_{n+1} = x_n - F'(x_n)^{-1} F(x_n), \quad \text{for each } n = 0, 1, 2, \ldots, \qquad (20.1.2)$$

where x_0 is an initial guess.

There exist several works related to the convergence of Newton's method (see, for example, [1–39]).

On the other hand, the convergence domain usually gets smaller as the order of convergence of the method increases. That is why it is important to enlarge the convergence domain as much as possible using the same conditions and constants as before. This is our main motivation for this chapter. In particular, we revisit the two-step Newton's methods defined for each $n = 0, 1, 2 \ldots$ by

$$\begin{aligned} y_n &= x_n - F'(x_n)^{-1} F(x_n), \\ x_{n+1} &= y_n - F'(x_n)^{-1} F(y_n), \end{aligned} \qquad (20.1.3)$$

and

$$\begin{aligned} y_n &= x_n - F'(x_n)^{-1} F(x_n), \\ x_{n+1} &= y_n - F'(y_n)^{-1} F(y_n). \end{aligned} \qquad (20.1.4)$$

Two-step Newton's methods (20.1.3) and (20.1.4) are of convergence order 3 and 4, respectively [4,6,9,10,24,26].

A Contemporary Study of Iterative Methods. DOI: 10.1016/B978-0-12-809214-9.00020-6

265

It is well known fact that if the Lipschitz condition

$$\|F'(x_0)^{-1}(F'(x) - F'(y))\| \le L\|x - y\| \text{ for each } x \text{ and } y \in \mathcal{D}, \quad (20.1.5)$$

as well as

$$\|F'(x_0)^{-1}F(x_0)\| \le v, \quad (20.1.6)$$

hold for some $L > 0$ and $v > 0$, then a sufficient semilocal convergence condition for both Newton's method (20.1.2) and two-step Newton's method (20.1.4) is given by the Newton–Kantorovich hypothesis [27]:

$$h = Lv \le \frac{1}{2}. \quad (20.1.7)$$

Hypothesis (20.1.7) is only sufficient for the convergence of Newton's method. That is why we challenged it in a series of works [4–10,13] by introducing the center-Lipschitz condition defined as:

$$\|F'(x_0)^{-1}(F'(x) - F'(x_0))\| \le L_0\|x - x_0\| \text{ for each } x \in \mathcal{D}. \quad (20.1.8)$$

It is clear that

$$L_0 \le L \quad (20.1.9)$$

holds in general and $\dfrac{L}{L_0}$ can be arbitrarily large [5,6,10,13]. Our sufficient convergence conditions are given by

$$h_1 = L_1 v \le \frac{1}{2}, \quad (20.1.10)$$

$$h_2 = L_2 v \le \frac{1}{2}, \quad (20.1.11)$$

and

$$h_3 = L_3 v \le \frac{1}{2}, \quad (20.1.12)$$

where

$$L_1 = \frac{L_0 + L}{2},$$

$$L_2 = \frac{1}{8}\left(L + 4L_0 + \sqrt{L^2 + 8L_0 L}\right), \quad (20.1.13)$$

and

$$L_3 = \frac{1}{8}\left(4L_0 + \sqrt{L_0 L} + \sqrt{L^2 + 8L_0 L}\right).$$

Moreover, it is worth noticing that

$$h \le \frac{1}{2} \Rightarrow h_1 \le \frac{1}{2} \Rightarrow h_2 \le \frac{1}{2} \Rightarrow h_3 \le \frac{1}{2}, \qquad (20.1.14)$$

but not necessarily vice versa, unless $L_0 = L$ and

$$\frac{h_1}{h} \to \frac{1}{2}, \; \frac{h_2}{h} \to \frac{1}{4}, \; \frac{h_2}{h_1} \to \frac{1}{2}, \; \frac{h_3}{h} \to 0, \; \frac{h_3}{h_1} \to 0, \; \text{and} \; \frac{h_3}{h_2} \to 0 \text{ as } \frac{L_0}{L} \to 0. \qquad (20.1.15)$$

As a consequence, the convergence domain for Newton's method (20.1.2) has been extended under the same computational cost, since in practice the computation of L requires the computation of L_0 as a special case. Furthermore, the information on the location of the solution is at least as precise, and the error estimates on the distances $\|x_{n+1} - x_n\|$ and $\|x_n - x^\star\|$ are more precise.

In the case of the two-step Newton's method (20.1.3) a sufficient convergence condition using only (20.1.5) is given by [10,24,26]

$$h_4 = L_4 v \le \frac{1}{2}, \qquad (20.1.16)$$

where

$$L_4 = \frac{4 + \sqrt{21}}{4} L. \qquad (20.1.17)$$

In the present chapter using (20.1.5) and (20.1.8) we show that (20.1.12) can be used as a sufficient convergence condition for two-step Newton's method (20.1.4). Moreover, we show that a sufficient convergence condition for (20.1.3) is given by

$$h_5 = L_5 v \le \frac{1}{2}, \qquad (20.1.18)$$

where

$$L_5 = \frac{1}{4} \left(3L_0 + L + \sqrt{(3L_0 + L)^2 + L(4L_0 + L)} \right). \qquad (20.1.19)$$

Notice that

$$h_4 \le \frac{1}{2} \Rightarrow h_5 \le \frac{1}{2}, \qquad (20.1.20)$$

but not necessarily vice versa, unless $L_0 = L$ and

$$\frac{h_5}{h_4} \to \frac{1 + \sqrt{2}}{4 + \sqrt{21}} < 1 \text{ as } \frac{L_0}{L} \to 0. \qquad (20.1.21)$$

Condition (20.1.18) can be weakened even further.

Related to the local convergence matter, using the Lipschitz condition

$$\|F'(x^\star)^{-1}(F'(x) - F'(y))\| \le l\|x - y\| \text{ for each } x \text{ and } y \in \mathcal{D} \text{ and some } l > 0, \tag{20.1.22}$$

is the convergence radius used in the literature (see Rheinboldt [33] and Traub [39]) for both Newton's method (20.1.2) and two-step Newton's method (20.1.4) and given by

$$R_0 = \frac{2}{3l}. \tag{20.1.23}$$

In this chapter, we use the center-Lipschitz condition,

$$\|F'(x^\star)^{-1}(F'(x) - F'(x^\star))\| \le l\|x - x^\star\| \text{ for each } x \in \mathcal{D} \text{ and some } l_0 > 0, \tag{20.1.24}$$

to show that the convergence radius for both Newton's method (20.1.2) and two-step Newton's method (20.1.4) is given by

$$R_0 = \frac{2}{2l_0 + l}. \tag{20.1.25}$$

Again we have that

$$l_0 \le l \tag{20.1.26}$$

holds in general and $\dfrac{l}{l_0}$ can be arbitrarily large [5,6,10]. We get

$$R_0 \le R \tag{20.1.27}$$

and

$$\frac{R}{R_0} \to 3 \text{ as } \frac{l_0}{l} \to 0. \tag{20.1.28}$$

The radius of convergence R was found by us in [5,6,10] only for Newton's method. Here we also have this result for two-step Newton's method (20.1.4). Moreover, in view of (20.1.22), there exists a constant $l_1 > 0$ such that

$$\|F'(x^\star)^{-1}(F'(x) - F'(x_0))\| \le l_1\|x - x_0\| \text{ for all } x \in \mathcal{D}. \tag{20.1.29}$$

Note that

$$l_1 \le l \tag{20.1.30}$$

holds and $\dfrac{l}{l_1}$ can be arbitrarily large. Although the convergence radius R does not change, the error bounds are more precise when using (20.1.29). Finally, the corresponding results for the two-step Newton's method (20.1.3) are presented with

$$R = \frac{2}{2l_0 + 5l}.$$ (20.1.31)

Many high convergence order iterative methods can be written as two-step methods [4,6,9,10,22–26,39,40]. Therefore, the technique of recurrent functions or the technique of simplified majorizing sequences given in this chapter can be used for other high convergence order iterative methods. As an example, we suggest the Chebyshev method or the method of tangent parabolas, defined by

$$x_{n+1} = x_n - (I - M_n)F'(x_n)^{-1}F(x_n) \text{ for each } n = 0, 1, 2, \ldots,$$ (20.1.32)

where x_0 is an initial point and

$$M_n = \frac{1}{2}F'(x_n)^{-1}F''(x_n)F'(x_n)^{-1}F(x_n) \text{ for each } n = 0, 1, 2, \ldots$$

Here $F''(x)$ denotes the second Fréchet-derivative of operator F [6,10,27, 39]. Chebyshev method can be written as a two-step method of the form

$$y_n = x_n - F'(x_n)^{-1}F(x_n),$$

$$x_{n+1} = y_n - \frac{1}{2}F'(x_n)^{-1}F''(x_n)(y_n - x_n)^2 \text{ for each } n = 0, 1, 2, \ldots$$
(20.1.33)

The chapter is organized as follows. The convergence results of the majorizing sequences for two-step Newton's method (20.1.4) are given in Section 20.2. Next the convergence results of the majorizing sequences for two-step Newton's method (20.1.3) are given in Section 20.3. The semilocal convergence analysis of two-step Newton's method (20.1.4) is presented in Section 20.4. Then the local convergence analysis of two-step Newton's method (20.1.3) is presented in Section 20.5. Finally, numerical examples are given in the concluding Section 20.6.

20.2 MAJORIZING SEQUENCES FOR TWO-STEP NEWTON'S METHOD (20.1.4)

In this section, we present sufficient convergence conditions and bounds on the limit points of majorizing sequences for two-step method (20.1.4).

Lemma 20.2.1. *Let $L_0 > 0$, $L \geq L_0$, and $v > 0$ be given parameters. Set*

$$\alpha = \frac{2L}{L + \sqrt{L^2 + 8L_0L}}.$$ (20.2.1)

Suppose that

$$h_1 = L_1 v \leq \frac{1}{2}, \tag{20.2.2}$$

where

$$L_1 = \frac{1}{8}(L + 4L_0 + \sqrt{L^2 + 8L_0 L}). \tag{20.2.3}$$

Then the scalar sequence $\{t_n\}$ given by

$$\begin{cases} t_0 = 0, \quad s_0 = v, \\[2mm] t_{n+1} = s_n + \dfrac{L(s_n - t_n)^2}{2(1 - L_0 s_n)}, \\[2mm] s_{n+1} = t_{n+1} + \dfrac{L(t_{n+1} - s_n)^2}{2(1 - L_0 t_{n+1})} \quad \text{for each } n = 0, 1, 2, \ldots \end{cases} \tag{20.2.4}$$

is well defined, increasing, bounded from above by

$$t^{**} = \frac{v}{1 - \alpha}, \tag{20.2.5}$$

and converges to its unique least upper bound t^* which satisfies

$$v \leq t^* \leq t^{**}. \tag{20.2.6}$$

Moreover, the following estimates hold:

$$t_{n+1} - s_n \leq \alpha(s_n - t_n) \leq \alpha^{2n+1} v, \tag{20.2.7}$$

$$s_n - t_n \leq \alpha(t_n - s_{n-1}) \leq \alpha^{2n} v, \tag{20.2.8}$$

$$t^* - s_n \leq \frac{\alpha^{2n} v}{1 - \alpha}, \tag{20.2.9}$$

and

$$t^* - t_n \leq \frac{\alpha^{2n} v}{1 - \alpha} + \alpha^{2n} v. \tag{20.2.10}$$

Proof. We first notice that $\alpha \in [\frac{1}{2}, 1)$ by (20.2.1). We shall show using mathematical induction that

$$\frac{L(s_k - t_k)}{2(1 - L_0 s_k)} \leq \alpha \tag{20.2.11}$$

and

$$\frac{L(t_{k+1} - s_k)}{2(1 - L_0 t_{k+1})} \leq \alpha. \tag{20.2.12}$$

If $k = 0$ in (20.2.11), we must have

$$\frac{L(s_0 - t_0)}{2(1 - L_0 s_0)} \le \alpha \text{ or } \frac{Lv}{2(1 - L_0 v)} \le \alpha. \tag{20.2.13}$$

Using the value of α in (20.2.13), we can show instead that

$$\left(\frac{L}{2} + \frac{2LL_0}{L + \sqrt{L + \sqrt{L^2 + 8L_0 L}}} \right) v \le \frac{2L}{L + \sqrt{L + \sqrt{L^2 + 8L_0 L}}},$$

which is (20.2.2). If $k = 0$ in (20.2.12), we must have

$$\frac{L(t_1 - s_0)}{2(1 - L_0 t_1)} \le \alpha \text{ or } (L^2 - 4L_0^2\alpha + 2L_0 L\alpha)v^2 + 8L_0\alpha v - 4\alpha \le 0. \tag{20.2.14}$$

Case 1. $L^2 - 4L_0^2\alpha + 2L_0 L\alpha \ge 0$
Then (20.2.14) is satisfied, provided that

$$v \le \frac{-8L_0\alpha + \sqrt{(8L_0\alpha)^2 + 16\alpha(L^2 - 4L_0^2\alpha + 2L_0 L\alpha)}}{2(L^2 - 4L_0^2\alpha + 2L_0 L\alpha)} \tag{20.2.15}$$

or

$$\frac{2L_0\alpha + \sqrt{\alpha L^2 + 2L_0 L\alpha^2}}{2\alpha} v \le 1. \tag{20.2.16}$$

In view of (20.2.2) and (20.2.16), we must show

$$\frac{2L_0\alpha + \sqrt{\alpha L^2 + 2L_0 L\alpha^2}}{2\alpha} \le \frac{1}{4}(L + 4L_0 + \sqrt{L^2 + 8L_0 L})$$

or

$$2\sqrt{\alpha L^2 + 2L_0 L\alpha^2} \le \alpha L + \alpha\sqrt{L^2 + 8L_0 L},$$

or

$$\alpha \ge \frac{2L}{L + \sqrt{L^2 + 8L_0 L}},$$

which is true as equality by (20.2.1).
Case 2. $L^2 - 4L_0^2\alpha + 2L_0 L\alpha < 0$
Again we show that (20.2.15) is satisfied, as was shown in Case 1.
Case 3. $L^2 - 4L_0^2\alpha + 2L_0 L\alpha = 0$
Inequality (20.2.14) reduces to $2L_0 v \le 1$, which is true by (20.2.2). Hence estimates (20.2.11) and (20.2.12) hold for $k = 0$. Let us assume they hold for $k \le n$. Then, using (20.2.4), (20.2.11), and (20.2.12), we have in turn that

$$t_{k+1} - s_k = \frac{L(s_k - t_k)}{2(1 - L_0 s_k)}(s_k - t_k) \le \alpha(s_k - t_k),$$

$$s_{k+1} - t_{k+1} = \frac{L(t_{k+1} - s_k)}{2(1 - L_0 t_{k+1})}(t_{k+1} - s_k) \leq \alpha(t_{k+1} - s_k),$$

leading to

$$t_{k+1} - s_k \leq \alpha(\alpha^2)^k v, \qquad (20.2.17)$$

$$s_{k+1} - t_{k+1} \leq (\alpha^2)^{k+1} v, \qquad (20.2.18)$$

$$t_{k+1} \leq s_k + \alpha(\alpha^2)^k v \leq t_k + \alpha^{2k} v + \alpha\alpha^{2k} v$$
$$\leq t_{k-1} + \alpha^{2(k-1)} v + \alpha\alpha^{2(k-1)} v + \alpha^{2k} v + \alpha\alpha^{2k} v$$
$$\leq \cdots \leq t_0 + [\alpha^{2\cdot 0} + \cdots + \alpha^{2k}] v + \alpha[\alpha^{2\cdot 0} + \cdots + \alpha^{2k}] v \qquad (20.2.19)$$
$$= (1 + \alpha)\frac{1 - \alpha^{2(k+1)}}{1 - \alpha^2} v < t^{**},$$

and

$$s_{k+1} \leq (1 + \alpha)\frac{1 - \alpha^{2(k+1)}}{1 - \alpha^2} v + \alpha^{2(k+1)} v. \qquad (20.2.20)$$

In view of (20.2.11), (20.2.17), (20.2.18), and (20.2.19), we must show

$$\frac{L}{2}(s_{k+1} - t_{k+1}) + L_0\alpha s_{k+1} - \alpha \leq 0$$

or $\qquad\qquad\qquad\qquad\qquad\qquad\qquad\qquad\qquad\qquad\qquad$ (20.2.21)

$$\frac{L}{2}\alpha^{2(k+1)} v + L_0\alpha[(1 + \alpha)\frac{1 - \alpha^{2(k+1)}}{1 - \alpha^2} + \alpha^{2(k+1)}] v - \alpha \leq 0.$$

Estimate (20.2.21) motivates us to define recurrent functions f_k on $[0, \alpha^2]$ by

$$f_k(t) = \frac{L}{2}t^{k+1} v + L_0\sqrt{t}\left[(1 + \sqrt{t})\frac{1 - t^{k+1}}{1 - t} + t^{k+1}\right] v - \sqrt{t}. \qquad (20.2.22)$$

We need to find a relationship between two consecutive functions f_k. Using (20.2.22) we get that

$$f_{k+1}(t) = f_k(t) + [\frac{L}{2}t + L_0\sqrt{t}t + L_0\sqrt{t}(1 + \sqrt{t})](t - 1)t^k v \leq f_k(t),$$

$$\qquad\qquad\qquad\qquad\qquad\qquad\qquad\qquad\qquad\qquad (20.2.23)$$

since $\alpha \in [0, 1)$ and the quantity in the brackets for $t = \alpha^2$ is nonnegative. Then, in view of (20.2.21)–(20.2.23), we must show that

$$f_0(\alpha^2) \leq 0$$

or

$$\left[\frac{L}{2}\alpha + L_0(1 + \alpha + \alpha^2)\right]v \le 1. \tag{20.2.24}$$

We have that α is the unique positive root of the equation

$$2L_0t^2 + Lt - L = 0. \tag{20.2.25}$$

It follows from (20.2.24) and (20.2.25) that we must show

$$\frac{1}{2}(L + 2L_0 + 2L_0\alpha)v \le 1 \tag{20.2.26}$$

or, in view of (20.2.2),

$$\frac{1}{2}(L + 2L_0 + 2L_0\alpha) \le \frac{1}{4}(L + 4L_0 + \sqrt{L^2 + 8L_0L}),$$

or

$$\alpha \le \frac{2L}{L + \sqrt{L^2 + 8L_0L}},$$

which is true as an equality. The induction for (20.2.11) is completed. Estimate (20.2.12) is satisfied if

$$\frac{L}{2}(t_{k+1} - s_k) + \alpha L_0 t_{k+1} - \alpha \le 0$$

or $\tag{20.2.27}$

$$\frac{L}{2}\alpha\alpha^{2k}v + \alpha L_0(1 + \alpha)\frac{1 - \alpha^{2(k+1)}}{1 - \alpha^2}v - \alpha \le 0.$$

Estimate (20.2.27) motivates us to define recurrent functions g_k on $[0, \alpha^2]$ by

$$g_k(t) = \frac{L}{2}\sqrt{t}t^{k+1}v + \sqrt{t}L_0(1 + \sqrt{t})\frac{1 - t^{k+1}}{1 - t}v - \sqrt{t}. \tag{20.2.28}$$

We have that

$$g_{k+1}(t) = g_k(t) + (\frac{L}{2}\sqrt{t}t + L_0\sqrt{t}(1 + \sqrt{t}))(t - 1)t^k v$$
$$\le g_k(t) \text{ for all } t \in [0, \alpha^2], \tag{20.2.29}$$

since $\alpha \in [0, 1)$.

Hence we have that $g_{k+1}(\alpha^2) \le g_k(\alpha^2) \le \cdots \le g_1(\alpha^2)$ and, in view of (20.2.28), estimate (20.2.27) holds if

$$g_1(\alpha^2) \le 0 \tag{20.2.30}$$

or

$$\frac{1}{2}(L\alpha^2 + 2L_0(1+\alpha)(1+\alpha^2))v \le 1. \tag{20.2.31}$$

We have by (20.2.25) that

$$L\alpha^2 + 2L_0(1+\alpha)(1+\alpha^2) = \frac{L(L-L\alpha)}{2L_0} + 2L_0(1+\alpha)(1+\frac{L-L\alpha}{2L_0})$$

$$= \frac{L^2 - L^2\alpha + 2L_0(1+\alpha)[2L_0 + L - L\alpha]}{2L_0}$$

$$= \frac{L^2 - L^2\alpha + 4L_0^2 + 4L_0^2\alpha + 2L_0L + 2L_0L\alpha - 2L_0L\alpha - 2L_0L\alpha^2}{2L_0}$$

$$= \frac{L(L - L\alpha - 2L_0\alpha^2) + 4L_0^2 + 4L_0^2\alpha + 2L_0L}{2L_0}$$

$$= \frac{2L_0(L + 2L_0 + 2\alpha L_0)}{2L_0} = L + 2L_0 + 2\alpha L_0.$$

So we must have

$$(\frac{L}{2} + (1+\alpha)L_0)v \le 1. \tag{20.2.32}$$

Then, in view of (20.2.2), if suffices to show that

$$\frac{L}{2} + (1+\alpha)L_0 \le \frac{1}{4}\left(L + 4L_0 + \sqrt{L^2 + 8L_0L}\right)$$

or

$$\alpha \le \frac{-L + \sqrt{L^2 + 8L_0L}}{4L_0} = \frac{2L}{L + \sqrt{L^2 + 8L_0L}},$$

which is true as an equality. The induction for (20.2.12) is complete. As a consequence, sequence $\{t_n\}$ is increasing, bounded from above by t^{**} given by (20.2.5), and as such it converges to its unique least upper bound t^* which satisfies (20.2.6). Moreover, using (20.2.17) and (20.2.18), we have that

$$t_{k+m} - s_k = t_{k+m} - s_{k+m} + s_{k+m} - s_k$$

and

$$s_{k+m} - s_k = (s_{k+m} - s_{k+m-1}) + (s_{k+m-1} - s_k) \le \cdots$$

$$\le \alpha\alpha^{2(k+m-1)}v + \alpha^{2(k+m-1)}v + \alpha\alpha^{2(k+m-2)}v + \alpha^{2(k+m-2)}v$$

$$+ \cdots + \alpha\alpha^{2k}v + \alpha^{2k}v$$

so

$$t_{k+m} - s_k \leq \alpha^{2(m+k)}v + \alpha\alpha^{2k}v(1 + \cdots + \alpha^{2(m-1)}) + \alpha^{2k}v(1 + \cdots + \alpha^{2(m-1)})$$
$$= \alpha^{2k}(1 + \cdots + \alpha^{2m})v + \alpha\alpha^{2k}(1 + \cdots + \alpha^{2(m-1)})v.$$

(20.2.33)

Letting $m \to \infty$ in (20.2.33), we obtain (20.2.9). Moreover, we have

$$s_{k+m} - t_k \leq s_{m+k} - s_{m+k-1} + s_{m+k-1} - t_k$$
$$\leq \alpha\alpha^{2(m+k-1)}v + \alpha^{2(k+m-1)}v + \cdots + s_k - t_k$$ (20.2.34)
$$\leq \alpha\alpha^{2k}(1 + \cdots + \alpha^{2(m-1)})v + \alpha^{2k}(1 + \cdots + \alpha^{2(m-1)})v + \alpha^{2k}v.$$

Letting $m \to \infty$ in (20.2.34), we obtain (20.2.10). This completes the proof of the lemma. $\square$

We can state the following remark

Remark 20.2.2. *Let us define sequence $\{\bar{t}_n\}$ by*

$$\begin{cases} \bar{t}_0 = 0, \ \bar{s}_0 = v, \ \bar{t}_1 = \bar{s}_0 + \dfrac{L_0(\bar{s}_0 - \bar{t}_0)^2}{2(1 - L_0\bar{s}_0)}, \\[3mm] \bar{t}_{n+1} = \bar{s}_n + \dfrac{L(\bar{s}_n - \bar{t}_n)^2}{2(1 - L_0\bar{s}_n)}, \\[3mm] \bar{s}_{n+1} = \bar{t}_{n+1} + \dfrac{L(\bar{t}_{n+1} - \bar{s}_n)^2}{2(1 - L_0\bar{t}_{n+1})} \ \text{for each } n = 0, 1, 2, \ldots \end{cases}$$ (20.2.35)

Clearly, $\{\bar{t}_n\}$ converges under (20.2.2) and is tighter than $\{t_n\}$. Using a simple inductive argument, we obtain that

$$\bar{t}_n \leq t_n,$$ (20.2.36)
$$\bar{s}_n \leq s_n,$$ (20.2.37)
$$\bar{t}_{n+1} - \bar{s}_n \leq t_{n+1} - s_n,$$ (20.2.38)
$$\bar{s}_{n+1} - \bar{t}_{n+1} \leq s_{n+1} - t_{n+1},$$ (20.2.39)

and

$$\bar{t}^* = \lim_{n \to \infty} \bar{t}_n \leq t^*.$$ (20.2.40)

Moreover, strict inequalities hold in (20.2.36)–(20.2.39) if $L_0 < L$ for $n \geq 1$. Note also that sequence $\{\bar{t}_n\}$ may converge under weaker hypothesis than (20.2.2) (see [13] and the lemmas that follow).

Next we present a different technique for defining sequence $\{t_n\}$. This technique is easier but it provides a less precise upper bound on t^* and t^{**}. We will first simplify sequence $\{t_n\}$. Let $L = bL_0$ for some $b \geq 1$, $r_n = L_0 t_n$, and $q_n = L_0 s_n$. Then we have that sequence $\{r_n\}$ is given by

$$
\begin{cases}
r_0 = 0, \quad q_0 = L_0 v, \\[2mm]
r_{n+1} = q_n + \dfrac{b(q_n - r_n)^2}{2(1 - q_n)}, \\[3mm]
q_{n+1} = r_{n+1} + \dfrac{b(r_{n+1} - q_n)^2}{2(1 - r_{n+1})}.
\end{cases}
$$

Then set $p_n = 1 - r_n$, $m_n = 1 - q_n$ to obtain sequence $\{p_n\}$ given by

$$
\begin{cases}
p_0 = 1, \quad m_0 = 1 - L_0 v, \\[2mm]
p_{n+1} = m_n - \dfrac{b(m_n - p_n)^2}{2m_n}, \\[3mm]
m_{n+1} = p_{n+1} - \dfrac{b(p_{n+1} - m_n)^2}{2p_{n+1}}.
\end{cases}
$$

Finally, set $\beta_n = 1 - \dfrac{p_n}{m_{n-1}}$ and $\alpha_n = 1 - \dfrac{m_n}{p_n}$ to obtain the sequence $\{\beta_n\}$ defined by

$$
\alpha_{n+1} = \frac{b}{2}\left(\frac{\beta_{n+1}}{1 - \beta_{n+1}}\right)^2, \tag{20.2.41}
$$

$$
\beta_{n+1} = \frac{b}{2}\left(\frac{\alpha_n}{1 - \alpha_n}\right)^2. \tag{20.2.42}
$$

We also have by substituting and eliminating β_{n+1} that

$$
\alpha_{n+1} = \frac{b^3}{8}\frac{\alpha_n^4}{(1 - \alpha_n)^4\left[1 - \dfrac{b}{2}\dfrac{\alpha_n^2}{(1 - \alpha_n)^2}\right]^2}.
$$

Furthermore, it follows from (20.2.41) and (20.2.42) that the convergence of the sequences $\{\alpha_n\}$ and $\{\beta_n\}$ (i.e., $\{r_n\}$ and $\{q_n\}$) is related to the equation

$$
x = \frac{b}{2}\frac{x^2}{(1 - x)^2}
$$

whose zeros are

$$
x = 0, \quad x = \frac{4L_0}{L + L_0 + \sqrt{L^2 + 8L_0 L}} \quad \text{and} \quad x = \frac{L + L_0 + \sqrt{L^2 + 8L_0 L}}{4L_0}.
$$

Consequently, we get the following lemma.

Lemma 20.2.3. *Suppose that* (20.2.2) *holds. Then sequence* $\{t_n\}$ *is increasing, bounded from above by* $\dfrac{1}{L_0}$, *and converges to its unique least upper bound* t^*, *which satisfies*

$$v \leq t^* \leq \frac{1}{L_0}.$$

Now we can give the following lemma.

Lemma 20.2.4. *Suppose that there exists* $N = 0, 1, 2 \ldots$ *such that*

$$t_0 < s_0 < t_1 < s_1 < \cdots < s_N < t_{N+1} < \frac{1}{L_0}$$

and (20.2.43)

$$h^N = L_2(s_N - t_N) \leq \frac{1}{2},$$

where L_2 *is given in* (20.2.3). *Then the scalar sequence* $\{t_n\}$ *given in* (20.2.4) *is well defined, increasing, bounded from above by*

$$t_N^{**} = \frac{s_N - t_N}{1 - \alpha},$$

and converges to its unique least upper bound t_N^* *which satisfies*

$$v \leq t_N^* \leq t_N^{**}.$$

Moreover, estimates (20.2.7)–(20.2.10) *hold with* $s_N - t_N$ *replacing n for* $n \geq N$. *Notice that if* $N = 0$, *we obtain* (20.1.11) *and for* $N = 1$ *we obtain* (20.1.12) *[13]*.

20.3 MAJORIZING SEQUENCES FOR TWO-STEP NEWTON'S METHOD (20.1.3)

In this section we present majorizing sequences for the two-step method (20.1.3).

Lemma 20.3.1. *Let* $L_0 > 0$, $L \geq L_0$, *and* $v > 0$ *be given parameters. Set*

$$\alpha = \frac{L}{2L_0 + L}.$$ (20.3.1)

Suppose that

$$h_5 = L_5 v \leq \frac{1}{2},$$ (20.3.2)

where

$$L_5 = \frac{1}{4}\left(L + 3L_0 + \sqrt{(L+3L_0)^2 + L(L+4L_0)}\right).$$ (20.3.3)

Then the scalar sequence $\{t_n\}$ given by

$$\begin{cases} t_0 = 0, \ s_0 = v, \\[2mm] t_{n+1} = s_n + \dfrac{L(s_n - t_n)^2}{2(1 - L_0 t_n)}, \\[3mm] s_{n+1} = t_{n+1} + \dfrac{L[(t_{n+1} - s_n) + 2(s_n - t_n)]}{2(1 - L_0 t_{n+1})}(t_{n+1} - s_n) \\[3mm] \text{for each } n = 0, 1, 2, \ldots \end{cases}$$ (20.3.4)

is well defined, increasing, bounded from above by

$$t^{**} = \frac{v}{1 - \alpha},$$ (20.3.5)

and converges to its unique least upper bound t^* which satisfies

$$v \le t^* \le t^{**}.$$ (20.3.6)

Moreover, the following estimates hold:

$$t_{n+1} - s_n \le \alpha(s_n - t_n) \le \alpha^{2n+1} v,$$ (20.3.7)

$$s_n - t_n \le \alpha(t_n - s_{n-1}) \le \alpha^{2n} v,$$ (20.3.8)

$$t^* - s_n \le \frac{\alpha^{2n} v}{1 - \alpha},$$ (20.3.9)

and

$$t^* - t_n \le \frac{\alpha^{2n} v}{1 - \alpha} + \alpha^{2n} v.$$ (20.3.10)

Proof. We first notice that $\alpha \in [\frac{1}{3}, 1)$ by (20.3.1). Now we shall show that

$$\frac{L(s_k - t_k)}{2(1 - L_0 s_k)} \le \alpha$$ (20.3.11)

and

$$\frac{L(t_{k+1} - s_k) + 2L(s_k - t_k)}{2(1 - L_0 t_{k+1})} \le \alpha.$$ (20.3.12)

If $k = 0$, (20.3.11) is satisfied if

$$\frac{1}{4}(2L_0 + L)v \le \frac{1}{2},$$

which is true, since $\dfrac{2L_0 + L}{4} \le L_2$. For $k = 0$, (20.3.12) becomes

$$\frac{\dfrac{L^2 v^2}{2} + 2Lv}{2(1 - L_0(v + \dfrac{Lv^2}{2}))} \le \frac{L}{2L_0 + L}$$

or

$$L(4L_0 + L)v^2 + 4(3L_0 + L)v - 4 \le 0, \tag{20.3.13}$$

which is true by (20.3.2). Hence estimates (20.3.11) and (20.3.12) hold for $k = 0$. Then assume they hold for all $k \le n$. Now we have

$$t_{k+1} - s_k \le \alpha^{2k+1} v, \tag{20.3.14}$$

$$s_{k+1} - t_{k+1} \le (\alpha^2)^{k+1} v, \tag{20.3.15}$$

$$t_{k+1} = (1 + \alpha)\frac{1 - \alpha^{2(k+1)}}{1 - \alpha^2} v < t^{**}, \tag{20.3.16}$$

and

$$s_{k+1} \le (1 + \alpha)\frac{1 - \alpha^{2(k+1)}}{1 - \alpha^2} v + \alpha^{2(k+1)} v. \tag{20.3.17}$$

In view of (20.3.14)–(20.3.16), estimate (20.3.11) is satisfied if

$$\frac{L}{2}\alpha^{2n} v + L_0\alpha(1 + \alpha)\frac{1 - \alpha^{2n}}{1 - \alpha^2} v - \alpha \le 0. \tag{20.3.18}$$

Estimate (20.3.18) motivates us to define recurrent functions f_k on $[0, \alpha^2]$ by

$$f_k(t) = \frac{L}{2}t^k v + L_0\sqrt{t}(1 + \sqrt{t})\frac{1 - t^k}{1 - t} v - \sqrt{t}. \tag{20.3.19}$$

Then we obtain

$$f_{k+1}(t) = f_k(t) + \frac{1}{2}g(t)(t - 1)t^{k-1} v \le f_k(t), \tag{20.3.20}$$

by the choice of α, where

$$g(t) = 2L_0\sqrt{t}(1 + \sqrt{t}) + Lt.$$

In view of (20.3.20) we have that for $t = \alpha^2$

$$f_{k+1}(\alpha^2) \le f_k(\alpha^2). \tag{20.3.21}$$

As a consequence, it follows from (20.3.21) that (20.3.18) is satisfied if

$$f_1(\alpha^2) \leq 0 \qquad (20.3.22)$$

or

$$\frac{2L_0 + L}{2}v \leq 1. \qquad (20.3.23)$$

But (20.3.23) is true by (20.3.2). Similarly, (20.3.12) is satisfied if

$$\frac{L}{2}\alpha^{2k+1}v + L\alpha^{2k}v + \alpha L_0(1+\alpha)\frac{1 - \alpha^{2(k+1)}}{1 - \alpha^2}v - \alpha \leq 0, \qquad (20.3.24)$$

leading to the introduction of functions f_k^1 on $[0, \alpha^2]$ by

$$f_k^1(t) = \frac{L}{2}\sqrt{t}t^k v + Lt^k v + \sqrt{t}L_0(1+\sqrt{t})\frac{1 - t^{k+1}}{1 - t}v - \sqrt{t}. \qquad (20.3.25)$$

Then we have

$$f_{k+1}^1(t) = f_k^1(t) + g^1(t)(t-1)t^k v \leq f_k^1(t), \qquad (20.3.26)$$

where

$$g^1(t) = \frac{L}{2}\sqrt{t} + L_0\sqrt{t}(1+\sqrt{t}) + L. \qquad (20.3.27)$$

Hence it follows from (20.3.26) that (20.3.24) is satisfied if $f_0^1(\alpha^2) \leq 0$ (since $f_k^1(\alpha^2) \leq f_{k-1}^1(\alpha^2) \leq \cdots \leq f_0^1(\alpha^2)$), which reduces to showing (20.3.13). The rest of the proof is identical to the proof of Lemma 20.2.1. □

Remark 20.3.2. *Let us define sequence* $\{\bar{t}_n\}$ *by*

$$\begin{cases} \bar{t}_0 = 0, \ \bar{s}_0 = v, \ \bar{t}_1 = \bar{s}_0 + \dfrac{L_0(\bar{s}_0 - \bar{t}_0)^2}{2(1 - L_0\bar{t}_0)}, \\[2mm] \bar{s}_1 = \bar{t}_1 + \dfrac{L(\bar{t}_1 - \bar{s}_0)^2 + 2L_0(\bar{s}_0 - \bar{t}_0)(\bar{t}_1 - \bar{s}_0)}{2(1 - L_0\bar{t}_1)}, \\[2mm] \bar{t}_{n+1} = \bar{s}_n + \dfrac{L(\bar{s}_n - \bar{t}_n)^2}{2(1 - L_0\bar{t}_n)}, \\[2mm] \bar{s}_{n+1} = \bar{t}_{n+1} + \dfrac{L[(\bar{t}_{n+1} - \bar{s}_n) + 2(\bar{s}_n - \bar{t}_n)](\bar{t}_{n+1} - \bar{s}_n)}{2(1 - L_0\bar{t}_{n+1})} \\[2mm] \textit{for each } n = 0, 1, 2, \ldots \end{cases} \qquad (20.3.28)$$

Then sequence $\{\bar{t}_n\}$ *is at least as tight as majorizing sequence* $\{t_n\}$.

Using a sequence of modifications of sequence $\{t_n\}$, we get

$$r_0 = 0, \quad q_0 = L_0 \nu,$$

$$r_{n+1} = q_n + \frac{b(q_n - r_n)^2}{2(1 - q_n)},$$

$$q_{n+1} = r_{n+1} + \frac{b[(r_{n+1} - q_n) + 2(q_n - r_n)](r_{n+1} - q_n)}{2(1 - q_{n+1})},$$

$$p_{n+1} = m_n - \frac{b(m_n - p_n)^2}{2m_n},$$

$$m_{n+1} = p_{n+1} - \frac{b[(p_{n+1} - m_n) - 2b(p_n - m_n)](p_{n+1} - m_n)}{2m_{n+1}},$$

$$\alpha_{n+1} = \frac{b\beta_{n+1}(1 - \alpha_n)(1 - \alpha_{n+1}) + 2b\alpha_n\beta_{n+1}}{2(1 - \beta_{n+1})(1 - \alpha_n)(1 - \beta_{n+1})}.$$

$$\beta_{n+1} = \frac{b}{2}\left(\frac{\alpha_n}{1 - \alpha_n}\right)^2.$$

As a consequence, we can state the following lemmas.

Lemma 20.3.3. *Suppose that* (20.3.2) *holds. Then the sequence* $\{t_n\}$ *is increasing, bounded from above by* $\dfrac{1}{L_0}$, *and converges to its unique least upper bound satisfying*

$$\nu \le t^* \le \frac{1}{L_0}.$$

Lemma 20.3.4. *Suppose that there exists* $N = 0, 1, 2, \ldots$ *such that*

$$t_0 < s_0 < t_1 < s_1 < \cdots < s_N < t_{N+1} < \frac{1}{L_0}$$

and (20.3.29)

$$h^N = L_5(s_N - t_N) \le \frac{1}{2},$$

where L_5 *is given in* (20.3.3). *Then the conclusions of Lemma 20.3.3 hold but with sequence* $\{t_n\}$ *given by* (20.3.4).

20.4   SEMILOCAL CONVERGENCE OF TWO-STEP NEWTON'S METHOD (20.1.4)

From now on $U(\omega, \rho)$ and $\bar{U}(\omega, \rho)$ respectively stand for the open and closed ball in X with center ω and radius $\rho > 0$.

First, for the semilocal convergence matter, we use (20.1.4) to obtain the identities

$$x_{n+1} - y_n$$

$$= [-F'(y_n)^{-1}F'(x_0)][F'(x_0)^{-1}\int_0^1 [F'(x_n + t(y_n - x_n)) \quad (20.4.1)$$

$$- F'(x_n)](y_n - x_n)dt],$$

$$y_{n+1} - x_{n+1}$$

$$= [-F'(x_{n+1})^{-1}F'(x_0)][F'(x_0)^{-1}\int_0^1 [F'(y_n + t(x_{n+1} - y_n)) \quad (20.4.2)$$

$$- F'(y_n)](x_{n+1} - y_n)dt].$$

Moreover, if $F(x^*) = F(y^*) = 0$, we obtain

$$0 = F(y^*) - F(x^*) = \int_0^1 F'(x^* + t(y^* - x^*))(y^* - x^*)dt. \quad (20.4.3)$$

Then we can give the following result.

Theorem 20.4.1. *Let $F : D \subset X \to Y$ be Fréchet differentiable. Suppose that there exists $x_0 \in D$ and parameters $L_0 > 0$, $L \geq L_0$, and $v \geq 0$ such that for each $x, y \in D$,*

$$F'(x_0)^{-1} \in (Y, X),$$

$$\|F'(x_0)^{-1}F(x_0)\| \leq v,$$

$$\|F'(x_0)^{-1}[F'(x) - F'(x_0)]\| \leq L_0\|x - x_0\|,$$

$$\|F'(x_0)^{-1}[F'(x) - F'(y)]\| \leq L\|x - y\|.$$

Moreover, suppose that hypotheses of lemmas in Section 20.2 hold and

$$\bar{U}(x_0, t^*) \subseteq D,$$

where t^* is given in Lemma 20.2.1. Then the sequence $\{x_n\}$ generated by two-step method (20.1.4) is well defined, remains in $\bar{U}(x_0, t^*)$ for each $n = 0, 1, 2, \ldots$, and converges to a solution $x^* \in \bar{U}(x_0, t^*)$ of equation $F(x) = 0$. Moreover, the following estimates hold for each $n = 0, 1, 2, \ldots$:

$$\|x_{n+1} - y_n\| \leq \bar{t}_{n+1} - \bar{s}_n,$$

$$\|y_n - x_n\| \leq \bar{s}_n - \bar{t}_n,$$

$$\|x_n - x^*\| \leq t^* - \bar{t}_n,$$

and

$$\|y_n - y^*\| \leq t^* - \bar{s}_n,$$

where the sequence $\{\bar{t}_n\}$ is given in (20.2.35). Furthermore, if there exists $r \geq t^*$ such that

$$\bar{U}(x_0, r) \subseteq \mathcal{D}$$

and

$$L_0(t^* + r) < 2,$$

then the limit point x^* is the unique solution of equation $F(x) = 0$ in $\bar{U}(x_0, r)$.

Secondly, for the local convergence we obtain the identities

$$
\begin{aligned}
y_n - x^* = & [-F'(x_n)^{-1} F'(x^*)][F'(x^*)^{-1} \int_0^1 [F'(x^* + t(x_n - x^*)) \\
& - F'(x_n)](x_n - x^*)dt]
\end{aligned}
\tag{20.4.4}
$$

and

$$
\begin{aligned}
x_{n+1} - x^* = & [-F'(y_n)^{-1} F'(x^*)][F'(x^*)^{-1} \int_0^1 [F'(x^* + t(y_n - x^*)) \\
& - F'(y_n)](y_n - x^*)dt].
\end{aligned}
\tag{20.4.5}
$$

We can arrive at [5,6,10,13]:

Theorem 20.4.2. *Let $F : \mathcal{D} \subset X \rightarrow Y$ be Fréchet differentiable. Suppose that there exists $x^* \in \mathcal{D}$ and parameters $l_0 > 0$, $l_1 > 0$, and $l > 0$ such that for each $x, y \in \mathcal{D}$,*

$$F(x^*) = 0,$$

$$F'(x^*)^{-1} \in (Y, X),$$

$$\|F'(x^*)^{-1}(F'(x) - F'(x^*))\| \leq l_0 \|x - x^*\|,$$

$$\|F'(x^*)^{-1}[F'(x) - F'(x_0)]\| \leq l_1 \|x - x_0\|,$$

$$\|F'(x^*)^{-1}[F'(x) - F'(y)]\| \leq l \|x - y\|,$$

and

$$\bar{U}(x^*, R) \subseteq \mathcal{D},$$

where

$$R = \frac{2}{2l_0 + l}.$$

Then the sequence $\{x_n\}$ generated by two-step method (20.1.4) is well-defined for each $n = 0, 1, 2, \ldots$, and converges to $x^* \in \bar{U}(x_0, R)$ provided that

$x_0 \in U(x^*, R)$. Moreover, the following estimates hold for each $n = 0, 1, 2, \ldots$:

$$\|y_n - x^*\| \le \frac{\bar{l}\|x_n - x^*\|^2}{2(1 - l_0\|x_n - x^*\|)}$$

and

$$\|x_{n+1} - x^*\| \le \frac{l\|y_n - x^*\|^2}{2(1 - l_0\|y_n - x^*\|)},$$

where

$$\bar{l} = \begin{cases} l_1 & \text{if } n = 0, \\ l & \text{if } n \ge 0. \end{cases}$$

Remark 20.4.3. If $l_1 = l = l_0$, the result reduces to [33,39] in the case of Newton's method. The radius is then given by $R_0 = \dfrac{2}{3l}$.

If $l_1 = l$, the result reduces to [5,6,10] in the case of Newton's method. The radius is again given by R. However, if $l_1 < l$, then the error bounds are finer (see $\bar{l}$ and $\|y_0 - x^*\|$).

20.5 LOCAL CONVERGENCE OF TWO-STEP NEWTON'S METHOD (20.1.4)

In this section we obtain the following identities for the semilocal convergence, but using (20.4.1), (20.4.3), and

$$y_{n+1} - x_{n+1} = [-F'(x_{n+1})^{-1} F'(x_0)][F'(x_0)^{-1} \int_0^1 [F'(y_n + t(x_{n+1} - y_n))$$
$$- F'(y_n)](x_{n+1} - y_n)dt].$$

$$(20.5.1)$$

Then, again, we arrive at

Theorem 20.5.1. Let $F : D \subset X \to Y$ be Fréchet differentiable. Suppose that there exists $x_0 \in D$ and parameters $L_0 > 0$, $L \ge L_0$, and $v \ge 0$ such that for each $x, y \in D$,

$$F'(x_0)^{-1} \in (Y, X),$$

$$\|F'(x_0)^{-1} F(x_0)\| \le v,$$

$$\|F'(x_0)^{-1}[F'(x) - F'(x_0)]\| \le L_0\|x - x_0\|,$$

$$\|F'(x_0)^{-1}[F'(x) - F'(y)]\| \le L\|x - y\|.$$

Moreover, suppose that hypotheses of Lemma 20.3.1, or Lemma 20.3.3, or Lemma 20.3.4 hold and

$$\bar{U}(x_0, t^*) \subseteq \mathcal{D},$$

where t^* is given in (20.3.5). Then the sequence $\{x_n\}$ generated by two-step method (20.1.3) is well defined, remains in $\bar{U}(x_0, t^*)$ for each $n = 0, 1, 2, \ldots$, and converges to a solution $x^* \in \bar{U}(x_0, t^*)$ of equation $F(x) = 0$. Furthermore, the following estimates hold for each $n = 0, 1, 2, \ldots$:

$$\|x_{n+1} - y_n\| \leq \bar{t}_{n+1} - \bar{s}_n,$$
$$\|y_n - x_n\| \leq \bar{s}_n - \bar{t}_n,$$
$$\|x_n - x^*\| \leq t^* - \bar{t}_n,$$

and

$$\|y_n - y^*\| \leq t^* - \bar{s}_n,$$

where the sequence $\{\bar{t}_n\}$ is given in (20.3.28). If there exists $r \geq t^*$ such that

$$\bar{U}(x_0, r) \subseteq \mathcal{D}$$

and

$$L_0(t^* + r) < 2,$$

then the limit point x^* is the unique solution of equation $F(x) = 0$ in $\bar{U}(x_0, r)$.

The identities for the local convergence case using (20.1.3) are (20.4.4) and

$$x_{n+1} - x^*$$
$$= [-F'(x_n)^{-1} F'(x^*)][F'(x^*)^{-1} \int_0^1 [F'(x^* + t(y_n - x^*))$$
$$- F'(y_n)](y_n - x^*)dt + (F'(y_n) - F'(x_n))(y_n - x^*)].$$

We then obtain

Theorem 20.5.2. *Let $F : \mathcal{D} \subset X \to Y$ be Fréchet differentiable. Suppose that there exists $x^* \in \mathcal{D}$ and parameters $l_0 > 0$, $l_1 > 0$, and $l > 0$ such that for each $x, y \in \mathcal{D}$,*

$$F(x^*) = 0,$$
$$F'(x^*)^{-1} \in (Y, X),$$
$$\|F'(x^*)^{-1}(F'(x) - F'(x^*))\| \leq l_0 \|x - x^*\|,$$
$$\|F'(x^*)^{-1}[F'(x) - F'(x_0)]\| \leq l_1 \|x - x_0\|,$$

$$\|F'(x^*)^{-1}[F'(x) - F'(y)]\| \leq l\|x - y\|,$$

and

$$\bar{U}(x^*, R) \subseteq \mathcal{D},$$

where

$$R = \frac{2}{2l_0 + 5l}.$$

Then the sequence $\{x_n\}$ generated by two-step method (20.1.3) is well defined for each $n = 0, 1, 2, \ldots$ and converges to $x^* \in \bar{U}(x_0, R)$ provided that $x_0 \in U(x^*, R)$. Moreover, the following estimates hold for each $n = 0, 1, 2, \ldots$:

$$\|y_n - x^*\| \leq \frac{\bar{l}\|x_n - x^*\|^2}{2(1 - l_0\|x_n - x^*\|)}$$

and

$$\|x_{n+1} - x^*\| \leq \frac{l[\|y_n - x^*\| + 2\|y_n - x_n\|]\|y_n - x^*\|}{2(1 - l_0\|y_n - x^*\|)},$$

where $\bar{l}$ is given in Theorem 20.4.2.

20.6 NUMERICAL EXAMPLES

Example 20.6.1. *In the following example we consider the equation*

$$x^3 - 0.49 = 0. \tag{20.6.1}$$

We take the starting point $x_0 = 1$ and consider the domain $\Omega = B(x_0, 0.5)$. In this case, we obtain

$$\nu = 0.17, \tag{20.6.2}$$

$$L = 3, \tag{20.6.3}$$

and

$$L_0 = 2.5. \tag{20.6.4}$$

Notice that Kantorovich hypothesis $L\nu \leq 0.5$ is not satisfied, but condition (2.2) in Lemma 20.3.1 is satisfied since

$$L_1 = 2.66333\ldots$$

and

$$h_1 = L_1\nu = 0.452766\ldots \leq 0.5.$$

So two-step Newton's method starting from $x_0 \in B(x_0, 0.5)$ converges to the solution of (20.6.1).

Example 20.6.2. *Let* $X = Y = C[0, 1]$, *the space of continuous functions defined on* $[0, 1]$, *be equipped with the max-norm. Let* $\Omega = \{x \in C[0, 1]; \|x\| \leq R\}$, *with* $R > 1$, *and* F *be defined on* Ω *by*

$$F(x)(s) = x(s) - f(s) - \lambda \int_0^1 G(s, t) x(t)^3 \, dt, \quad x \in C[0, 1], \ s \in [0, 1],$$

where $f \in C[0, 1]$ *is a given function,* λ *is a real constant, and the kernel* G *is the Green's function*

$$G(s, t) = \begin{cases} (1 - s)t, & t \leq s, \\ s(1 - t), & s \leq t. \end{cases}$$

In this case, for each $x \in \Omega$, $F'(x)$ *is a linear operator defined on* Ω *by the following expression:*

$$[F'(x)(v)](s) = v(s) - 3\lambda \int_0^1 G(s, t) x(t)^2 v(t) \, dt, \quad v \in C[0, 1], \ s \in [0, 1].$$

If we choose $x_0(s) = f(s) = 1$, *it follows that*

$$\|I - F'(x_0)\| \leq 3|\lambda|/8.$$

Thus if

$$|\lambda| < 8/3,$$

$F'(x_0)^{-1}$ *is defined and*

$$\|F'(x_0)^{-1}\| \leq \frac{8}{8 - 3|\lambda|}.$$

Moreover,

$$\|F(x_0)\| \leq \frac{|\lambda|}{8},$$

$$\|F'(x_0)^{-1} F(x_0)\| \leq \frac{|\lambda|}{8 - 3|\lambda|}.$$

On the other hand, for $x, y \in \Omega$ *we have*

$$[(F'(x) - F'(y))v](s) = 3\lambda \int_0^1 G(s, t)(x(t)^2 - y^2(t))v(t) \, dt.$$

Consequently,

$$\|F'(x) - F'(y)\| \leq \|x - y\| \frac{3|\lambda|(\|x\| + \|y\|)}{8}$$

$$\leq \|x - y\| \frac{6R|\lambda|}{8},$$

$$\|F'(x) - F'(1)\| \le \|x - 1\| \frac{1 + 3|\lambda|(\|x\| + 1)}{8}$$
$$\le \|x - 1\| \frac{1 + 3(1 + R)|\lambda|}{8}.$$

Choosing $\lambda = 1$ and $R = 2.6$, we have

$$\nu = \frac{1}{5},$$
$$L = 3.12,$$

and

$$L_0 = 2.16.$$

Hence, condition (20.1.7), namely $2L\nu = 1.248 \le 1$, is not satisfied, but condition (2.2), i.e., $L_1\nu = 0.970685 \le 1$, is satisfied. We can ensure the convergence of $\{x_n\}$.

Example 20.6.3. *Let $X = Y = C[0, 1]$ be equipped with the max-norm. Consider the following nonlinear boundary value problem:*

$$\begin{cases} u'' = -u^3 - \gamma u^2, \\ u(0) = 0, \quad u(1) = 1. \end{cases}$$

It is well known that this problem can be formulated as the integral equation

$$u(s) = s + \int_0^1 Q(s, t) \, (u^3(t) + \gamma \, u^2(t)) \, dt \tag{20.6.5}$$

where Q is the Green function:

$$Q(s, t) = \begin{cases} t\,(1 - s), & t \le s, \\ s\,(1 - t), & s < t. \end{cases}$$

We observe that

$$\max_{0 \le s \le 1} \int_0^1 |Q(s, t)| \, dt = \frac{1}{8}.$$

Then problem (20.6.5) is in the form (20.1.1), where $F : D \longrightarrow Y$ is defined as

$$[F(x)](s) = x(s) - s - \int_0^1 Q(s, t) \, (x^3(t) + \gamma \, x^2(t)) \, dt.$$

Set $u_0(s) = s$ and $D = U(u_0, R_0)$. It is easy to verify that $U(u_0, R_0) \subset U(0, R_0 + 1)$ since $\|u_0\| = 1$. If $2\gamma < 5$, the operator F' satisfies conditions *of Theorem 20.4.1 with*

$$v = \frac{1 + \gamma}{5 - 2\gamma},$$

$$L = \frac{\gamma + 6 R_0 + 3}{4(5 - 2\gamma)},$$

and

$$L_0 = \frac{2\gamma + 3 R_0 + 6}{8(5 - 2\gamma)}.$$

Note that $L_0 < L$. Choosing $R_0 = 1$ and $\gamma = 0.6$, condition (20.1.16), i.e.,

$$\frac{4 + \sqrt{21}}{4} Lv = 0.570587 \cdots \leq 0.5,$$

is not satisfied, but condition (20.3.2) is satisfied as

$$\frac{1}{4}\left(3L_0 + L + \sqrt{(3L_0 + L)^2 + L(4L_0 + L)}\right)v = 0.381116 \cdots \leq 0.5.$$

So we can ensure the convergence of $\{x_n\}$.

Example 20.6.4. *Let $\mathcal{X} = [-1, 1]$, $\mathcal{Y} = \mathbb{R}$, $x_0 = 0$ and let $F : \mathcal{X} \to \mathcal{Y}$ be the polynomial*

$$F(x) = \frac{1}{6}x^3 + \frac{1}{6}x^2 - \frac{5}{6}x + \frac{1}{9}.$$

In this case, since

$$\|F'(0)^{-1} F(0)\| \leq 0.13333 \cdots = v,$$

$$L = \frac{22}{10},$$

and

$$L_0 = \frac{13}{10},$$

condition (20.1.16), namely

$$\frac{4 + \sqrt{21}}{4} Lv = 0.629389 \cdots \leq 0.5,$$

is not satisfied, but condition (20.3.2), i.e.,

$$\frac{1}{4}\left(L + 3L_0 + \sqrt{(L + 3L_0)^2 + L(L + 4L_0)}\right)v = 0.447123 \cdots \leq 0.5,$$

is satisfied. Hence the sequence $\{x_n\}$ generated by two step Newton's method (20.1.3) is well defined and converges to a solution x^* of $F(x) = 0$.

Example 20.6.5. *Let*

$$X = Y = \mathbb{R}^3,$$
$$D = \bar{U}(0, 1),$$

and

$$x^* = \begin{pmatrix} 0 \\ 0 \\ 0 \end{pmatrix}.$$

Define function F on D for

$$w = \begin{pmatrix} x \\ y \\ z \end{pmatrix}$$

by

$$F(w) = \begin{pmatrix} e^x - 1 \\ \dfrac{e-1}{2}y^2 + y \\ z \end{pmatrix}.$$

Then the Fréchet-derivative is given by

$$F'(v) = \begin{bmatrix} e^x & 0 & 0 \\ 0 & (e-1)y+1 & 0 \\ 0 & 0 & 1 \end{bmatrix}.$$

    Using the norm of the maximum of the rows and the Lipschitz conditions of Theorem 20.5.1, we see that since $F'(x^*) = diag\{1, 1, 1\}$, we can define parameters for Newton's method by

$$l = l_1 = e \tag{20.6.6}$$

and

$$l_0 = 2. \tag{20.6.7}$$

    Then the two-step Newton's method (20.1.4) starting from $x_0 \in B(x^*, R^*)$ converges to a solution of $F(x) = 0$. Note that this radius is greater than given by Rheinboldt or Traub [39] namely $R^*_{TR} = \dfrac{2}{3e} < \dfrac{2}{4+e} = R^*$. So we can ensure the convergence to a solution. Note that again $l_0 < l$. Then, the two-step Newton method (20.1.3) starting from $x_0 \in B(x^*, R)$, where $R = \dfrac{2}{2l_0 + 5l} = \dfrac{2}{4+5e}$, converges to x^*.

Example 20.6.6. *Let* $\mathcal{X} = \mathcal{Y} = \mathcal{C}[0, 1]$, *the space of continuous functions defined on* $[0, 1]$, *equipped with the max norm and* $\mathcal{D} = \overline{U}(0, 1)$. *Define the function F on* $\mathcal{D}$ *given by*

$$F(h)(x) = h(x) - 5 \int_0^1 x\,\theta\,h(\theta)^3\,d\theta. \qquad (20.6.8)$$

Then, we have:

$$F'(h[u])(x) = u(x) - 15 \int_0^1 x\,\theta\,h(\theta)^2\,u(\theta)\,d\theta \quad \text{for all } u \in \mathcal{D}.$$

Using (20.6.8), hypotheses of Theorem 20.5.2 hold for $x^\star(x) = 0$ *(*$x \in [0, 1]$*),* $l = l_1 = 15$ *and* $l_0 = 7.5$.

Then the two-step Newton method (20.1.4) starting from $x_0 \in B(x^*, R^*)$ *converges to a solution. Note that radius* R^* *is bigger than that of Rheinboldt or Traub [39], given by* $R^*_{TR} = \dfrac{2}{45} < \dfrac{1}{15} = R^*$. *Moreover, we can ensure the convergence to a solution. Note that again* $l_0 < l$. *Then the two-step Newton's method (20.1.3) starting from* $x_0 \in B(x^*, R)$, *where* $R = \dfrac{2}{2l_0 + 5l} = \dfrac{1}{45}$, *converges to* x^*.

REFERENCES

[1] S. Amat, S. Busquier, C. Bermúdez, Á.A. Magreñán, Expanding the applicability of a third order Newton-type method free of bilinear operators, Algorithms 8 (3) (2015) 669–679.

[2] S. Amat, S. Busquier, C. Bermúdez, Á.A. Magreñán, On the election of the damped parameter of a two-step relaxed Newton-type method, Nonlinear Dynam. 84 (1) (2016) 9–18.

[3] S. Amat, Á.A. Magreñán, N. Romero, On a two-step relaxed Newton-type method, Appl. Math. Comput. 219 (24) (2013) 11341–11347.

[4] S. Amat, S. Busquier, J.M. Gutiérrez, Third-order iterative methods with applications to Hammerstein equation: a unified approach, J. Comput. Appl. Math. 235 (2011) 2936–2943.

[5] I.K. Argyros, A unifying local-semilocal convergence analysis and applications for two-point Newton-like methods in Banach space, J. Math. Anal. Appl. 298 (2004) 374–397.

[6] I.K. Argyros, Computational theory of iterative methods, in: C.K. Chui, L. Wuytack (Eds.), Stud. Comput. Math., vol. 15, Elsevier Publ. Co., New York, USA, 2007.

[7] I.K. Argyros, Y.J. Cho, S. Hilout, On the midpoint method for solving equations, Appl. Math. Comput. 216 (2010) 2321–2332.

[8] I.K. Argyros, A semilocal convergence analysis for directional Newton methods, Math. Comput. 80 (2011) 327–343.

[9] I.K. Argyros, J.A. Ezquerro, J.M. Gutiérrez, M.Á. Hernández, S. Hilout, On the semilocal convergence of efficient Chebyshev-secant-type methods, J. Comput. Appl. Math. 235 (2011) 3195–3206.

[10] I.K. Argyros, Y.J. Cho, S. Hilout, Numerical Method for Equations and Its Applications, CRC Press/Taylor and Francis, New York, 2012.

[11] I.K. Argyros, S. González, Á.A. Magreñán, Majorizing sequences for Newton's method under centred conditions for the derivative, Int. J. Comput. Math. 91 (12) (2014) 2568–2583.

[12] I.K. Argyros, J.M. Gutiérrez, Á.A. Magreñán, N. Romero, Convergence of the relaxed Newton's method, J. Korean Math. Soc. 51 (1) (2014) 137–162.

[13] I.K. Argyros, S. Hilout, Weaker conditions for the convergence of Newton's method, J. Complexity 28 (2012) 364–387.

[14] I.K. Argyros, Á.A. Magreñán, Extended convergence results for the Newton–Kantorovich iteration, J. Comput. Appl. Math. 286 (2015) 54–67.

[15] I.K. Argyros, Á.A. Magreñán, L. Orcos, J.A. Sicilia, Local convergence of a relaxed two-step Newton like method with applications, J. Math. Chem. 55 (7) (2017) 1427–1442, https://www.scopus.com/inward/record.uri?eid=2-.

[16] I.K. Argyros, Á.A. Magreñán, J.A. Sicilia, Improving the domain of parameters for Newton's method with applications, J. Comput. Appl. Math. 318 (2017) 124–135.

[17] W. Bi, Q. Wu, H. Ren, Convergence ball and error analysis of Ostrowski–Traub's method, Appl. Math. J. Chinese Univ. Ser. B 25 (2010) 374–378.

[18] E. Cătinaş, The inexact, inexact perturbed, and quasi-Newton methods are equivalent models, Math. Comput. 74 (249) (2005) 291–301.

[19] J. Chen, I.K. Argyros, R.P. Agarwal, Majorizing functions and two-point Newton-type methods, J. Comput. Appl. Math. 234 (2010) 1473–1484.

[20] F. Cianciaruso, Convergence of Newton–Kantorovich approximations to an approximate zero, Numer. Funct. Anal. Optim. 28 (5–6) (2007) 631–645.

[21] P. Deuflhard, Newton Methods for Nonlinear Problems: Affine Invariance and Adaptive Algorithms, Springer-Verlag, Berlin, Heidelberg, 2004.

[22] J.A. Ezquerro, M.Á. Hernández, M.A. Salanova, Recurrent relations for the midpoint method, Tamkang J. Math. 31 (2000) 33–42.

[23] J.A. Ezquerro, M.Á. Hernández, On the R-order of the Halley method, J. Math. Anal. Appl. 303 (2005) 591–601.

[24] J.A. Ezquerro, M.Á. Hernández, N. Romero, Newton-type methods of high order and domains of semilocal and global convergence, Appl. Math. Comput. 214 (1) (2009) 142–154.

[25] J.M. Gutiérrez, M.Á. Hernández, Recurrent relations for the super-Halley method, Comput. Math. Appl. 36 (1998) 1–8.

[26] M.Á. Hernández, N. Romero, On a characterization of some Newton-like methods of R-order at least three, J. Comput. Appl. Math. 183 (2005) 53–66.

[27] L.V. Kantorovich, G.P. Akilov, Functional Analysis, Pergamon Press, Oxford, 1982.

[28] Á.A. Magreñán, I.K. Argyros, On the local convergence and the dynamics of Chebyshev–Halley methods with six and eight order of convergence, J. Comput. Appl. Math. 298 (2016) 236–251.

[29] Á.A. Magreñán, Different anomalies in a Jarratt family of iterative root-finding methods, Appl. Math. Comput. 233 (2014) 29–38.

[30] Á.A. Magreñán, A new tool to study real dynamics: the convergence plane, Appl. Math. Comput. 248 (2014) 215–224.

[31] Á.A. Magreñán, I.K. Argyros, Optimizing the applicability of a theorem by F. Potra for Newton-like methods, Appl. Math. Comput. 242 (2014) 612–623.

[32] Á.A. Magreñán, I.K. Argyros, J.A. Sicilia, New improved convergence analysis for Newton-like methods with applications, J. Math. Chem. 55 (7) (2017) 1505–1520.

[33] J.M. Ortega, W.C. Rheinboldt, Iterative Solution of Nonlinear Equation in Several Variables, Society for Industrial and App. Math. Philadelpia, PA, USA, 1970, 598 pp.

[34] P.K. Parida, D.K. Gupta, Recurrence relations for a Newton-like method in Banach spaces, J. Comput. Appl. Math. 206 (2007) 873–887.

[35] P.K. Parida, D.K. Gupta, Semilocal convergence of a family of third-order Chebyshev-type methods under a mild differentiable condition, Int. J. Comput. Math. 87 (2010) 3405–3419.

[36] F.A. Potra, On the convergence of a class of Newton-like methods, in: Iterative Solution of Nonlinear Systems of Equations, Oberwolfach, 1982, in: Lecture Notes in Math., vol. 953, Springer, Berlin, New York, 1982, pp. 125–137.

[37] F.A. Potra, On Q-order and R-order of convergence, J. Optim. Theory Appl. 63 (1989) 415–431.

[38] P.D. Proinov, New general convergence theory for iterative processes and its applications to Newton–Kantorovich type theorems, J. Complexity 26 (2010) 3–42.
[39] J.F. Traub, Iterative Method for Solutions of Equations, Prentice-Hall, New Jersey, 1964.
[40] X. Wang, C. Gu, J. Kou, Semilocal convergence of a multipoint fourth-order super-Halley method in Banach spaces, Num. Alg. 56 (2010) 497–516.

Chapter 21

Introduction to complex dynamics

21.1 BASIC DYNAMICAL CONCEPTS

In this chapter some concepts of complex dynamics that will be used in this book are shown. One of the most frequent problems in mathematics is solving a nonlinear equation $f(z) = 0$, with $f : \mathbb{C} \to \mathbb{C}$. The solutions of these equations cannot be solved in a direct way, except in very special cases. That is why most of the methods for solving these equations are iterative and also the reason why we study the complex dynamics associated with iterative methods in this chapter.

From now on we define a rational function $R : \hat{\mathbb{C}} \to \hat{\mathbb{C}}$, where $\hat{\mathbb{C}}$ is the Riemann sphere. Notice that $R(z) = \frac{P(z)}{Q(z)}$ where $P(z)$ and $Q(z)$ are polynomials with complex coefficients without common factors. Moreover, the degree of a rational function $R(z)$ is defined as the highest degree of $P(z)$ and $Q(z)$. We begin with the definition of the orbit of a point.

Definition 21.1.1. *The orbit of a point $z_0 \in \hat{\mathbb{C}}$ is defined as*

$$\mathcal{O}(z_0) = \{z_0, \, R(z_0), \, R^2(z_0), \ldots, R^n(z_0), \ldots\}.$$

We will analyze the phase plane of the map R by classifying the starting points based on the asymptotic behavior of their orbits. In the dynamical study of rational functions, one of the most commonly found problems is studying the behavior of the orbits of a point $z_0 \in \hat{\mathbb{C}}$. If the orbit converges to some value, it will be a fixed point of the rational function $R(z)$.

Definition 21.1.2. *A point $z_0 \in \bar{C}$ is called a* fixed point *of $R(z)$ if it satisfies* $R(z) = z$.

There exist different types of a fixed point z_0 depending on its associated multiplier $\mu = |R'(z_0)|$.

Definition 21.1.3. *Taking the associated multiplier into account, a fixed point z_0 is called:*

- superattractor *if $|R'(z_0)| = 0$,*
- attractor *if $|R'(z_0)| < 1$,*

A Contemporary Study of Iterative Methods. DOI: 10.1016/B978-0-12-809214-9.00021-8
Copyright © 2018 Elsevier Inc. All rights reserved.

- repulsor *if $|R'(z_0)| > 1$,*
- *and* parabolic *if $|R'(z_0)| = 1$.*

The name associated to the different fixed point is representative of its behavior. If z_0 is an attracting fixed point of $R(z)$, that is, $|R'(z_0)| < 1$, the iterations of every point in a neighborhood of z_0 will converge to the fixed point. Then we have that for any z sufficiently close to z_0 there exists $\beta < 1$ such that

$$\frac{|R(z) - R(z_0)|}{|z - z_0|} < \beta < 1,$$

and as z_0 is a fixed point, we obtain $R(z_0) = z_0$. Moreover,

$$|R(z) - z_0| < \beta|z - z_0|,$$

consequently, $R(z)$ is closer to z_0 than z. Now, iterating k times, we obtain that

$$|R^k(z) - z_0| < \beta^k|z - z_0|,$$

and taking into account that $\beta < 1$, we obtain

$$\lim_{n \to \infty} R^k(z) = z_0.$$

Following an analogous process, it is easy to see that every point in a neighborhood of a repelling fixed point moves away from it.

Special attention must be given to the point ∞ as a fixed point.

Definition 21.1.4. *∞ is a fixed point of a rational function $R(z)$ if and only if $z = 0$ is a fixed point of the function*

$$F : z \to \frac{1}{R(\frac{1}{z})}.$$

Moreover, if ∞ is a fixed point of $R(z)$, its associated multiplier is $\mu = F'(0)$.

Consequently, ∞ can be an attractor or even superattractor as can be seen in the following example.

Example 21.1.5. *We want to calculate the behavior of the fixed points of the rational equation*

$$R(z) = \frac{z^2 + z}{2}.$$

It is easy to see that $z = 0$ is an attracting fixed point since $R'(0) = 0.5 < 1$ and $z = 1$ is a repelling fixed point since $R'(1) = 1.5 > 1$, but what happens with ∞? To answer that question, we must calculate first the function

$$F(z) = \frac{1}{R(\frac{1}{z})} = \frac{2z^2}{1+z}.$$

As $F(0) = 0$ and $F'(0) = 0$, ∞ is a superattracting fixed point of $R(z)$.

On the other hand, we must distinguish between simple and multiple zeros of an equation.

Definition 21.1.6. *Let z_0 be a fixed point of a rational function $R(z)$. We say that z_0 has multiplicity $m \geq 1$ if z_0 is a root of multiplicity m of the equation $G(z) = R(z) - z = 0$, that is, $G^{(j)}(z_0) = 0$ for each $j = 0, 1, \ldots, m-1$ and $G^{(m)} \neq 0$. z_0 is simple if $m = 1$, that is, $G(z_0) = 0$ and $G'(z_0) \neq 0$.*

The maximum number of fixed points that a rational function can have is stated in the following result.

Theorem 21.1.7. *A rational function $R(z)$ of degree $d \geq 1$ has exactly $d + 1$ fixed points reckoned with multiplicity.*

The fixed points of a rational function are special cases of periodic points which are defined as follows.

Definition 21.1.8. *A point $z_0 \in \hat{\mathbb{C}}$ is called a periodic point of period p if $R^p(z_0) = z_0$ and $R^n(z_0) \neq z_0$ for each $n < p$.*

Notice that the orbit associated to a periodic point z_0 of period n has only n different terms.

Definition 21.1.9. *The orbit associated to a periodic point of period n is called an n-cycle.*

The multiplier associated to an n-cycle is the same for every point of the cycle

$$|(R^n)'(z_0)| = \cdots = |(R^n)'(z_n)| = |R'(z_0)||R'(z_1)| \cdots |R'(z_n)|.$$

A fixed point of a cycle can be classified by means of the value of the multiplier as

- *superattractor* if $|(R^n)'(z_0)| = 0$,
- *attractor* if $|(R^n)'(z_0)| < 1$,
- *repulsor* if $|(R^n)'(z_0)| > 1$,
- and *parabolic* if $|(R^n)'(z_0)| = 1$.

Due to the form of the method, some fixed points which are not roots of the function $f(z)$ can be introduced as will be shown in the following example. These points are called *strange fixed points*.

Example 21.1.10. *The famous Chebyshev method defined as*

$$C(z) = z - \left(1 + \frac{1}{2}\frac{f(z)f''(z)}{f'(z)^2}\right)\frac{f(z)}{f'(z)}$$

applied to the polynomial $p(z) = z^2 - z$ *has two fixed points at* $z = 0$ *and* $z = 1$ *and two strange fixed points at* $z = \frac{5-\sqrt{5}}{10}$ *and* $z = \frac{5+\sqrt{5}}{10}$.

In the study of iterative methods one of the most important concepts is the notion of the basin of attraction.

Definition 21.1.11. The basin of attraction *of an attracting fixed point* α *is defined as*

$$\mathcal{A}(\alpha) = \{z_0 \in \hat{\mathbb{C}} \; : \; R^n(z_0) \to \alpha, \; n \to \infty\}.$$

Moreover, the immediate basin of attraction of an attracting fixed point α, *denoted by* $\mathcal{A}^*(\alpha)$, *is the connected component of* $\mathcal{A}(\alpha)$ *that contains* α. *If* z_0 *is an attracting periodic point of* $R(z)$ *of period n, the basin of attraction of the orbit* $\mathcal{O}(z_0)$ *is the set*

$$\mathcal{A}(\mathcal{O}(z_0)) = \bigcup_{j=0}^{n-1} R^j(\mathcal{A}(z_0)),$$

where $\mathcal{A}(z_0)$ *is the basin of attraction of* z_0, *which is a fixed point of* R^n. *Furthermore, the immediate basin of attraction is the set*

$$\mathcal{A}^*(\mathcal{O}(z_0)) = \bigcup_{j=0}^{n-1} R^j(\mathcal{A}^*(z_0)),$$

In Fig. 21.1, we see the basins of attraction associated with the quadratic polynomial $p(z) = z^2 - 1$, the difference between the basins of attraction is a straight line. Moreover, the basins correspond with the half-planes

$$\mathbb{C}^- = \{z | \Re(z) < 0\}$$

and

$$\mathbb{C}^+ = \{z | \Re(z) > 0\},$$

where $\Re(z)$ denotes the real part of z.

Each point of $\mathbb{C}^-$ converges to the negative root $z = -1$, and the iteration of every point in the half-plane $\mathbb{C}^+$ converges to the positive one $z = 1$. Furthermore, on the imaginary axis which is the separation between both half-planes, it is well know that the method has chaotic behavior.

FIGURE 21.1 Basins of attraction associated to the fixed points of Newton's method applied to polynomial $p(z) = z^2 - 1$. The basin of $z = 1$ appears in dark gray, while the basin of $z = -1$ is shown in light gray.

In Fig. 21.2 we present the basins of attraction associated to the fixed points of the Newton's method, $N(z)$, applied to $p(z) = z^3 - 1$, where

$$N(z) = z - \frac{p(z)}{p'(z)} = z - \frac{z^3 - 1}{3z^2} = \frac{2z^3 + 1}{3z^2}.$$

On the other hand, important is also the notion of a critical point defined as follows.

Definition 21.1.12. *Let $R(z)$ be a rational function of degree d. A point $w \in \hat{\mathbb{C}}$ for which the cardinality of $R^{-1}(w)$ is lower than d is called a critical value of $R(z)$. A point $z \in R^{-1}(w)$ which is a root of $R(z) - w$ of multiplicity greater than 1 is called a critical point of $R(z)$. Moreover, ∞ is a critical point of $R(z)$ if 0 is a critical point of $F(z)$. The multiplicity of ∞ as critical point of $R(z)$ is the same as the multiplicity of 0 as critical point of $F(z)$.*

Remark 21.1.13. *A point z is a critical point of a holomorphic function $p(z)$ if $p'(z) = 0$.*

Theorem 21.1.14. *Let $C = C(R)$ be the set of all critical points of a rational function $R(z)$. Then*

• *The set of the critical points of $R^n(z)$ is*

$$C(R^n) = C \cup R^{-1}(C) \cup \cdots \cup R^{-n}(C);$$

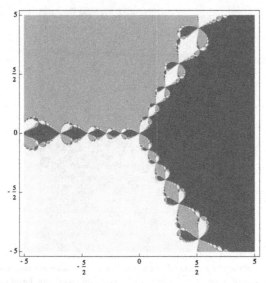

FIGURE 21.2 Basins of attraction associated to the fixed points of Newton's method applied to polynomial $p(z) = z^3 - 1$. In dark gray appears the basin of $z = 1$, in light gray the basin of $z = e^{2\pi i/3}$ and in white the basin of $z = e^{4\pi i/3}$.

- *The set of the critical values of $R^n(z)$ is*

$$R(C) = C \cup R^2(C) \cup \cdots \cup R^n(C).$$

Critical points play a very important role in the study of the dynamics of a rational function $R(z)$ due to the fact that they are points for which there exist a neighborhood where the function $R(z)$ is not a local homeomorphism. But the real importance of the critical points resides in the fact that the topology of the Julia and Fatou sets associated to a function $R(z)$ is related with the orbits of the critical points, as it is indicated in the following theorems. All these theorems can be extended to rationally indifferent cycles. A cycle is called rationally indifferent if its multiplier is $\mu = e^{pz/q} = 1$.

Theorem 21.1.15. *Let $R(z)$ be a rational function of degree greater or equal to two, then the immediate basin of attraction of each attracting cycle of $R(z)$ contains at least one critical point of $R(z)$.*

The following result is really important since is states that we have to study the orbits of the critical points to determinate the (super)attracting cycles.

Theorem 21.1.16 (Fatou, Julia). *The immediate basin of attraction of an attracting or superattracting cycle contains at least one critical point.*

Theorem 21.1.17 (Shishikura 1987). *Let $R(z)$ be a rational function of degree greater or equal to two, then $R(z)$ has at most $2d - 2$ critical points. Consequently, it has at most $2d - 2$ attracting or rationally indifferent periodic orbits.*

21.2 JULIA AND FATOU SETS

This section will be dedicated to the properties of the Julia [14] and Fatou sets [13]. These sets are dedicated to the mathematicians who discovered them, namely Gaston Julia and Pierre Fatou. Both authors fought in 1918 to win the *le grand prix des sciences mathématiques* with their sets. Both sets have complementary behavior. There exist several ways of introducing both theories but in this book we will choose the one that Fatou developed in [13]. First of all, we need the notion of a normal family given by Montel in [24].

Definition 21.2.1. *A family τ of meromorphic functions defined on a domain* $U \subseteq \hat{\mathbb{C}}$

$$\tau = \{f_i : U \to \hat{\mathbb{C}}; \quad f_i \text{ is meromorphic}\}$$

is normal if each sequence $(f_n)_{n \in \mathbb{N}}$ of elements of τ has a subsequence $(f_{n_k})_{k \in \mathbb{N}}$ which converges uniformly on each compact subset of U.

Once the notion of a normal family has been defined, we will focus our attention on the family of iteration of the rational function $R(z)$ where equicontinuity means that the iteration of near points does not diverge.

Definition 21.2.2. *A point z_0 belongs to the Fatou set, denoted by $\mathbb{F}(R)$ (also known as the normality or stability domain), if there exists a neighborhood U of z_0 such that the family*

$$\tau = \{R^n : U \to \hat{\mathbb{C}} : \quad n = 0, 1, 2, \ldots\}$$

is normal in U.

This definition is very technical so we will use the following more intuitive and colloquial definition.

Definition 21.2.3. *The* Fatou set *of the rational function R, $\mathcal{F}(R)$, is the set of points $z \in \hat{\mathbb{C}}$ whose orbits tend to an attractor (fixed point, periodic orbit, or infinity). Its complement in $\hat{\mathbb{C}}$ is the* Julia set, $\mathcal{J}(R)$. *This means that the basin of attraction of any fixed point belongs to the Fatou set and the boundaries of these basins of attraction belong to the Julia set.*

The different properties of these sets have been studied by many authors such as Fagella et al. (see [12]) or Peitgen et al. (see [25]). A brief list of them is shown in this section.

Let $R(z)$ be a rational function and $\mathcal{J}(R)$ the Julia set associated to $R(z)$. Then the following assertions hold:

  (i) $\mathcal{J}(R) = \mathcal{J}(R^n)$, for each $n \in \mathbb{N}$.

 (ii) If $\mathcal{J}(R)$ has nonempty interior, then $\mathcal{J}(R) = \hat{\mathbb{C}}$.

(iii) $\mathcal{J}(R) \neq \emptyset$ and $\mathcal{J}(R)$ is dense itself (is perfect).

(iv) For each point $z \in \mathcal{J}(R)$ the set of preimages of z is dense in $\mathcal{J}(R)$.

(v) $R(\mathcal{J}(R)) = R^{-1}(\mathcal{J}(R)) = \mathcal{J}(R)$, that is, $\mathcal{J}(R)$ is completely invariant.

(vi) If P_R is the set of repelling periodic points of $R(z)$, then $P_R \subset \mathcal{J}(R)$ and $\bar{P}_R = \mathcal{J}(R)$.

(vii) If ϑ is an attracting cycle of $R(z)$, then the basin of attraction of ϑ belongs to $\mathcal{F}(R)$. Moreover, its boundary is contained in $\mathcal{J}(R)$.

(viii) $R(z)$ is topologically transitive in $\mathcal{J}(R)$.

(ix) If $z_0 \in \mathcal{J}(R)$, then the closure of $\{z \in \hat{\mathbb{C}} | R^n(z) = z_0\} = \mathcal{J}(R)$.

(x) $R^{-1}(\mathcal{F}(R)) = R(\mathcal{F}(R)) = \mathcal{F}(R)$ and $R^{-1}(\mathcal{J}(R)) = R(\mathcal{J}(R)) = \mathcal{J}(R)$, that is, both sets are completely invariant.

The component of a Fatou set can be one of the following:

1. Basin of attraction of the attracting fixed points.
2. Basin of periodic cycles.
3. Basin of attraction of the indifferent fixed points.
4. Siegel disks.
5. Herman rings.
6. Baker domains.
7. Errant domains.

For a complete description of each Fatou component we refer the reader to [27]. In the iteration of polynomials we will only find the first three kinds. Moreover, we have the following results which link the critical points with the Fatou components.

Theorem 21.2.4. *The number of Fatou components of a rational function $R(z)$ can be 0, 1, 2, or ∞.*

Theorem 21.2.5. *If the orbits of the critical points of a rational function $R(z)$ are finite, then the Lebesgue measure of the Julia set of $R(z)$ is 0.*

To end this section we will distinguish what kind of points belong to each set.

Remark 21.2.6. *The fixed points*

- *Superattractors and attractors belong to the Fatou set.*
- *Repulsors and indifferent points belong to the Julia set.*

21.3   TOPOLOGICAL CONJUGATIONS

This section will focus attention to the topological conjugations which will help us in the study of the dynamics of iterative methods. Before introducing this concept we need the definition of a Möbius transformation.

Definition 21.3.1. *A Möbius transformation of parameters $a, b, c, d \in \mathbb{C}$ is a function Γ with the following form*

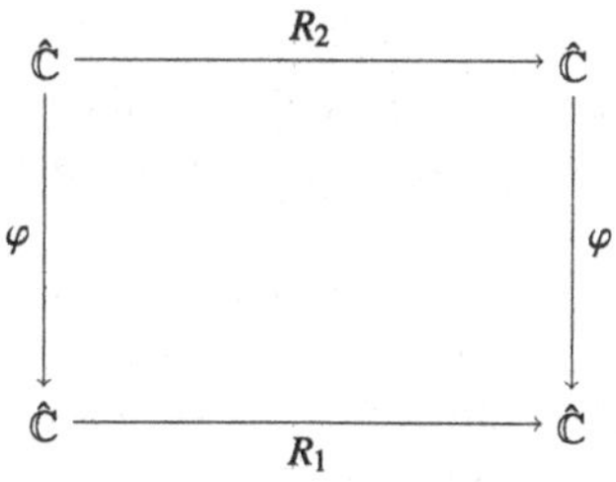

FIGURE 21.3 Diagram of a topological conjugation.

$$\Gamma(z) = \frac{az+b}{cz+d}, \quad \forall z \in \hat{\mathbb{C}}$$

where a, b, c, d are such that $ad - bc \neq 0$.

Definition 21.3.2. *Let $R_1, R_2 : \hat{\mathbb{C}} \to \hat{\mathbb{C}}$ be two rational functions. We say that they are conjugated if and only if there exists a Möbius transformation $\varphi : \hat{\mathbb{C}} \to \hat{\mathbb{C}}$ such that $\varphi \circ R_1 \circ \varphi^{-1}(z) = R_2(z)$ for each z.*

If R_1 and R_2 are conjugated by M, then it is clear that

$$M(\mathcal{J}(R_1)) = \mathcal{J}(R_2)$$

and

$$M(\mathcal{F}(R_1)) = \mathcal{F}(R_2).$$

The properties of the topological conjugations have been briefly studied by many authors, e.g., Beardon [8] and Fagella et al. [12]. In the following result a short resumé of these properties is shown (see Fig. 21.3).

Theorem 21.3.3. *Let $R_1, R_2 : \hat{\mathbb{C}} \to \hat{\mathbb{C}}$ be two rational functions and let $\varphi : \hat{\mathbb{C}} \to \hat{\mathbb{C}}$ be a topological conjugation between $R_1(z)$ and $R_2(z)$. Then the following assertions hold:*

(i) $\varphi^{-1} : \hat{\mathbb{C}} \to \hat{\mathbb{C}}$ *is also a topological conjugation between $R_1(z)$ and $R_2(z)$.*
(ii) $\varphi \circ R_1^n(z) = R_2^n \circ \varphi(z)$, *for each $n \in \mathbb{N}$.*
(iii) p *is a periodic point of $R_1(z)$ if and only if $\varphi(p)$ is a periodic point of $R_2(z)$. Moreover, both periodic points have the same period.*
(iv) *If p is a periodic point of $R_1(z)$ and $\varphi'(z) \neq 0$ for each $z \in \mathcal{O}(p)$, then p and $\varphi(p)$ have the same behavior.*
(v) *If p is a periodic point of $R_1(z)$ with basin of attraction $\mathcal{A}(p)$, then the basin of attraction of $\varphi(p)$ is $\varphi(\mathcal{A}(p))$.*

21.4 PARAMETER PLANES

In this section, the parameter spaces associated to an iterative method which has a parameter (it could come from the function or even from the iterative method) will be shown (similar studies can be seen in [1–7,9–11,15–18,4,19–26,28,29]. In order to show the applicability of this tool we will apply it to the damped Newton's method applied to the following polynomial $p(z) = (z - a)^2(z - b)$ where $a \neq b$.

It is well known that there is at least one critical point associated with each invariant Fatou component. In practice, the critical points of a family are the solutions of the derivative of the iterative function. Moreover, there exists critical point no related to the roots of the function, these points are called free critical points. In order to find the best members of the family in terms of stability the parameter spaces will be shown.

The study of the orbits of the critical points gives rise about the dynamical behavior of an iterative method. In concrete, to determinate if there exist any attracting periodic orbit different to the roots of the polynomial $p(z)$, the following question must be answered: For which values of the parameter, the orbits of the free critical points are attracting periodic orbits? One way of giving response to that questions is drawing the parameter spaces of the convergence of that free critical points. There will exist open regions of the parameter space for which the iteration of the free critical points does not converge to any of the roots of the polynomial $p(z)$, that is, these regions converges to attracting cycles, to an strange fixed point or even to infinity.

If we apply damped Newton's method to polynomial $p(z)$, we obtain:

$$N_{\lambda, p}(z) = z - \lambda \frac{(z - a)^2(z - b)}{3z^2 - (4a + 2b)z + a^2 + ab}.$$

It is also important in the study of the dynamics the Möbius transformation

$$M(z) = 1 + \frac{2(z - a)}{a - b}. \tag{21.4.1}$$

This transformation carries root a to $z = 1$ and root b to $z = -1$. Now applying this transformation we obtain the

$$R_\lambda(z) = \frac{(3 - \lambda)z^2 + z + \lambda}{3z + 1}$$

and we are going to draw the parameter planes of this rational function.

The free critical points associated to $R_\lambda(z)$ are:

- If $\lambda = 3$, $R_3'(z) = \frac{-8}{(3z-1)^2}$, and so, there are no critical points.
- If $\lambda \neq 3$, we obtain two free critical points

$$cp_1 = \frac{3 - \lambda + 2\sqrt{2}\sqrt{3\lambda - \lambda^2}}{3\lambda - 9}$$

and

$$cp_2 = \frac{3 - \lambda - 2\sqrt{2}\sqrt{3\lambda - \lambda^2}}{3\lambda - 9}.$$

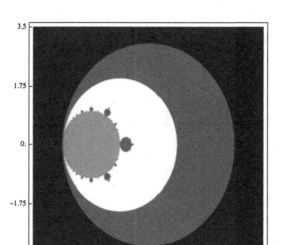

FIGURE 21.4 Parameter plane associated to cp_1.

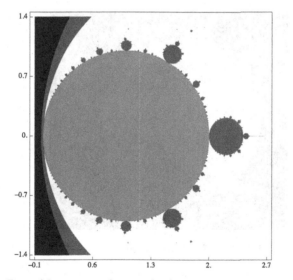

FIGURE 21.5 Zoom of the parameter plane associated to cp_1.

A point is painted in light gray if the iteration of the method starting in $z_0 = cp_1$ converges to the fixed point -1 (related to root b), in white if it converges to 1 (related to root a) and in black if the iteration diverges to ∞. Moreover, the regions in dark gray correspond to zones of convergence to different cycles or other "bad" behavior. As a consequence, every point of the plane which is

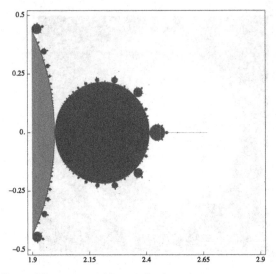

FIGURE 21.6 Zoom of the parameter plane associated to cp_1.

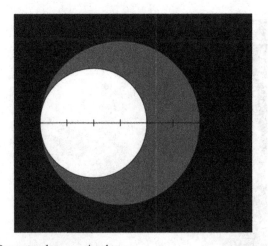

FIGURE 21.7 Parameter plane associated to cp_2.

neither white nor light gray is not a good choice of λ in terms of numerical behavior.

In Figs. 21.4 and 21.7 parameter planes associated to cp_1 and cp_2 are respectively shown. Moreover, in Figs. 21.5 and 21.6, zoom of the parameter plane associated to cp_1 are shown.

Finally, in Galleries I (Fig. 21.8) and II (Fig. 21.9), some basins associated to the behaviors found in the parameter planes are shown.

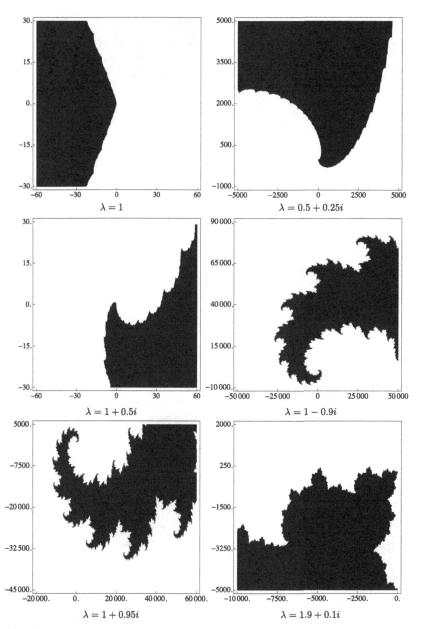

FIGURE 21.8 Gallery *I*. Basins of attraction associated to the fixed points of $R_\lambda(z)$. The basin of $z = -1$ appears in black and the basin of $z = 1$ appears in white.

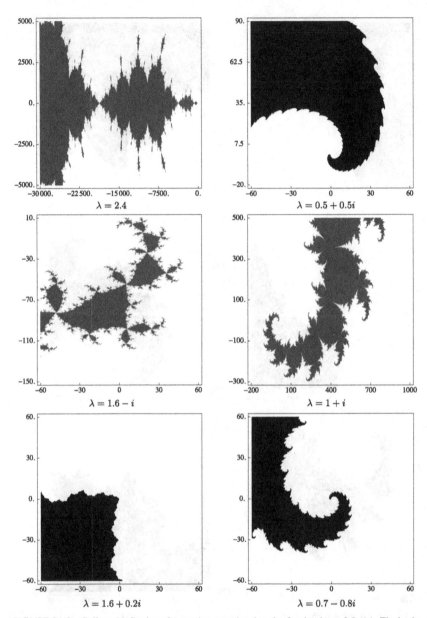

FIGURE 21.9 Gallery *11*. Basins of attraction associated to the fixed points of $R_\lambda(z)$. The basin of $z = -1$ appears in black and the basin of $z = 1$ appears in white. Moreover, nonconvergent zones appear in gray.

REFERENCES

[1] S. Amat, I.K. Argyros, S. Busquier, Á.A. Magreñán, Local convergence and the dynamics of a two-point four parameter Jarratt-like method under weak conditions, Numer. Algorithms 74 (2) (2017) 371–391.
[2] S. Amat, S. Busquier, C. Bermúdez, Á.A. Magreñán, On the election of the damped parameter of a two-step relaxed Newton-type method, Nonlinear Dynam. 84 (1) (2016) 9–18.
[3] I.K. Argyros, A. Cordero, Á.A. Magreñán, J.R. Torregrosa, On the convergence of a higher order family of methods and its dynamics, J. Comput. Appl. Math. 309 (2017) 542–562.
[4] I.K. Argyros, Á.A. Magreñán, A study on the local convergence and the dynamics of Chebyshev–Halley-type methods free from second derivative, Numer. Algorithms 71 (1) (2016) 1–23.
[5] I.K. Argyros, Á.A. Magreñán, On the convergence of an optimal fourth-order family of methods and its dynamics, App. Math. Comput. 252 (2015) 336–346.
[6] I.K. Argyros, Á.A. Magreñán, L. Orcos, Local convergence and a chemical application of derivative free root finding methods with one parameter based on interpolation, J. Math. Chem. 54 (7) (2016) 1404–1416.
[7] I.K. Argyros, Á.A. Magreñán, Local convergence and the dynamics of a two-step Newton-like method, Internat. J. Bifur. Chaos. 26 (5) (2016) 1630012.
[8] A.F. Beardon, Iteration of Rational Functions, Springer-Verlag, New York, 1991.
[9] A. Cordero, J.M. Gutiérrez, Á.A. Magreñán, J.R. Torregrosa, Stability analysis of a parametric family of iterative methods for solving nonlinear models, Appl. Math. Comput. 285 (2016) 26–40.
[10] A. Cordero, Á.A. Magreñán, C. Quemada, J.R. Torregrosa, Stability study of eighth-order iterative methods for solving nonlinear equations, J. Comput. Appl. Math. 291 (2016) 348–357.
[11] A. Cordero, L. Feng, Á.A. Magreñán, J.R. Torregrosa, A new fourth-order family for solving nonlinear problems and its dynamics, J. Math. Chem. 53 (3) (2014) 893–910.
[12] N. Fagella, X. Jarque, Iteración compleja y fractales, Vicens Vives, Barcelona, 2007.
[13] P. Fatou, Sur les équations fonctionelles, Bull. Soc. Math. France 47 (1919) 161–271.
[14] G. Julia, Memoire sur l'iteration des fonctions rationelles, J. Math. Pures. Appl. 8 (1) (1918) 47–246.
[15] Y.H. Geum, Y.I. Kim, Á.A. Magreñán, A biparametric extension of King's fourth-order methods and their dynamics, Appl. Math. Comput. 282 (2016) 254–275.
[16] J.M. Gutiérrez, Á.A. Magreñán, N. Romero, Dynamic aspects of damped Newton's method, in: Civil-Comp Proceedings, vol. 100, 2012.
[17] J.M. Gutiérrez, Á.A. Magreñán, J.L. Varona, Fractal dimension of the universal Julia sets for the Chebyshev–Halley family of methods, in: AIP Conf. Proc., vol. 1389, 2011, pp. 1061–1064.
[18] J.M. Gutiérrez, Á.A. Magreñán, J.L. Varona, The "Gauss–Seidelization" of iterative methods for solving nonlinear equations in the complex plane, Appl. Math. Comput. 218 (6) (2011) 2467–2479.
[19] T. Lotfi, Á.A. Magreñán, K. Mahdiani, J.J. Rainer, A variant of Steffensen–King's type family with accelerated sixth-order convergence and high efficiency index: dynamic study and approach, Appl. Math. Comput. 252 (2015) 347–353.
[20] Á.A. Magreñán, Different anomalies in a Jarratt family of iterative root-finding methods, Appl. Math. Comput. 233 (2014) 29–38.
[21] Á.A. Magreñán, A new tool to study real dynamics: the convergence plane, Appl. Math. Comput. 248 (2014) 215–224.
[22] Á.A. Magreñán, I.K. Argyros, On the local convergence and the dynamics of Chebyshev–Halley methods with six and eight order of convergence, J. Comput. Appl. Math. 298 (2016) 236–251.
[23] J. Milnor, Dynamics in One Complex Variable: Introductory Lectures, third edition, Princeton University Press, Princeton, New Jersey, 2006.
[24] P. Montel, Leçons sur les familles normales des fonctions analytiques et leurs applications, Gauthier-Villars, Paris, 1927.

[25] H.-O. Peitgen, D. Saupe, F. von Haeseler, Cayley's problem and Julia sets, Math. Intelligencer 6 (2) (1984) 11–20.

[26] G. Saunder, Iteration of Rational Functions in One Complex Variable and Basins of Attractive Points, PhD thesis, University of California, Berkeley, 1984.

[27] D. Sullivan, Quasiconformal homeomorphisms and dynamics I. Solution of Fatou–Julia problem wandering domains, Ann. Math. 122 (2) (1985) 401–418.

[28] J.F. Traub, Iterative Methods for the Solution of Equations, Prentice Hall, Englewood Cliffs, New Jersey, 1964.

[29] J.L. Varona, Graphic and numerical comparison between iterative methods, Math. Intelligencer 24 (1) (2002) 37–46.

Chapter 22

Convergence and the dynamics of Chebyshev–Halley type methods

22.1 INTRODUCTION

In this chapter we are concerned with the problem of approximating a locally unique solution x^* of the equation

$$F(x) = 0, \tag{22.1.1}$$

where F is a differentiable function defined on a convex subset D of S with values in S, where S is $\mathbb{R}$ or $\mathbb{C}$.

Several problems from applied mathematics can be solved by means of finding the solutions of equations like (22.1.1) using mathematical modeling [4,10, 33,36]. To find the solutions of equation (22.1.1), the most used methods are iterative. A very important problem in the study of iterative procedures is the convergence domain. In general, the radius of convergence is small. Therefore, it is important to enlarge the radius.

The dynamical properties related to an iterative method applied to polynomials give important information about its stability and reliability. In recently studies, several recognized authors [1–39] have found interesting dynamical planes, including periodical behavior and other anomalies. One of our main interests in this chapter is the study of the parameter spaces associated to a family of iterative methods, which allow us to distinguish between the good and bad methods in terms of their numerical properties. These are our study objectives in this chapter.

Recently, J. Sharma [36] studied the local convergence of the method defined for each $n = 0, 1, 2, \ldots$ by

$$
\begin{aligned}
y_n &= x_n - F'(x_n)^{-1} F(x_n), \\
z_n &= x_n - (1 + (F(x_n) - 2\beta F(y_n))^{-1} F(y_n)) F'(x_n)^{-1} F(x_n), \\
x_{n+1} &= x_n - A_n^{-1} F(z_n),
\end{aligned}
$$

$$\tag{22.1.2}$$

A Contemporary Study of Iterative Methods. DOI: 10.1016/B978-0-12-809214-9.00022-X
Copyright © 2018 Elsevier Inc. All rights reserved.

where x_0 is an initial point, $\beta \in S$ a given parameter, and

$$A_n = [x_n, y_n; F] + [z_n, y_n, x_n; F](z_n - y_n)$$
$$+ [z_n, y_n, x_n, x_n; F](z_n - y_n)(z_n - x_n),$$

with $[x_n, y_n; F]$, $[z_n, y_n, x_n; F]$, and $[z_n, y_n, x_n, x_n; F]$ being divided differences of order one, two, three, respectively [4,10,33,37].

The order of convergence was shown to be at least six, and if $\beta = 1$ then the order of convergence is eight.

This method includes the modifications of Chebyshev's method ($\beta = 0$), Halley's method ($\beta = 1/2$), and super-Halley method ($\beta = 1$). Method (22.1.2) is a useful alternative to the third order Chebyshev–Halley-methods [20–26] defined for each $n = 0, 1, 2, \ldots$ by

$$x_{n+1} = x_n - (1 + \frac{1}{2}(1 - \beta K_F(x_n))^{-1} K_F(x_n)) F'(x_n)^{-1} F(x_n), \qquad (22.1.3)$$

where

$$K_F(x_n) = F'(x_n)^{-1} F''(x_n) F'(x_n)^{-1} F(x_n),$$

since the computation of $F''(x_n)$ is avoided.

Method (22.1.2) can also be used instead of a method due to D. Li, P. Liu, and J. Kou [29] defined for each $n = 0, 1, 2, \ldots$ by

$$
\begin{aligned}
y_n &= x_n - F'(x_n)^{-1} F(x_n), \\
z_n &= x_n - (1 + (F(x_n) - 2\beta F(y_n))^{-1} F(y_n)) F'(x_n)^{-1} F(x_n), \\
x_{n+1} &= z_n - B_n^{-1} F(z_n),
\end{aligned}
$$

$$(22.1.4)$$

where

$$B_n = F'(x_n) + \bar{F}''(x_n)(z_n - x_n)$$

and

$$\bar{F}''(x_n) = 2F(y_n) F'(x_n)^2 F(x_n)^{-2}.$$

In the present chapter we present a local convergence analysis for method (22.1.2) using hypotheses only on the first Fréchet-derivative, contractive techniques, and Lipschitz constants. Moreover, we provide a radius of convergence and computable error estimates on the distances $\|x_n - x^*\|$ not given in [36] and other studies.

The dynamics of this family applied to an arbitrary quadratic polynomial $p(z) = (z - a)(z - b)$ will also be analyzed. The study of the dynamics of families of iterative methods has grown in the last years due to the fact that this study allows us to know the best choices of the parameter in terms of stability and to find the values of the parameter for which anomalies appear, such us convergence to cycles, divergence to infinity, etc.

The rest of the chapter is organized as follows: in Section 22.2 we present the local convergence analysis of method (22.1.2). The dynamics of method (22.1.2) are given in Section 22.3. Finally, the numerical examples are presented in the concluding Section 22.4.

22.2 LOCAL CONVERGENCE

The analysis is based on Lipschitz constants and contraction considerations. Let $L_0 > 0$, $L > 0$, $M \geq 1$, $l_1 > 0$, $l_2 > 0$, $a_0 > 0$, $a_1 > 0$, $a_2 > 0$, and $\beta \in S$ be constants. Define scalar function g_1 on the interval $[0, \frac{1}{L_0})$ by

$$g_1(t) = \frac{Lt}{2(1 - L_0 t)}$$

and parameter r_1 by

$$r_1 = \frac{2}{2L_0 + L} < \frac{1}{L_0}. \tag{22.2.1}$$

Notice that $g_1(r_1) = 1$. Define two more functions on the interval $[0, \frac{1}{L_0})$ by

$$g_2(t) = \frac{L_0 t}{2} + \frac{|\beta| M L}{1 - L_0 t}$$

and

$$h_2(t) = g_2(t) - 1.$$

Suppose that

$$|\beta| M L < 1. \tag{22.2.2}$$

Then we have by (22.2.2) that $h_2(0) = |\beta| M L - 1 < 0$ and $h_2(t) \to \infty$ as $t \to (\frac{1}{L_0})^-$. The intermediate value theorem asserts that function h_2 has a zero in the interval $(0, \frac{1}{L_0})$. Denote by r_2 the smallest such zero. Moreover, define additional functions on the interval $(0, r_2]$ by

$$g_3(t) = g_1(t) \left[1 + \frac{2M^2}{2(1 - L_0 t) - L_0 t(1 - L_0 t) - 2|\beta| M L} \right]$$

and

$$h_3(t) = g_3(t) - 1.$$

Then we have that $h_3(0) = -1$ and $h_3(t) \to \infty$ as $t \to r_2^-$. Hence function h_3 has a zero in $(0, r_2)$. Denote by r_3 the smallest such zero.

Now define further functions on $[0, r_3)$ by

$$g_4(t) = \left[l_1 g_3(t) + l_2 g_1(t) + a_1(g_1(t) + g_3(t))\right.$$
$$\left. + a_2(g_1(t) + g_3(t))(1 + g_3(t))t\right]t,$$

$$g_5(t) = (a_0 + a_1(g_1(t) + g_3(t))t + a_2(g_1(t) + g_3(t))(1 + g_3(t))t^2)M$$
$$+ (1 + L_0 t)M g_3(t),$$

$$g_6(t) = g_1(t) + \frac{g_5(t)}{(1 - L_0 t)(1 - g_4(t))},$$

and one more function on $[0, r_3]$ by

$$h_4(t) = g_4(t) - 1.$$

We have that $h_4(0) = -1$ and $h_4(t) \to +\infty$ as $t \to r_3^-$. Hence, function h_4 has a zero in the interval $(0, r_3)$. Denote by r_4 the smallest such zero.

Next we suppose that

$$a_0 M < 1 \qquad (22.2.3)$$

and define function h_6 on $[0, \frac{1}{L_0})$ by

$$h_6(t) = g_6(t) - 1.$$

We have that $h_6(0) = a_0 M - 1 < 0$ (by (22.2.2)) and $h_6(t) \to \infty$ as $t \to \frac{1}{L_0}^-$. Hence, function h_6 has a zero in the interval $(0, \frac{1}{L_0})$. Denote by r_6 the smallest such zero.

Set

$$r = \min\{r_4, r_6\} < \frac{1}{L_0}. \qquad (22.2.4)$$

Then, for each $t \in [0, r)$,

$$0 \le g_1(t) < 1, \qquad (22.2.5)$$

$$0 \le g_2(t)t < 1, \qquad (22.2.6)$$

$$0 \le g_3(t)t < 1, \qquad (22.2.7)$$

$$0 \le g_4(t) < 1, \qquad (22.2.8)$$

and

$$0 \le g_6(t) < 1. \qquad (22.2.9)$$

From now on, let us denote by $U(w, \rho)$, $\bar{U}(w, \rho)$ the open and closed balls in S, respectively, with center $v \in S$ and radius $\rho > 0$.

Now we present our main convergence result.

Theorem 22.2.1. *Let $F : \mathcal{D} \subset S \to S$ be a differentiable function and $[\cdot, \cdot; F]$, $[\cdot, \cdot, \cdot; F]$, $[\cdot, \cdot, \cdot, \cdot; F]$, $[\cdot, \cdot, \cdot, \cdot, \cdot; F]$ be divided differences of order one, two, three and four, respectively, for F. Let $l_1 > 0$, $l_2 > 0$, $a_0 > 0$, $a_1 > 0$, $a_2 > 0$, $L_0 > 0$, $L > 0$, $M > 0$, and $\beta \in S$ be given parameters. Suppose that there exists $x^* \in \mathcal{D}$ such that for all $x, y, z \in D$ the following hold:*

$$F(x^*) = 0, \quad F'(x^*)^{-1} \in L(S, S), \tag{22.2.10}$$

$$|\beta|ML < 1,$$

$$a_0 M < 1,$$

$$\|F'(x^*)^{-1}(F'(x) - F'(x^*))\| \le L_0\|x - x^*\|, \tag{22.2.11}$$

$$\|F'(x^*)^{-1}(F'(x) - F'(y))\| \le L\|x - y\|, \tag{22.2.12}$$

$$\|F'(x^*)^{-1}F'(x)\| \le M, \tag{22.2.13}$$

$$\|F'(x^*)^{-1}([x, y, z; F] - F'(x^*))\| \le l_1\|x - x^*\| + l_2\|y - x^*\|, \tag{22.2.14}$$

$$\|F'(x^*)^{-1}[z, y; F]\| \le a_0, \tag{22.2.15}$$

$$\|F'(x^*)^{-1}[z, y, x; F]\| \le a_1, \tag{22.2.16}$$

$$\|F'(x^*)^{-1}[z, y, x, x; F]\| \le a_2, \tag{22.2.17}$$

and

$$\bar{U}(x^*, r) \subseteq \mathcal{D}, \tag{22.2.18}$$

where the radius r is given in (22.2.4). Then the iteration $\{x_n\}$ generated by method (22.1.2) for $x_0 \in U(x^*, r) \setminus \{x^*\}$ is well defined, remains in $\bar{U}(x^*, r)$ for each $n = 0, 1, 2, \ldots$, and converges to x^*. Moreover, the following error bounds are true:

$$\|y_n - x^*\| \le g_1(\|x_n - x^*\|)\|x_n - x^*\| < \|x_n - x^*\| < r, \tag{22.2.19}$$

$$\|z_n - x^*\| \le g_3(\|x_n - x^*\|)\|x_n - x^*\| < \|x_n - x^*\|, \tag{22.2.20}$$

and

$$\|x_{n+1} - x^*\| \le g_6(\|x_n - x^*\|)\|x_n - x^*\| < \|x_n - x^*\|, \tag{22.2.21}$$

where the "g" functions are defined previously. Furthermore, for $R \in [r, \frac{2}{L_0})$ the limit point x^* is the unique zero of function F in $\bar{U}(x^*, R) \cap D$.

Proof. We prove (22.2.20)–(22.2.22) by induction. In view of (22.2.11), for r and $x_0 \in U(x^*, r)$, we get

$$\|F'(x^*)^{-1}(F'(x_0) - F'(x^*))\| \le L_0\|x_0 - x^*\| < L_0 r < 1.$$

Banach lemma on invertible functions [4,10,33] implies that $F'(x_0)^{-1} \in L(S, S)$ and

$$\|F'(x_0)^{-1}F'(x^*)\| \le \frac{1}{1 - L_0\|x_0 - x^*\|} \le \frac{1}{1 - L_0 r}. \tag{22.2.22}$$

As a consequence, y_0 is well defined by the first substep of the method (22.1.2) for $n = 0$. By the first substep of method (22.2.1) for $n = 0$, we get the identity

$$
\begin{aligned}
y_0 - x^* &= x_0 - x^* - F'(x_0)^{-1}F(x_0) \\
&= -F'(x_0)^{-1}F'(x^*)\int_0^1 \left[F'(x^* + \theta(x_0 - x^*))\right. \\
&\quad \left. - F'(x_0)\right](x_0 - x^*)d\theta.
\end{aligned} \tag{22.2.23}
$$

Using (22.2.4), (22.2.5), (22.2.12), (22.2.22), and (22.2.23), we get

$$
\begin{aligned}
\|y_0 - x^*\| &\le \|F'(x_0)^{-1}F'(x^*)\| \| \int_0^1 F'(x^*)^{-1}[F'(x^* + \theta(x_0 - x^*)) \\
&\quad - F'(x_0)]\|d\theta\|x_0 - x^*\| \\
&\le \frac{L\|x_0 - x^*\|^2}{2(1 - L_0\|x_0 - x^*\|)} = g_1(\|x_0 - x^*\|)\|x_0 - x^*\| \\
&< \|x_0 - x^*\| < r,
\end{aligned}
$$

which shows (22.2.19) for $n = 0$ and $y_0 \in U(x^*, r)$.

In view of (22.2.4), (22.2.6), (22.2.12), (22.2.13), (22.2.19), (22.2.22), and (22.2.23), we obtain

$$
\begin{aligned}
&\|(F'(x^*)(x_0 - x^*))^{-1}\left[F(x_0) - F(x^*) - 2\beta F(y_0) - F'(x^*)(x_0 - x^*)\right]\| \\
&\le \frac{1}{\|x_0 - x^*\|}\| \int_0^1 F'(x^*)^{-1}(F'(x^* + \theta(x_0 - x^*)) - F'(x^*))(x_0 - x^*)d\theta\| \\
&\quad + 2|\beta|\| \int_0^1 F'(x^*)^{-1}F'(x^* + \theta(x_0 - x^*))d\theta \\
&\quad F'(x_0)^{-1}F'(x^*)\int_0^1 F'(x^*)^{-1}[F'(x^* + \theta(x_0 - x^*)) - F'(x_0)](x_0 - x^*)d\theta\| \\
&\le \frac{L_0\|x_0 - x^*\|}{2} + \frac{|\beta|ML}{1 - L_0\|x_0 - x^*\|} = g_2(\|x_0 - x^*\|) \\
&< g_2(r) < 1,
\end{aligned} \tag{22.2.24}
$$

where we used that

$$
\begin{aligned}
F'(x^*)^{-1}F(y_0) &= F'(x^*)^{-1}(F(y_0) - F(x^*)) \\
&= \int_0^1 F'(x^*)^{-1}F'(x^* + \theta(y_0 - x^*))(y_0 - x^*)d\theta,
\end{aligned} \tag{22.2.25}
$$

so

$$\| F'(x^*)^{-1} F(y_0) \| \le M \| y_0 - x^* \| \le M g_1(\| x_0 - x^* \|) \| x_0 - x^* \|$$

and

$$\| x^* + \theta(y_0 - x^*) - x^* \| = \theta \| y_0 - x^* \| \le \| y_0 - x^* \| < r.$$

As a consequence, $(F(x_0) - 2\beta F(y_0))^{-1} \in L(S, S)$ and

$$\| (F(x_0) - 2\beta F(y_0))^{-1} F'(x^*) \| \le \frac{1}{\| x_0 - x^* \| (1 - g_2(\| x_0 - x^* \|))}. \quad (22.2.26)$$

Hence z_0 is well defined. Using the second substep of method (22.1.2) for $n = 0$, (22.2.4), (22.2.6), (22.2.19), (22.2.22), (22.2.25) (for $y_0 = x_0$), and (22.2.26), we obtain that

$$\| z_0 - x^* \| \le \| x_0 - x^* - F'(x_0)^{-1} F(x_0) \|$$
$$+ \| F'(x^*)^{-1} F(y_0) \| \| F'(x^*)^{-1} F(x_0) \| \| F'(x_0)^{-1} F'(x^*) \|$$
$$\times \| (F(x_0) - 2\beta F(y_0))^{-1} F'(x^*) \|$$
$$\le g_1(\| x_0 - x^* \|) \| x_0 - x^* \|$$
$$+ \frac{M^2 \| x_0 - x^* \| \| y_0 - x^* \|}{(1 - L_0 \| x_0 - x^* \|) \| x_0 - x^* \| [1 - \frac{1}{2}(L_0 \| x_0 - x^* \| + \frac{2|\beta| M L}{1 - L_0 \| x_0 - x^* \|})]}$$
$$\le g_1(\| x_0 - x^* \|) \| x_0 - x^* \|$$
$$\times \left[1 + \frac{2M^2}{2(1 - L_0 \| x_0 - x^* \|) - L_0 \| x_0 - x^* \| (1 - L_0 \| x_0 - x^* \|) - 2\beta M L} \right]$$
$$= g_3(\| x_0 - x^* \|) \| x_0 - x^* \| < \| x_0 - x^* \| < r,$$

which shows (22.2.20) for $n = 0$ and $z_0 \in U(x^*, r)$.

We also have by (22.2.14)–(22.2.17), (22.2.19), and (22.2.20) that

$$\| F'(x^*)^{-1} \left([z_0, y_0; F] - F'(x^*) + [z_0, y_0, x_0; F](z_0 - y_0) \right.$$
$$\left. + [z_0, y_0, x_0, x_0; F](z_0 - y_0)(z_0 - z_0) \right)$$
$$\le \| F'(x^*)^{-1} \left([z_0, y_0; F] - F'(x^*) \right) \| + \| F'(x^*)^{-1} [z_0, y_0, x_0; F] \| \| z_0 - y_0 \| +$$
$$\| F'(x^*)^{-1} [z_0, y_0, x_0, x_0; F] \| \| z_0 - y_0 \| \| z_0 - x_0 \|$$
$$\le l_1 \| z_0 - x^* \| + l_2 \| y_0 - x^* \| + a_1 \| z_0 - y_0 \| + a_2 \| z_0 - y_0 \| \| z_0 - x_0 \|$$
$$\le l_1 g_3(\| x_0 - x^* \|) + l_2 g_1(\| x_0 - x^* \|) \| x_0 - x^* \| + a_1(\| z_0 - x^* \| + \| y_0 - x^* \|)$$
$$+ a_2(\| z_0 - x^* \| + \| x_0 - x^* \|)(\| z_0 - x^* \| + \| x_0 - x^* \|)$$
$$\le l_1 g_3(\| x_0 - x^* \|) + l_2 g_1(\| x_0 - x^* \|) \| x_0 - x^* \| + a_1(g_1(\| x_0 - x^* \|)$$
$$+ g_3(\| x_0 - x^* \|)) \| x_0 - x^* \|$$
$$+ a_2(g_1(\| x_0 - x^* \|) + g_3(\| x_0 - x^* \|)) \| x_0 - x^* \|^2$$

$$= g_4(\|x_0 - x^*\|) < g_4(r) < 1.$$
(22.2.27)

It follows from (22.2.27) that $([z_0, y_0; F] + [z_0, y_0, x_0; F](z_0 - y_0) + [z_0, y_0, x_0, x_0; F](z_0 - y_0)(z_0 - x_0))$ is invertible and

$$\|([z_0, y_0; F] + [z_0, y_0, x_0; F](z_0 - y_0)$$
$$+ [z_0, y_0, x_0, x_0; F](z_0 - y_0)(z_0 - x_0))^{-1} F'(x^*)\|$$
(22.2.28)
$$\leq \frac{1}{1 - g_4(\|x_0 - x^*\|)}.$$

Now, using (22.2.4), (22.2.9), (22.2.15), (22.2.16), (22.2.17), (22.2.19), and (22.2.20), we obtain

$$\| ([z_0, y_0; F] + [z_0, y_0, x_0; F](z_0 - y_0)$$
$$+ [z_0, y_0, x_0, x_0; F](z_0 - y_0)(z_0 - x_0)) F(x_0) - F'(x_0)F(z_0)\|$$
$$\leq \left[\|F'(x^*)^{-1}[z_0, y_0; F]\| + \|F'(x^*)^{-1}[z_0, y_0, x_0; F]\| \|z_0 - y_0\| \right.$$
$$+ \|F'(x^*)^{-1}[z_0, y_0, x_0, x_0; F]\|$$
$$\times \|z_0 - y_0\| \|z_0 - x_0\| \Big] \int_0^1 F'(x^*)^{-1} F'(x^* + \theta(x_0 - x^*) d\theta \|x_0 - x^*\|$$
$$+ \left(\|F'(x^*)^{-1}(F'(x_0) - F'(x^*))\| + \|F'(x^*)^{-1}F'(x^*)\| \right)$$
$$\times \int_0^1 \|F'(x^*)^{-1} F'(x^* + \theta(x_0 - x^*)\| d\theta \|z_0 - x^*\|$$
$$\leq (a_0 + a_1(\|z_0 - x^*\| + \|y_0 - x^*\|) + a_2(\|z_0 - x^*\| + \|y_0 - x^*\|)(z_0 - x^*\|$$
$$+ \|x_0 - x^*\|))M\|x_0 - x^*\|$$
$$+ (1 + L_0\|x_0 - x^*\|)M\|z_0 - z^*\|$$
$$\leq (a_0 + a_1(g_1(\|x_0 - x^*\|) + g_3(\|x_0 - x^*\|))\|x_0 - x^*\| + a_2(g_1(\|x_0 - x^*\|)$$
$$+ g_3(\|x_0 - x^*\|))$$
$$\times (1 + g_3(\|x_0 - x^*\|))\|x_0 - x^*\|^2)M\|x_0 - x^*\|$$
$$+ (1 + L_0\|x_0 - x^*\|)Mg_3(\|x_0 - x^*\|)\|x_0 - x^*\|$$
$$= g_5(\|x_0 - x^*\|)\|x_0 - x^*\|.$$
(22.2.29)

By method (22.1.2) we get that

$$x_1 - x^* = x_0 - x^*$$
$$- \frac{F(z_0)}{[z_0, y_0; F] + [z_0, y_0, x_0; F](z_0 - y_0) + [x_0, y_0, x_0, x_0; F](z_0 - y_0)(z_0 - x_0)}$$
$$= y_0 - x^*$$

$$+\frac{([z_0,y_0;F]+[z_0,y_0,x_0;F](z_0-y_0)+[z_0,y_0,x_0,x_0;F](z_0-y_0)(z_0-x_0))F(x_0)-F'(x_0)F(z_0)}{F'(x_0)([z_0,y_0;F]+[z_0,y_0,x_0;F](z_0-y_0)+[z_0,y_0,x_0,x_0;F](z_0-y_0)(z_0-x_0))}.$$

$$(22.2.30)$$

Next, using (22.2.4), (22.2.9), (22.2.19), (22.2.28), (22.2.29), and (22.2.30), we obtain

$$\|x_1-x^*\| \le g_1(\|x_0-x^*\|)\|x_0-x^*\|$$
$$+\frac{g_5(\|x_0-x^*\|)\|x_0-x^*\|}{(1-L_0\|x_0-x^*\|)(1-g_4(\|x_0-x^*\|))}$$
$$=g_6(\|x_0-x^*\|)\|x_0-x^*\| < \|x_0-x^*\|,$$

which shows (22.2.21) for $n=0$ and $x_1 \in U(x^*,r)$. The induction is completed if x_0, y_0, z_0, x_1 are replaced by x_k, y_k, z_k, x_{k+1} in the previous estimates, leading to (22.2.19)–(22.2.21). Using the estimates $\|x_{k+1}-x^*\| \le c\|x_k-x^*\| < r$, $c=g_6(r) \in [0,1)$, we deduce that $\lim_{k\to\infty} x_k = x^*$ and $x_{k+1} \in U(x^*,r)$. Finally, let $T=\int_0^1 F'(y^*+\theta(x^*-y^*))d\theta$ for some $y^* \in \bar{U}(x^*,R)$ with $F(y^*)=0$. By (22.2.11), we obtain

$$\|F'(x^*)^{-1}(T-F'(x^*))\| = \|\int_0^1 F'(x^*)^{-1}(F'(y^*+\theta(x^*-y^*))-F'(x^*))d\theta\|$$
$$\le \|\int_0^1 L_0\|y^*+\theta(x^*-y^*)-x^*\|d\theta = L_0\int_0^1(1-\theta)\|y^*-x^*\| \le \frac{L_0}{2}R < 1.$$

$$(22.2.31)$$

Next by (22.2.31) we get $T^{-1} \in L(S,S)$. Then, from the identity $0=F(y^*)-F(x^*)=T(y^*-x^*)$, we deduce that $x^*=y^*$. □

Now, we can state the following remark.

Remark 22.2.2. *(i) Notice that by (22.2.11) and since*

$$\|F'(x^*)^{-1}F'(x)\| = \|F'(x^*)^{-1}(F'(x)-F'(x^*))+I\|$$
$$\le 1+\|F'(x^*)^{-1}(F'(x)-F'(x^*))\| \quad (22.2.32)$$
$$\le 1+L_0\|x-x^*\|,$$

condition (22.2.13) can be removed and M can be replaced by

$$M(t)=1+L_0t, \text{ or simply by } M(t)=M=2, \text{ since } t \in [0,\frac{1}{L_0}).$$

Moreover, condition (22.2.12) can be replaced by

$$\|F'(x^*)^{-1}(F'(x^*+\theta(x-x^*))-F'(x))\| \le L(1-\theta)\|x-x^*\|$$
for each $x \in D$ and $\theta \in [0,1]$.

$$(22.2.33)$$

(ii) *Our results find applications for operators F satisfying the autonomous differential equation [4,10] given by*

$$F'(x) = P(F(x)),$$

where P is a known continuous operator. Notice that $F'(x^*) = P(F(x^*)) = P(0)$, so we can apply the results without actually knowing the solution x^*. If, e.g., $F(x) = e^x - 1$, then we can choose $P(x) = x + 1$.

(iii) *The radius r_1 was shown in [4], [10] to be the convergence radius for Newton's method under conditions (2.11) and (2.30), i.e.,*

$$x_{n+1} = x_n - F'(x_n)^{-1}F(x_n), \quad \text{for each } n = 0, 1, 2, \ldots \quad (22.2.34)$$

By (22.2.7) and the definition of r_1, we get that the convergence radius r of the method (22.1.2) is smaller than the convergence radius r_1 of the second order Newton's method (22.2.34). In [4,10] we showed that r_1 is at least as large as the convergence ball radius given by Rheinboldt [33] and Traub [34], namely

$$r_R = \tfrac{2}{3L}. \quad (22.2.35)$$

If $L_0 < L$, we get that

$$r_R < r_1$$

and

$$\frac{r_R}{r_1} \to \frac{1}{3} \quad as \quad \frac{L_0}{L} \to 0.$$

As a consequence, our convergence ball radius r_1 is at most three times larger than r_R.

Remark 22.2.3. *The results obtained can be extended to hold for methods defined for each $n = 0, 1, 2, \ldots$ by*

$$y_n = x_n - F'(x_n)^{-1}F(x_n),$$
$$z_n = x_n - (1 + (F(x_n) - 2\beta F(y_n)))^{-1}F(y_n)F'(x_n)^{-1}F(x_n), \quad (22.2.36)$$
$$x_{n+1} = y_n - A_n^{-1}F(z_n),$$

or

$$y_n = x_n - F'(x_n)^{-1}F(x_n),$$
$$z_n = x_n - (1 + (F(x_n) - 2\beta F(y_n)))^{-1}F(y_n)F'(x_n)^{-1}F(x_n), \quad (22.2.37)$$
$$x_{n+1} = z_n - A_n^{-1}F(z_n).$$

These methods are faster than method (22.1.2) since they use y_n and z_n, respectively, in the place of x_n in the third substep of method (22.1.2). In view of the *proof of Theorem 22.2.1, define updated functions g_5 and g_6 as follows:*

$$g_5^1(t) = Mg_3(t),$$

$$g_6^1(t) = g_1(t) + \frac{g_5^1(t)}{1 - g_4(t)},$$

and

$$g_6^2(t) = g_3(t) + \frac{g_5^1(t)}{1 - g_4(t)}.$$

Using these changes we can state the following result.

Theorem 22.2.4. *Suppose that the hypotheses of previous Theorem 22.2.1 hold, except for condition (22.2.2). Then the conclusions of Theorem 22.2.1 hold for methods (22.2.36) and (22.2.37), but estimate (22.2.21) holds with g_6^1 replacing g_6 for method (22.2.36) and g_6^2 replacing g_6 for method (22.2.37).*

It is worth noticing that condition (22.2.2) is not needed for the convergence of method (22.2.36) or method (22.2.37).

22.3 DYNAMICAL STUDY OF THE METHOD (22.1.2)

In this section we are going to study the complex dynamics of the method (22.1.2). By applying this operator on a quadratic polynomial with two different roots A and B, $p(z) = (z - A)(z - B)$. Using the Möebius map $h(z) = \frac{z-A}{z-B}$, which carries root A to 0, root B to ∞ and ∞ to 1, we obtain the rational operator associated to the family of iterative schemes is finally

$$G(z, \beta) = \frac{z^6(2 + z - 2\beta)^2}{(-1 - 2z + 2z\beta)^2}. \tag{22.3.1}$$

It is clear that $z = 0$ and $z = \infty$ are fixed points of $G(z, \beta)$. Moreover, there exist some strange fixed which are:

- $z = 1$ related to divergence to ∞
- The roots of

$$p(z) = 1 + 5z + 9z^2 + 9z^3 + 9z^4 + 5z^5 + z^6 - 4z\beta - 12z^2\beta - 12z^3\beta$$
$$- 12z^4\beta - 4z^5\beta + 4z^2\beta^2 + 4z^3\beta^2 + 4z^4\beta^2$$

These solutions of this polynomial depend on the value of the parameter β.

In Fig. 22.1 the bifurcation diagram of the fixed points is shown.

On the other hand, it is a well-known fact that there is at least one critical point associated with each invariant Fatou component. The critical points of the family are the solutions of is $G'(z, \beta) = 0$, where

$$G'(z, \beta) = \frac{4z^5(2 + z - 2\beta)\left(-3 - 6z - 3z^2 + 3\beta + 8z\beta + 3z^2\beta - 4z\beta^2\right)}{(-1 - 2z + 2z\beta)^3}.$$

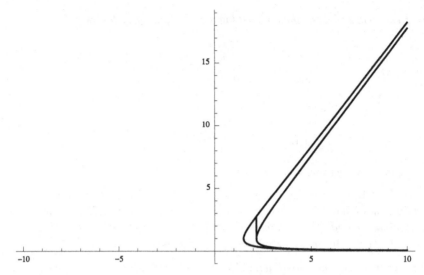

FIGURE 22.1   Bifurcation diagram of the fixed points of $G(z, \beta)$.

It is clear that $z = 0$ and $z = \infty$ are critical points. Furthermore, the free critical points are:

$$cp_1(\beta) = 2(-1 + \beta)$$

$$cp_2(\beta) = \frac{3 - 4\beta + 2\beta^2 - \sqrt{-6\beta + 19\beta^2 - 16\beta^3 + 4\beta^4}}{3(-1 + \beta)}$$

$$cp_3(\beta) = \frac{3 - 4\beta + 2\beta^2 + \sqrt{-6\beta + 19\beta^2 - 16\beta^3 + 4\beta^4}}{3(-1 + \beta)}$$

We state the relations between the free critical points in the following lemma.

Lemma 22.3.1. *a) If $\beta = \frac{1}{2}$*
  (i) $cp_1(\beta) = cp_2(\beta) = cp_3(\beta) = -1$.
b) If $\beta = \frac{3}{2}$
  (i) $cp_1(\beta) = cp_2(\beta) = cp_3(\beta) = 1$.
c) If $\beta = 0$
  (i) $cp_1(\beta) = -2$ and $cp_2(\beta) = cp_3(\beta) = -1$.
d) If $\beta = 2$
  (i) $cp_1(\beta) = 2$ and $cp_2(\beta) = cp_3(\beta) = 1$.
It is also clear that for every value of β $cr2(\beta) = \frac{1}{cp_3(\beta)}$

So, there are only three independent free critical points, without loss of generality, we consider in this chapter the free critical point $cp_2(\beta)$. Now, we are going to look for the best members of the family by means of using the parameter space associated to the free critical points.

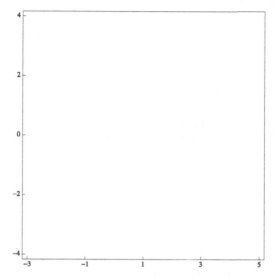

FIGURE 22.2   Parameter space associated to the free critical point $cp_1(\beta)$.

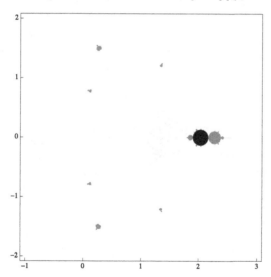

FIGURE 22.3   Parameter space associated to the free critical point $cp_2(\beta)$.

In Fig. 22.2, the parameter space associated to $cp_1(\beta)$ is shown and in Figs. 22.3–22.8 the parameter spaces associated to $cp_2(\beta)$ are shown. A point is painted in white if the iteration of the method starting in $z_0 = cp_1(\beta)$ converges to the fixed point 0 (related to root A) or if it converges to ∞ (related to root B) and in black if the iteration converges to 1 (related to ∞). Moreover, it appears in gray the convergence, after a maximum of 2000 iterations and with a tolerance of 10^{-6}, to any of the strange fixed points (distinct to 1), cycles or other

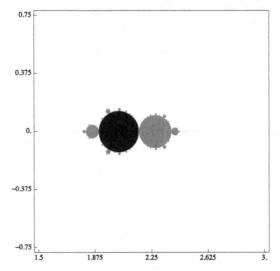

FIGURE 22.4 Details of the parameter space associated to the free critical point $cp_2(\beta)$.

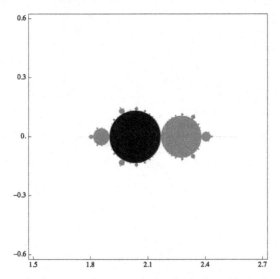

FIGURE 22.5 Details of the parameter space associated to the free critical point $cp_2(\beta)$.

"bad" behavior. As a consequence, every point of the plane which is not white is not a good choice of β in terms of numerical behavior.

In these dynamical planes we have painted in gray the convergence to 0, in white the convergence to ∞ and in black the zones with no convergence to the roots.

Then, focusing the attention in the region shown in Fig. 22.5 it is evident that there exist members of the family with complicated behavior. In Fig. 22.9,

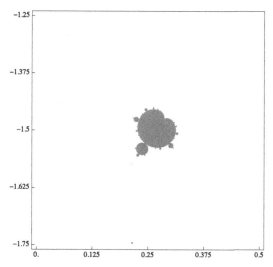

FIGURE 22.6 Details of the parameter space associated to the free critical point $cp_2(\beta)$.

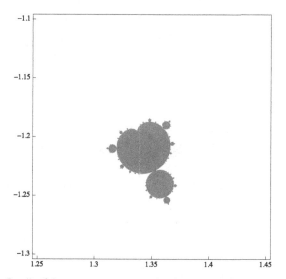

FIGURE 22.7 Details of the parameter space associated to the free critical point $cp_2(\beta)$.

the dynamical planes of a member of the family with regions of convergence to any of the strange fixed points is shown

In Figs. 22.10, 22.11 and 22.12 dynamical planes of members of the family with regions of convergence to an attracting cycle is shown.

On the other hand, in Fig. 22.13, the dynamical plane of a member of the family with regions of convergence to $z = 1$, related to ∞ is presented.

Other special cases are shown in Figs. 22.14, 22.15, 22.16 and 22.17.

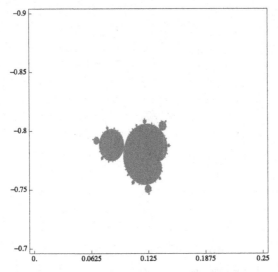

FIGURE 22.8 Details of the parameter space associated to the free critical point $cp_2(\beta)$.

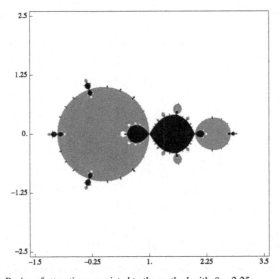

FIGURE 22.9 Basins of attraction associated to the method with $\beta = 2.25$.

22.4 NUMERICAL EXAMPLE

Example. *Let* $\mathcal{X} = [-1, 1]$, $\mathcal{Y} = \mathbb{R}$, $x_0 = 0$, *and let* $F : \mathcal{X} \to \mathcal{Y}$ *be the polynomial*

$$F(x) = \frac{1}{6}x^3 + \frac{1}{6}x^2 - \frac{5}{6}x + \frac{1}{3}.$$

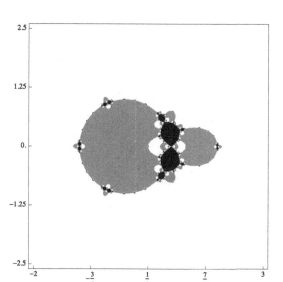

FIGURE 22.10 Basins of attraction associated to the method with $\beta = 1.85$.

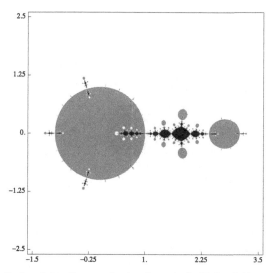

FIGURE 22.11 Basins of attraction associated to the method with $\beta = 2.41$.

Then, choosing

$$\beta = 0.7,$$

we get

$$L_0 = 1.42385\ldots,$$

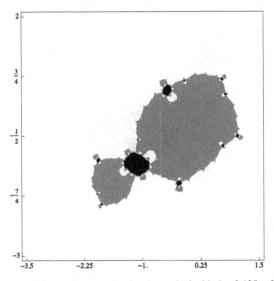

FIGURE 22.12   Basins of attraction associated to the method with $\beta = 0.125 - 0.77i$.

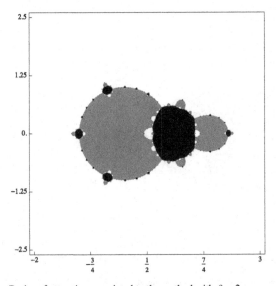

FIGURE 22.13   Basins of attraction associated to the method with $\beta = 2$.

$$L = 1.45653\ldots,$$
$$M = 0.946746\ldots,$$
$$l_1 = 2.9252\ldots,$$
$$l_2 = 1.9626\ldots,$$

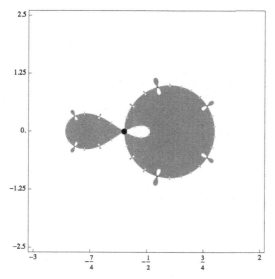

FIGURE 22.14 Basins of attraction associated to the method with $\beta = 0$.

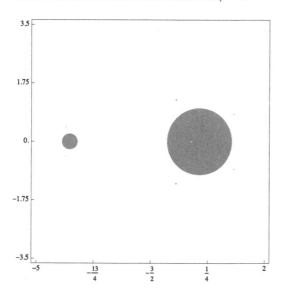

FIGURE 22.15 Basins of attraction associated to the method with $\beta = -1$.

$$a_0 = 0.946746\ldots,$$
$$a_1 = 0.72826\ldots,$$

and

$$a_2 = 0.29130\ldots$$

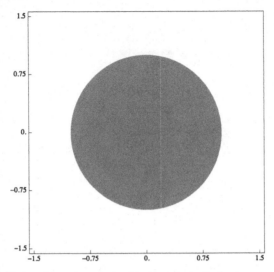

FIGURE 22.16   Basins of attraction associated to the method with $\beta = 1$.

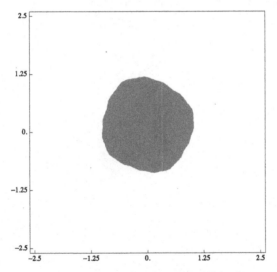

FIGURE 22.17   Basins of attraction associated to the method with $\beta = 3i$.

Then, by the definition of the "g" functions, we obtain

$$r_4 = 0.0161533\ldots, \quad r_6 = 0.00386947\ldots,$$

and as a consequence

$$r = r_6 = 0.00386947\ldots$$

So we can ensure the convergence of the method (22.1.2) with $\beta = 0.7$.

REFERENCES

[1] S. Amat, I.K. Argyros, S. Busquier, Á.A. Magreñán, Local convergence and the dynamics of a two-point four parameter Jarratt-like method under weak conditions, Numer. Algorithms 74 (2) (2017) 371–391.

[2] S. Amat, S. Busquier, C. Bermúdez, Á.A. Magreñán, On the election of the damped parameter of a two-step relaxed Newton-type method, Nonlinear Dynam. 84 (1) (2016) 9–18.

[3] S. Amat, M.A. Hernández, N. Romero, A modified Chebyshev's iterative method with at least sixth order of convergence, Appl. Math. Comput. 206 (1) (2008) 164–174.

[4] I.K. Argyros, Convergence and Application of Newton-Type Iterations, Springer, 2008.

[5] I.K. Argyros, A. Cordero, Á.A. Magreñán, J.R. Torregrosa, On the convergence of a higher order family of methods and its dynamics, J. Comput. Appl. Math. 309 (2017) 542–562.

[6] I.K. Argyros, Á.A. Magreñán, A study on the local convergence and the dynamics of Chebyshev–Halley-type methods free from second derivative, Numer. Algorithms 71 (1) (2016) 1–23.

[7] I.K. Argyros, Á.A. Magreñán, On the convergence of an optimal fourth-order family of methods and its dynamics, Appl. Math. Comput. 252 (2015) 336–346.

[8] I.K. Argyros, Á.A. Magreñán, L. Orcos, Local convergence and a chemical application of derivative free root finding methods with one parameter based on interpolation, J. Math. Chem. 54 (7) (2016) 1404–1416.

[9] I.K. Argyros, Á.A. Magreñán, Local convergence and the dynamics of a two-step Newton-like method, Internat J. Bifur. Chaos. 26 (5) (2016) 1630012.

[10] I.K. Argyros, S. Hilout, Numerical methods in Nonlinear Analysis, World Scientific Publ. Comp., New Jersey, 2013.

[11] R. Behl, Development and Analysis of Some New Iterative Methods for Numerical Solutions of Nonlinear Equations, PhD thesis, Punjab University, 2013.

[12] D.D. Bruns, J.E. Bailey, Nonlinear feedback control for operating a nonisothermal CSTR near an unstable steady state, Chem. Eng. Sci. 32 (1977) 257–264.

[13] V. Candela, A. Marquina, Recurrence relations for rational cubic methods I: the Halley method, Computing 44 (1990) 169–184.

[14] V. Candela, A. Marquina, Recurrence relations for rational cubic methods II: the Chebyshev method, Computing 45 (4) (1990) 355–367.

[15] F. Chicharro, A. Cordero, J.R. Torregrosa, Drawing dynamical and parameters planes of iterative families and methods, Sci. World J. (2013) 780153.

[16] C. Chun, Some improvements of Jarratt's method with sixth-order convergence, Appl. Math. Comput. 190 (2) (1990) 1432–1437.

[17] A. Cordero, J.M. Gutiérrez, Á.A. Magreñán, J.R. Torregrosa, Stability analysis of a parametric family of iterative methods for solving nonlinear models, Appl. Math. Comput. 285 (2016) 26–40.

[18] A. Cordero, Á.A. Magreñán, C. Quemada, J.R. Torregrosa, Stability study of eighth-order iterative methods for solving nonlinear equations, J. Comput. Appl. Math. 291 (2016) 348–357.

[19] A. Cordero, L. Feng, Á.A. Magreñán, J.R. Torregrosa, A new fourth-order family for solving nonlinear problems and its dynamics, J. Math. Chem. 53 (3) (2014) 893–910.

[20] J.A. Ezquerro, M.A. Hernández, Recurrence relations for Chebyshev-type methods, Appl. Math. Optim. 41 (2) (2000) 227–236.

[21] J.A. Ezquerro, M.A. Hernández, New iterations of R-order four with reduced computational cost, BIT Numer. Math. 49 (2009) 325–342.

[22] J.A. Ezquerro, M.A. Hernández, On the R-order of the Halley method, J. Math. Anal. Appl. 303 (2005) 591–601.

[23] J.M. Gutiérrez, M.A. Hernández, Recurrence relations for the super-Halley method, Computers Math. Applic. 36 (7) (1998) 1–8.

[24] M. Ganesh, M.C. Joshi, Numerical solvability of Hammerstein integral equations of mixed type, IMA J. Numer. Anal. 11 (1991) 21–31.

[25] M.A. Hernández, Chebyshev's approximation algorithms and applications, Computers Math. Applic. 41 (3–4) (2001) 433–455.

[26] M.A. Hernández, M.A. Salanova, Sufficient conditions for semilocal convergence of a fourth order multipoint iterative method for solving equations in Banach spaces, Southwest J. Pure Appl. Math. 1 (1999) 29–40.

[27] P. Jarratt, Some fourth order multipoint methods for solving equations, Math. Comput. 20 (95) (1966) 434–437.

[28] J. Kou, Y. Li, An improvement of the Jarratt method, Appl. Math. Comput. 189 (2007) 1816–1821.

[29] D. Li, P. Liu, J. Kou, An improvement of the Chebyshev–Halley methods free from second derivative, Appl. Math. Comput. 235 (2014) 221–225.

[30] Á.A. Magreñán, Different anomalies in a Jarratt family of iterative root-finding methods, Appl. Math. Comput. 233 (2014) 29–38.

[31] Á.A. Magreñán, A new tool to study real dynamics: the convergence plane, Appl. Math. Comput. 248 (2014) 215–224.

[32] S.K. Parhi, D.K. Gupta, Recurrence relations for a Newton-like method in Banach spaces, J. Comput. Appl. Math. 206 (2) (2007) 873–887.

[33] L.B. Rall, Computational Solution of Nonlinear Operator Equations, Robert E. Krieger, New York, 1979.

[34] H. Ren, Q. Wu, W. Bi, New variants of Jarratt method with sixth-order convergence, Numer. Algorithms 52 (4) (2009) 585–603.

[35] W.C. Rheinboldt, An adaptive continuation process for solving systems of nonlinear equations, in: Banach Ctr. Publ., vol. 3, Polish Academy of Science, 1978, pp. 129–142.

[36] J.R. Sharma, Improved Chebyshev–Halley methods with six and eight order of convergence, Appl. Math. Comput. 256 (2015) 119–124.

[37] J.F. Traub, Iterative Methods for the Solution of Equations, Prentice-Hall Ser. Automat. Comput., Prentice-Hall, Englewood Cliffs, NJ, 1964.

[38] X. Wang, J. Kou, C. Gu, Semilocal convergence of a sixth-order Jarratt method in Banach spaces, Numer. Algorithms 57 (2011) 441–456.

[39] J. Kou, X. Wang, Semilocal convergence of a modified multi-point Jarratt method in Banach spaces under general continuity conditions, Numer. Algorithms 60 (2012) 369–390.

Chapter 23

Convergence planes of iterative methods

23.1 INTRODUCTION

In this chapter we are concerned with the problem of approximating a locally unique solution x^* of equation

$$F(x) = 0, \qquad\qquad (23.1.1)$$

where F is a differentiable function defined on a convex subset D of S with values in S, where S is $\mathbb{R}$ or $\mathbb{C}$.

Many problems from applied mathematics, including engineering, can be solved by means of finding the solutions of equations like (23.1.1) using mathematical modeling [3–5,25,27,28,31].

The convergence analysis of iterative methods is usually divided into two categories: semilocal and local convergence analysis. The semilocal convergence matter is, based on the information around an initial point, to give criteria ensuring the convergence of iteration procedures. A very important problem in the study of iterative procedures is the convergence domain. In general the convergence domain is small. Therefore, it is important to enlarge the convergence domain without additional hypothesis. Another important problem is to find more precise error estimates on the distances $\|x_{n+1} - x_n\|$ and $\|x_n - x^*\|$.

The dynamical properties related to an iterative method applied to polynomials give important information about its stability and reliability. In recently studies, different authors [1,2,5,7,6,8–24,29–33] have found interesting behaviors. One of our main interests in this chapter is the study of the parameter spaces associated to a family of iterative methods, which allow us to distinguish between the good and bad methods in terms of their numerical properties, but, taking into account that there are more than one parameter, we are going to use the convergence planes introduced by Magreñán [26].

Recently, Cordero et al. [24] studied the local convergence of the method defined for each $n = 0, 1, 2, \ldots$ by

$$
\begin{aligned}
y_n &= x_n - a\frac{F(x_n)}{F'(x_n)}, \\[2mm]
x_{n+1} &= x_n - \frac{F(x_n)^2}{bF(x_n)^2 + cF(y_n)^2}\frac{F(x_n)}{F'(x_n)},
\end{aligned}
\qquad (23.1.2)
$$

A Contemporary Study of Iterative Methods. DOI: 10.1016/B978-0-12-809214-9.00023-1

333

where x_0 is an initial point and a, b, c are suitable parameters. They compared the region of convergence of method (23.1.2) to the corresponding ones given by

- Newton's method [4,5]:

$$x_{n+1} = x_n - \frac{F(x_n)}{F'(x_n)}, \qquad (23.1.3)$$

which has second order of convergence;
- Traub's method [31]:

$$y_n = x_n - \frac{F(x_n)}{F'(x_n)}, \quad x_{n+1} = x_n - \frac{F(x_n) + F(y_n)}{F'(x_n)}, \qquad (23.1.4)$$

which has third order of convergence;
- Ostrowski's method [31]:

$$y_n = x_n - \frac{F(x_n)}{F'(x_n)}, \quad x_{n+1} = x_n - \frac{F(x_n)}{F(x_n) - 2F(y_n)} \frac{F(y_n)}{F'(x_n)}, \qquad (23.1.5)$$

whose order of convergence is four.

In this chapter we present the local convergence analysis of method (23.1.2) using hypotheses only on the first derivative of function F. We also provide a computable radius of convergence and error estimates based on Lipschitz constants.

In this chapter some convergence planes of this family applied to the quadratic polynomial $p(z) = z^2 - 1$ will also be analyzed. The study of the dynamics of families of iterative methods has grown in the last years due to the fact that this study allows us to know the best choices of the parameter in terms of stability and to find the values of the parameter for which anomalies appear, such as convergence to cycles, divergence to infinity, etc. But there is a problem when a family has more than one parameter. This problem was solved by the second author in [26] by means of introducing a new graphic tool.

The rest of the chapter is organized as follows: in Section 23.2 we present the local convergence analysis of method (23.1.2). The convergence planes associated to method (23.1.2) are given in Section 23.3. Finally, the numerical examples are given in the concluding Section 23.4.

23.2 LOCAL CONVERGENCE

We present the local convergence analysis of method (23.1.2) in this section. It is convenient for the local convergence analysis that follows to introduce some functions and parameters. Let $L_0 > 0$, $L > 0$, $M \geq 1$, and $a, b, c \in S$ be given parameters with $L_0 \leq L$ and $b \neq 0$. Let us define function g_1 on the interval $[0, \frac{1}{L_0})$ by

$$g_1(t) = \frac{1}{2(1 - L_0 t)} [Lt + 2M|1 - \alpha|]$$

and parameter r_1 by

$$r_1 = \frac{2(1 - M|1 - a|)}{2L_0 + L}.$$

Suppose that

$$M|1 - a| < 1. \tag{23.2.1}$$

Then it follows from (23.2.1) that

$$r_1 < r_A = \frac{2}{2L_0 + L} \leq r_{TR} = \frac{2}{3L}. \tag{23.2.2}$$

Define functions g_0 and h_0 on the interval $[0, \frac{1}{L_0})$ by

$$g_0(t) = \frac{M^2|c|g_1^2(t)}{|b|(1 - \frac{L_0}{2}t)^2}$$

and

$$h_0(t) = g_0(t) - 1.$$

Moreover, suppose that

$$\frac{|c|}{|b|} M^4 |1 - \alpha|^2 < 1. \tag{23.2.3}$$

Next we have by (23.2.3) that $h_0(0) = g_0(0) - 1 = \frac{M^4|c||1 - \alpha|^2}{|b|} - 1 < 0$ and $h_0(t) \to \infty$ as $t \to (\frac{1}{L_0})^-$. It follows from the intermediate value theorem that function h_0 has a zero in the interval $(0, \frac{1}{L_0})$. Denote by r_0 the smallest such zero. Moreover, define functions g_2 and h_2 on the interval $[0, r_0)$ by

$$g_2(t) = \frac{Lt}{2(1 - L_0 t)} + \frac{M^3(|1 - b| + |c|g_1^2(t))}{|b|(1 - L_0 t)(1 - g_0(t))(1 - \frac{L_0}{2}t)^2}$$

and set

$$h_2(t) = g_2(t) - 1.$$

Moreover, suppose that

$$\frac{M^3}{|b|} \left[|1 - b| + M|c||1 - \alpha|^2 + |c|M^2|1 - \alpha|^2 \right] < 1. \tag{23.2.4}$$

Next we have by (23.2.4) that $h_2(0) < 0$ and $h_2(t) \to \infty$ as $t \to r_0^-$. Consequently, function h_2 has a zero in the interval $(0, r_0)$. Denote by r_2 the smallest such zero. Set

$$r = \min\{r_1, r_2\}. \tag{23.2.5}$$

As a consequence, we have for each $t \in [0, r)$ that

$$0 \le g_0(t) < 1, \quad 0 \le g_1(t) < 1, \text{ and } 0 \le g_2(t) < 1. \tag{23.2.6}$$

From now on, let $U(v, \rho)$, $\bar{U}(v, \rho)$ stand for the open and closed balls in S, respectively, with center $v \in S$ and radius $\rho > 0$.

Now we present the following convergence result.

Theorem 23.2.1. *Let $F : \mathcal{D} \subset S \to S$ be a differentiable function. Suppose that there exist $x^* \in \mathcal{D}$, and given parameters $L_0 > 0$, $L > 0$, $M \ge 1$, $a, b, c \in S$ with $b \neq 0$ such that (23.2.1) and (23.2.4) hold, and for each $x, y \in \mathcal{D}$ the following are true:*

$$F(x^*) = 0, \quad F'(x^*)^{-1} \in L(S, S), \tag{23.2.7}$$

$$\|F'(x^*)^{-1}(F'(x) - F'(x^*))\| \le L_0 \|x - x^*\|, \tag{23.2.8}$$

$$\|F'(x^*)^{-1}(F'(x) - F'(y))\| \le L \|x - y\|, \tag{23.2.9}$$

$$\|F'(x^*)^{-1}F'(x)\| \le M, \tag{23.2.10}$$

and

$$\bar{U}(x^*, r) \subseteq \mathcal{D}, \tag{23.2.11}$$

where r is defined by (23.2.5). Then the sequence $\{x_n\}$ generated by method (23.1.2) for $x_0 \in U(x^*, r) \setminus \{x^*\}$ is well defined, remains in $U(x^*, r)$ for each $n = 0, 1, 2, \ldots$, and converges to x^*. Moreover, the following estimates hold:

$$\|y_n - x^*\| \le g_1(\|x_n - x^*\|)\|x_n - x^*\| < \|x_n - x^*\| < r \tag{23.2.12}$$

and

$$\|x_{n+1} - x^*\| \le g_2(\|x_n - x^*\|)\|x_n - x^*\| < \|x_n - x^*\|, \tag{23.2.13}$$

where the "g" functions are defined previously. Furthermore, if there exists $R \in [r, \frac{2}{L_0})$ such that $\bar{U}(x^*, R) \subseteq \mathcal{D}$, then the limit point x^* is the only solution of equation $F(x) = 0$ in $\bar{U}(x^*, R)$.

Proof. We will use mathematical induction. It follows from the definition of r, the hypothesis $x_0 \in U(x^*, r)$, and (23.2.8) that

$$\|F'(x^*)^{-1}(F'(x_0) - F'(x^*))\| \le L_0 \|x_0 - x^*\| < L_0 r < 1. \tag{23.2.14}$$

By (23.2.14) and Banach lemma on invertible functions [4,5,28], we have that $F'(x_0)^{-1} \in L(S, S)$ and

$$\|F'(x_0)^{-1}F'(x^*)\| \leq \frac{1}{1 - L_0\|x_0 - x^*\|} \leq \frac{1}{1 - L_0 r}. \tag{23.2.15}$$

Hence, y_0 is well defined by the first substep of the method (23.1.2) for $n = 0$. By the first substep of method (23.2.1) for $n = 0$ and (23.2.7), we obtain the approximation

$$y_0 - x^*$$
$$= x_0 - x^* - F'(x_0)^{-1}F(x_0) + (1-a)F'(x_0)^{-1}F - (x^*)F'(x^*)^{-1}F(x_0)$$
$$= -F'(x_0)^{-1}F'(x^*)\int_0^1 \left[F'(x^* + \theta(x_0 - x^*)) - F'(x_0)\right](x_0 - x^*)d\theta$$
$$+ (1-a)F'(x_0)^{-1}F'(x^*)F'(x^*)^{-1}F(x_0). \tag{23.2.16}$$

Then, by (23.2.5), (23.2.6), (23.2.9), (23.2.10), and (23.2.16), we get in turn that

$$\|y_0 - x^*\| \leq \|F'(x_0)^{-1}F'(x^*)\| \|\int_0^1 F'(x^*)^{-1}[F'(x^* + \theta(x_0 - x^*))$$
$$- F'(x_0)]\|d\theta\|x_0 - x^*\|$$
$$\leq \frac{L\|x_0 - x^*\|^2}{2(1 - L_0\|x_0 - x^*\|)} + \frac{M|1 - a|M\|x_0 - x^*\|}{1 - L_0\|x_0 - x^*\|}$$
$$= g_1(\|x_0 - x^*\|)\|x_0 - x^*\| < \|x_0 - x^*\| < r,$$

which shows (23.2.27) for $n = 0$ and $y_0 \in U(x^*, r)$, where we also used the estimates

$$F(x_0) = F(x_0) - F(x^*) = \int_0^1 F'(x^*)^{-1}F'(x^* + \theta(x_0 - x^*))(x_0 - x^*)d\theta \tag{23.2.17}$$

and

$$\|F'(x^*)^{-1}F(x_0)\| = \|\int_0^1 F'(x^*)^{-1}F'(x^* + \theta(x_0 - x^*))(x_0 - x^*)d\theta\|$$
$$\leq M\|x_0 - x^*\|. \tag{23.2.18}$$

We need an estimate on the norm of

$$(b-1)F(x_0)^2 + cF(y_0)^2.$$

We have by (23.2.7), (23.2.27), and (23.2.18) that

$$\|F'(x^*)^{-2}((b-1)F(x_0)^2 + cF(y_0)^2)\|$$
$$\leq |1-b|\|F'(x^*)^{-1}\|(F(x_0) - F(x^*))\|^2 + |c|\|F'(x^*)^{-1}(F(y_0) - F(x^*))\|^2$$
$$\leq |1-b|M^2\|x_0 - x^*\|^2 + |c|M^2\|y_0 - x^*\|^2$$
$$\leq M^2(|1-b|\|x_0 - x^*\|^2 + |c|\|y_0 - x^*\|^2)$$
$$\leq M^2(|1-b| + cg_1^2(\|x_0 - x^*\|))\|x_0 - x^*\|^2.$$

$$(23.2.19)$$

Next we shall show that $bF(x_0)^2 + cF(y_0)^2$ is invertible. We have by (23.2.5), (23.2.6), (23.2.7), (23.2.27), (23.2.18), and $x_0 \neq x^*$ that

$$\|(F'(x^*)(x_0 - x^*))^{-1}\left[F(x_0) - F(x^*) - F'(x_0)(x_0 - x^*)\right]\|$$
$$\leq \|x_0 - x^*\|^{-1}\frac{L_0}{2}\|x_0 - x^*\|^2 = \frac{L_0}{2}\|x_0 - x^*\| < L_0 r < 1.$$

It follows that $F(x_0) \neq 0$ and

$$\|F(x_0)^{-1}F'(x^*)\| \leq \frac{1}{\|x_0 - x^*\|(1 - \frac{L_0}{2}\|x_0 - x^*\|)}.$$

We also have

$$\|\frac{c}{b}F(x_0)^{-2}F(y_0)^2\|$$
$$\leq \left|\frac{c}{b}\right| \frac{M^2\|y_0 - x^*\|^2}{\|x_0 - x^*\|^2(1 - \frac{L_0}{2}\|x_0 - x^*\|)^2} \leq \left|\frac{c}{b}\right| \frac{M^2 g_1^2(\|x_0 - x^*\|)}{(1 - \frac{L_0}{2}\|x_0 - x^*\|)^2}$$
$$= g_0(\|x_0 - x^*\|) < g_0(t) < 1.$$

$$(23.2.20)$$

Consequently, $1 + \frac{c}{b}F(x_0)^{-2}F(y_0)^2$ is invertible and

$$\|(1 + \frac{c}{b}F(x_0)^{-2}F(y_0)^2)^{-1}\| \leq \frac{1}{1 - g_0(\|x_0 - x^*\|)}.$$

It follows that $bF^2(x_0) + cF(y_0)^2 = bF(x_0)^2\left[1 + \frac{c}{b}F(x_0)^{-2}F(y_0)^2\right]$ is invertible and

$$\|(bF^2(x_0) + cF(y_0)^2)^{-1}(F'(x^*))^2\|$$
$$\leq \frac{1}{2|b|\|x_0 - x^*\|^2(1 - g_0(\|x_0 - x^*\|))(1 - \frac{L_0}{2}\|x_0 - x^*\|)^2}.$$
$$(23.2.21)$$

Therefore, x_1 is well defined by the second substep of method (23.1.2) for $n = 0$. Then, we have by method (23.1.2) for $n = 0$, (23.2.5), (23.2.6), (23.2.27), (23.2.15), (23.2.18), (23.2.19), and (23.2.21) that

$$
x_1 - x^* = x_0 - x^* - F'(x_0)^{-1} F(x_0)
$$
$$
+ \left[1 - \frac{F^2(x_0)}{bF(x_0)^2 + cF(y_0)^2} \right] \left[F'(x_0)^{-1} F'(x^*) \right] \left[F'(x^*)^{-1} F(x_0) \right]
$$

so

$$
\|x_1 - x^*\| \le \frac{L\|x_0 - x^*\|^2}{2(1 - L_0\|x_0 - x^*\|)}
$$
$$
+ \frac{\|F'(x^*)^{-2}((b-1)F(x_0)^2 + cF(y_0)^2)\|}{\|F'(x^*)^{-2}(bF(x_0)^2 + cF(y_0)^2)\|} \|F'(x_0)^{-1} F'(x^*)\| \|F'(x^*)^{-1} F(x_0)\|
$$
$$
\le \left[\frac{L\|x_0 - x^*\|^2}{2(1 - L_0\|x_0 - x^*\|)} \right.
$$
$$
+ \frac{M^3(|1 - b|\|x_0 - x^*\|^2 + |c|\|y_0 - x^*\|^2)}{|b|\|x_0 - x^*\|^2(1 - L_0\|x_0 - x^*\|)(1 - \frac{L_0}{2}\|x_0 - x^*\|)^2(1 - g_0(\|x_0 - x^*\|))} \right]
$$
$$
\times \|x_0 - x^*\|
$$
$$
\le \left[\frac{L\|x_0 - x^*\|^2}{2(1 - L_0\|x_0 - x^*\|)} \right.
$$
$$
\left. + \frac{M^3(|1 - b| + |c|g_1^2(\|x_0 - x^*\|))}{|b|(1 - \frac{L_0}{2}\|x_0 - x^*\|)^2(1 - L_0\|x_0 - x^*\|)(1 - g_0(\|x_0 - x^*\|))} \right] \|x_0 - x^*\|
$$
$$
= g_2(\|x_0 - x^*\|)\|x_0 - x^*\| < \|x_0 - x^*\| < r,
$$

which shows (23.2.28) for $n = 0$ and $x_1 \in U(x^*, r)$. By simply replacing x_0, y_0, x_1 with x_k, y_k, x_{k+1} in the preceding estimates, we arrive at (23.2.14)–(23.2.16). Using the estimates $\|x_{k+1} - x^*\| < \|x_k - x^*\| < r$, we deduce that $\lim_{k \to \infty} x_k = x^*$ and $x_{k+1} \in U(x^*, r)$. Finally, to show the uniqueness part, let $T = \int_0^1 F'(y^* + \theta(x^* - y^*))d\theta$ for some $y^* \in \bar{U}(x^*, R)$ with $F(y^*) = 0$. In view of (23.2.10), we obtain

$$
\|F'(x^*)^{-1}(T - F'(x^*))\| \le \| \int_0^1 L_0 \|y^* + \theta(x^* - y^*)\|d\theta
$$
$$
= L_0 \int_0^1 (1 - \theta)\|y^* - x^*\| \le \frac{L_0}{2} R < 1.
$$
(23.2.22)

It follows from (23.2.22) that $T^{-1} \in L(S, S)$. Then from the identity $0 = F(y^*) - F(x^*) = T(y^* - x^*)$, we conclude that $x^* = y^*$. $\qquad\square$

Next, we can present the following remark.

Remark 23.2.2. *1.* *In view of (23.2.8) and the estimate*

$$
\begin{aligned}
\|F'(x^*)^{-1}F'(x)\| &= \|F'(x^*)^{-1}(F'(x) - F'(x^*)) + I\| \\
&\leq 1 + \|F'(x^*)^{-1}(F'(x) - F'(x^*))\| \\
&\leq 1 + L_0\|x_0 - x^*\|,
\end{aligned}
$$

condition (23.2.8) can be dropped and M can be replaced by

$$
M(t) = 1 + L_0 t.
$$

2. The results obtained here can be used for operators F satisfying the autonomous differential equation [4,5] of the form

$$
F'(x) = P(F(x)),
$$

where P is a known continuous operator. Since $F'(x^*) = P(F(x^*)) = P(0)$, we can apply the results without actually knowing the solution x^*. Let as an example $F(x) = e^x - 1$. Then we can choose $P(x) = x + 1$.

3. The radius r_A was shown in [4], [5] to be the convergence radius for Newton's method (23.1.3) under conditions (23.2.8) and (23.2.9). It follows from (23.2.6) and the definition of r_1 that the convergence radius r of the method (23.1.2) cannot be larger than the convergence radius r_A of the second order Newton's method (23.1.3). As already noted, r_A is at least as large as the convergence ball given by Rheinboldt [30], namely

$$
r_R = \tfrac{2}{3L}. \tag{23.2.23}
$$

In particular, for $L_0 < L$ we have that

$$
r_R < r_A
$$

and

$$
\frac{r_R}{r_A} \to \frac{1}{3} \quad as \quad \frac{L_0}{L} \to 0.
$$

That is, our convergence ball r_1 is at most three times larger than Rheinboldt's. The same value for r_R was given by Traub [31].

4. Notice that condition (23.2.4) implies condition (23.2.3), but it does not necessarily imply condition (23.2.1). Moreover, condition (23.2.1) can be dropped as follows: Define parameter r_3 by

$$
r_3 = g_1(r_2)r_2 \tag{23.2.24}
$$

and set

$$
r^* = \max\{r_2, r_3\}. \tag{23.2.25}
$$

Next we can present the following result.

Theorem 23.2.3. *Suppose that the hypotheses of previous Theorem 23.2.1 hold except for (23.2.1) and (23.2.26). Moreover, suppose that*

$$\bar{U}(x^*, r^*) \subseteq \mathcal{D}, \qquad (23.2.26)$$

where the radius r^* is defined by (23.2.25). Then the sequence $\{x_n\}$ generated by method (23.1.2) for $x_0 \in U(x^*, r_2) \setminus \{x^*\}$ is well defined, remains in $U(x^*, r_2)$ for each $n = 0, 1, 2, \ldots$, and converges to x^*. Moreover, the following estimates hold:

$$\|y_n - x^*\| \le g_1(\|x_n - x^*\|)\|x_n - x^*\| \qquad (23.2.27)$$

and

$$\|x_{n+1} - x^*\| \le g_2(\|x_n - x^*\|)\|x_n - x^*\| < \|x_n - x^*\| < r_2, \qquad (23.2.28)$$

where the radius r_2 and the "g" functions are given previously. Furthermore, if there exists $R \in [r_2, \frac{2}{L_0})$ such that $\bar{U}(x^*, R) \subseteq \mathcal{D}$, then the limit point x^* is the only solution of equation $F(x) = 0$ in $\bar{U}(x^*, R)$.

Notice that it is not necessary for the iterates y_n to be in $U(x^*, r_2)$. This way condition (23.2.1) is not needed. However, the iterates y_n must be in $\bar{U}(x^*, r^*) \subseteq \mathcal{D}$.

23.3 CONVERGENCE PLANES OF THE METHOD (23.1.2) APPLIED TO THE QUADRATIC POLYNOMIAL $p(z) = z^2 - 1$

In this section, using the tool developed in [26], we will show the convergence planes of the method (23.1.2) when it is applied to the polynomial $p(x) = x^2 - 4$ for different values of the three parameters a, b and c. In concrete, we are going to choose three different values of the starting point x_0 and then we are going to compare how the convergence planes change.

The convergence plane is obtained by associating each point of the plane with a value of two of the parameters. That is, the tool is based on taking the vertical axis as the value of one parameter and the horizontal axis as the value of other parameter, so every point in the plane represents a value for both parameters. Once The convergence plane has been computed it is easy to distinguish the pairs of parameters for which the iteration of the starting point is convergent to any of the roots. So this tool provides a global vision about what points converges and shows what are the best choices of the parameters to ensure the convergence of the greatest set of starting points.

A point is painted, after a maximum of 1000 iterations and with a tolerance of 10^{-6}, in white if the iteration of the method starting in x_0 chosen converges

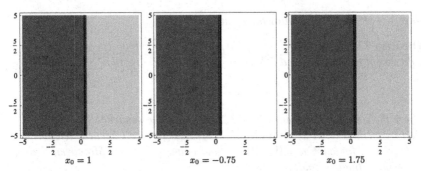

FIGURE 23.1   Convergence planes associated to the method (23.1.2) with $c = 0$ different starting points applied to the quadratic polynomial $p(z) = z^2 - 4$ in the region of the plane $(b, a) \in [-5, 5] \times [-5, 5]$.

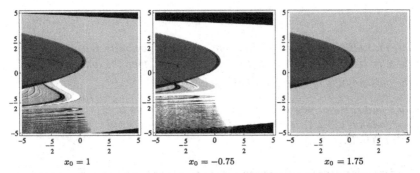

FIGURE 23.2   Convergence planes associated to the method (23.1.2) with $c = 1$ different starting points applied to the quadratic polynomial $p(z) = z^2 - 4$ in the region of the plane $(b, a) \in [-5, 5] \times [-5, 5]$.

to the root -2, in light gray if it converges to the root 2 and in dark gray if the iteration diverges to ∞. The regions in black correspond to zones of no convergence to any of the roots neither divergence to infinity. As a consequence, every point of the plane which is neither cyan nor magenta is not a good choice of the parameters for that starting point x_0 in terms of numerical behavior.

In Figs. 23.1–23.6 we see different convergence planes associated to the method (23.1.2) applied to the quadratic polynomial $p(z) = z^2 - 4$.

23.4   NUMERICAL EXAMPLES

Example 23.4.1. *Let $S = \mathbb{R}$, $D = [-1, 1]$, $x^* = 0$, and define function F on D by*

$$F(x) = e^x - 1. \qquad (23.4.1)$$

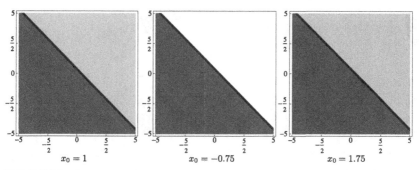

FIGURE 23.3 Convergence planes associated to the method (23.1.2) with $a = 0$ different starting points applied to the quadratic polynomial $p(z) = z^2 - 4$ in the region of the plane $(b, c) \in [-5, 5] \times [-5, 5]$.

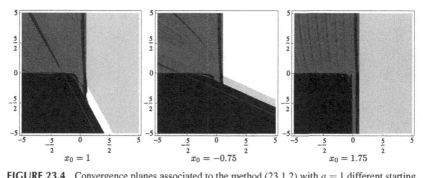

FIGURE 23.4 Convergence planes associated to the method (23.1.2) with $a = 1$ different starting points applied to the quadratic polynomial $p(z) = z^2 - 4$ in the region of the plane $(b, c) \in [-5, 5] \times [-5, 5]$.

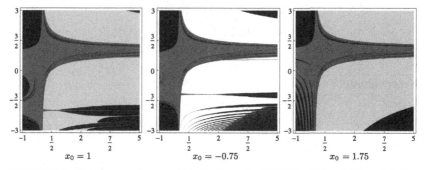

FIGURE 23.5 Convergence planes associated to the method (23.1.2) with $b = 0$ different starting points applied to the quadratic polynomial $p(z) = z^2 - 4$ in the region of the plane $(c, a) \in [-1, 5] \times [-3, 3]$.

Then, choosing

$$a = 1.25,$$
$$b = 0.96,$$

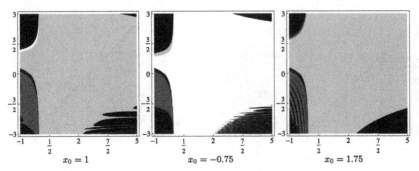

FIGURE 23.6 Convergence planes associated to the method (23.1.2) with $b = 1$ different starting points applied to the quadratic polynomial $p(z) = z^2 - 4$ in the region of the plane $(c, a) \in [-1, 5] \times [-3, 3]$.

and

$$c = 0.01,$$

we get

$$L_0 = e - 1,$$
$$L = e,$$

and

$$M = e.$$

Then, by the definition of the "g" functions, we obtain

$$r_1 = 0.104123\ldots,$$
$$r_2 = 0.005473\ldots,$$

and as a consequence

$$r = r_2 = 0.005473\ldots$$

So we can ensure the convergence of the method 23.1.2 with $a = 1.25$, $b = 0.96$, and $c = 0.01$.

REFERENCES

[1] S. Amat, S. Busquier, S. Plaza, Dynamics of the King and Jarratt iterations, Aequationes Math. 69 (3) (2005) 212–223.
[2] S. Amat, S. Busquier, S. Plaza, Chaotic dynamics of a third-order Newton-type method, J. Math. Anal. Appl. 366 (1) (2010) 24–32.
[3] S. Amat, M.A. Hernández, N. Romero, A modified Chebyshev's iterative method with at least sixth order of convergence, Appl. Math. Comput. 206 (1) (2008) 164–174.
[4] I.K. Argyros, Convergence and Application of Newton-Type Iterations, Springer, 2008.

[5] I.K. Argyros, S. Hilout, Numerical Methods in Nonlinear Analysis, World Scientific Publ. Comp., New Jersey, 2013.

[6] D.D. Bruns, J.E. Bailey, Nonlinear feedback control for operating a nonisothermal CSTR near an unstable steady state, Chem. Eng. Sci. 32 (1977) 257–264.

[7] D. Budzko, A. Cordero, J.R. Torregrosa, A new family of iterative methods widening areas of convergence, Appl. Math. Comput. 252 (2015) 405–417.

[8] V. Candela, A. Marquina, Recurrence relations for rational cubic methods I: the Halley method, Computing 44 (1990) 169–184.

[9] V. Candela, A. Marquina, Recurrence relations for rational cubic methods II: the Chebyshev method, Computing 45 (4) (1990) 355–367.

[10] F. Chicharro, A. Cordero, J.R. Torregrosa, Drawing dynamical and parameters planes of iterative families and methods, Sci. World J. (2013) 780153.

[11] C. Chun, Some improvements of Jarratt's method with sixth-order convergence, Appl. Math. Comput. 190 (2) (1990) 1432–1437.

[12] A. Cordero, J. García-Maimó, J.R. Torregrosa, M.P. Vassileva, P. Vindel, Chaos in King's iterative family, Appl. Math. Lett. 26 (2013) 842–848.

[13] A. Cordero, J.R. Torregrosa, P. Vindel, Dynamics of a family of Chebyshev–Halley type methods, Appl. Math. Comput. 219 (2013) 8568–8583.

[14] A. Cordero, J.R. Torregrosa, Variants of Newton's method using fifth-order quadrature formulas, Appl. Math. Comput. 190 (2007) 686–698.

[15] J.A. Ezquerro, M.A. Hernández, Recurrence relations for Chebyshev-type methods, Appl. Math. Optim. 41 (2) (2000) 227–236.

[16] J.A. Ezquerro, M.A. Hernández, New iterations of R-order four with reduced computational cost, BIT Numer. Math. 49 (2009) 325–342.

[17] J.A. Ezquerro, M.A. Hernández, On the R-order of the Halley method, J. Math. Anal. Appl. 303 (2005) 591–601.

[18] J.M. Gutiérrez, M.A. Hernández, Recurrence relations for the super-Halley method, Computers Math. Applic. 36 (7) (1998) 1–8.

[19] M. Ganesh, M.C. Joshi, Numerical solvability of Hammerstein integral equations of mixed type, IMA J. Numer. Anal. 11 (1991) 21–31.

[20] M.A. Hernández, Chebyshev's approximation algorithms and applications, Computers Math. Applic. 41 (3–4) (2001) 433–455.

[21] M.A. Hernández, M.A. Salanova, Sufficient conditions for semilocal convergence of a fourth order multipoint iterative method for solving equations in Banach spaces, Southwest J. Pure Appl. Math. 1 (1999) 29–40.

[22] P. Jarratt, Some fourth order multipoint methods for solving equations, Math. Comput. 20 (95) (1966) 434–437.

[23] J. Kou, Y. Li, An improvement of the Jarratt method, Appl. Math. Comput. 189 (2007) 1816–1821.

[24] D. Li, P. Liu, J. Kou, An improvement of the Chebyshev–Halley methods free from second derivative, Appl. Math. Comput. 235 (2014) 221–225.

[25] Á.A. Magreñán, Different anomalies in a Jarratt family of iterative root-finding methods, Appl. Math. Comput. 233 (2014) 29–38.

[26] Á.A. Magreñán, A new tool to study real dynamics: the convergence plane, Appl. Math. Comput. 248 (2014) 215–224.

[27] M. Petković, B. Neta, L. Petković, J. Džunić, Multipoint Methods for Solving Nonlinear Equations, Elsevier, 2013.

[28] L.B. Rall, Computational Solution of Nonlinear Operator Equations, Robert E. Krieger, New York, 1979.

[29] H. Ren, Q. Wu, W. Bi, New variants of Jarratt method with sixth-order convergence, Numer. Algorithms 52 (4) (2009) 585–603.

[30] W.C. Rheinboldt, An adaptive continuation process for solving systems of nonlinear equations, in: Banach Ctr. Publ., vol. 3, Polish Academy of Science, 1978, pp. 129–142.

[31] J.F. Traub, Iterative Methods for the Solution of Equations, Prentice-Hall Ser. Automat. Comput., Prentice-Hall, Englewood Cliffs, NJ, 1964.

[32] X. Wang, J. Kou, C. Gu, Semilocal convergence of a sixth-order Jarratt method in Banach spaces, Numer. Algorithms 57 (2011) 441–456.

[33] J. Kou, X. Wang, Semilocal convergence of a modified multi-point Jarratt method in Banach spaces under general continuity conditions, Numer. Algorithms 60 (2012) 369–390.

Chapter 24

Convergence and dynamics of a higher order family of iterative methods

24.1 INTRODUCTION

In this chapter we are concerned with the problem of approximating a locally unique solution x^* of equation

$$F(x) = 0, \qquad (24.1.1)$$

where F is a nonlinear operator defined on a convex subset D of a normed space X with values in a normed space Y.

A vast number of problems from applied sciences can be solved by means of finding the solutions equations like (24.1.1) using mathematical modeling [3, 31].

The convergence analysis of iterative methods is usually divided into two categories: semilocal and local convergence analysis. The semilocal convergence matter is, based on the information around an initial point, to give criteria ensuring the convergence of iteration procedures. A very important problem in the study of iterative procedures is the convergence domain. In general the convergence domain is small. Therefore, it is important to enlarge the convergence domain without additional hypothesis. Another important problem is to find more precise error estimates on the distances $\|x_{n+1} - x_n\|$ and $\|x_n - x^*\|$. These are our study objectives in this chapter regarding the dynamical behavior.

The dynamical properties related to an iterative method applied to polynomials give important information about its stability and reliability. In recently studies, recognized authors [1–4,18–21,23,31,32] have found interesting dynamical planes, including periodical behavior and others anomalies.

In this chapter, we consider the following optimal fourth-order family of methods presented in [13]:

$$v_n = x_n - \frac{2}{3} \frac{F(x_n)}{F'(x_n)},$$

$$x_{n+1} = x_n - \frac{[(\beta^2 - 22\beta - 27)F'(x_n) + 3(\beta^2 + 10\beta + 5)F'(v_n)]F(x_n)}{2(\beta F'(x_n) + 3F'(v_n))(3(\beta + 1)F'(v_n) - (\beta + 5)F'(x_n))},$$

$$(24.1.2)$$

A Contemporary Study of Iterative Methods. DOI: 10.1016/B978-0-12-809214-9.00024-3

where β is a complex parameter. Notice that every member of this family is an optimal fourth-order iterative method. In this chapter, the dynamics of this family applied to an arbitrary quadratic polynomial $p(z) = (z - a)(z - b)$ will be analyzed, characterizing the stability of all the fixed points. The graphic tool used to obtain the parameter space and the different dynamical planes have been introduced by Magreñán in [28,29].

The rest of the chapter is organized as follows: in Section 24.2 the study of the semilocal convergence is presented, in Section 24.3 the local convergence is studied, and in Section 24.4 some of the basic dynamical concepts related to the complex plane are presented, the stability of the fixed points of the family and the dynamical behavior of the family is analyzed, where the parameter space and some selected dynamical planes are presented.

24.2 SEMILOCAL CONVERGENCE

The semilocal convergence analysis of method (24.1.2) is presented in this section by treating it as a Newton-like method [7,33]. There is a plethora of semilocal convergence results for Newton-like methods under various Lipschitz-type conditions. In this section, we present only one such result. We refer the reader to [7] and the references therein for other results, where the idea of this section can also be used to generate semilocal convergence results for the method (24.1.2).

In this section, we assume that X and Y are Banach spaces. Denote by $U(w, \rho)$, $\bar{U}(w, \rho)$ the open and closed balls in X, respectively, with center $w \in X$ and radius $\rho > 0$.

Next we present a general semilocal convergence results for approximating a locally unique solution x^* of a nonlinear equation (24.1.1).

Theorem 24.2.1. *[7,33] Let $F : D \subset X \to Y$ be Fréchet-differentiable and let $A(x) \in L(X, Y)$ be an approximation of $F'(x)$. Suppose that there exist an open convex subset D_0 of D, $x_0 \in D_0$, $A(x_0)^{-1} \in L(Y, X)$, and constants $\eta, K > 0$, $L, \mu, l \geq 0$ such that for each $x, y \in D_0$ the following conditions hold:*

- $\|A(x_0)^{-1} F(x_0)\| \leq \eta$,
- $\|A(x_0)^{-1} \left(F'(x) - F'(y) \right)\| \leq K \|x - y\|$,
- $\|A(x_0)^{-1} \left(F'(x) - F'(x_0) \right)\| \leq M \|x - x_0\| + \mu$,
- $\|A(x_0)^{-1} \left(A(x) - A(x_0) \right)\| \leq L \|x - x_0\| + l$,
- $a = \mu + l < 1$,
- $h = \sigma \eta \leq \frac{1}{2}(1 - a)^2$, $\quad \sigma = \max\{K, M + L\}$,
  and
- $\bar{S} = \bar{U}(x_0, t^*) \subseteq D_0$, *where*

$$t^* = \frac{1 - a - \sqrt{(1 - a)^2 - 2h}}{\sigma}.$$

Then sequence $\{x_n\}$ generated by Newton-like method

$$x_{n+1} = x_n - A(x_n)^{-1} F(x_n) \quad \text{for each } n = 0, 1, 2, \ldots, \tag{24.2.1}$$

where x_0 is an initial point, is well defined, remains in $\bar{S}$ for each $n = 0, 1, 2, \ldots$, and converges to a solution $x^* \in \bar{S}$ of equation $F(x) = 0$. Moreover, the equation $F(x) = 0$ has a unique solution x^* in $\bar{U}(x_0, t^*) \cap D_0$ if $h = \frac{1}{2}(1 - a)^2$, or in $\bar{U}(x_0, t^{**}) \cap D_0$ if $h < \frac{1}{2}(1 - a)^2$, where

$$t^{**} = \frac{1 - a + \sqrt{(1 - a)^2 - 2h}}{\sigma}.$$

Furthermore, for $\phi(t) = \frac{\sigma}{2}t^2 - (1 - a)t + \eta$ and $\psi(t) = 1 - Lt - l$, the following estimates hold for each $n = 0, 1, 2, \ldots$:

$$\|x_{n+1} - x_n\| \leq t_{n+1} - t_n$$

and

$$\|x_n - x^*\| \leq t_n - t^*,$$

where the scalar sequence $\{t_n\}$ is defined by

$$t_0 = 0, \quad t_{n+1} = t_n + \frac{\phi(t_n)}{\psi(t_n)}.$$

Now define map Λ by

$$\Lambda(x) = B(x)^{-1} C(x),$$

where

$$B(x) = (\beta^2 - 22\beta - 27)F'(x) + 3(\beta^2 + 10\beta + 5)F'(x - \frac{2}{3}F'(x)^{-1}F(x))$$

and

$$C(x) = 2(\beta F'(x) + 3F'(x - \frac{2}{3}F'(x)^{-1}F(x))$$
$$\times (3(\beta + 1)F'(x - \frac{2}{3}F'(x)^{-1}F(x)) - (\beta + 5)F'(x)).$$

Then, with the above special choice of operator Λ, we obtain the following two corollaries. Notice that the second corollary in particular involves the convergence of method (24.1.2).

Corollary 24.2.2. *Suppose that the hypotheses of Theorem 24.2.1 hold with the mapping Λ replacing A. Then the conclusions of Theorem 24.2.1 for equation (24.1.1) and Newton-like method (24.2.1) hold with mapping Λ replacing A.*

Next the preceding corollary specializes even further in the case of method (24.1.2) when $X = Y = \mathbb{R}$ or $X = Y = \mathbb{C}$.

Corollary 24.2.3. *Suppose that the hypotheses of Corollary 24.2.2 hold and $X = Y = \mathbb{R}$ or $X = Y = \mathbb{C}$. Then the conclusions of Corollary 24.2.2 for equation (24.1.1) and the method (24.1.2) hold.*

Now we can present the following remark.

Remark 24.2.4. *The local convergence analysis of methods (24.2.1) and (24.1.2) is analogously obtained in the next section.*

24.3 LOCAL CONVERGENCE

We assume again that X and Y are Banach spaces. Notice that if the iterates in (24.2.1) exist and $x^* \in D$ is such that $F(x^*) = 0$, then we get the identity

$$x^* - x_{n+1} = A(x_n)^{-1} A(x^*)$$

$$\times \left[\int_0^1 A(x^*)^{-1} \left(F'(x_n + \theta(x^* - x_n)) - F'(x_n) \right) + A(x^*)^{-1} \left(F'(x_n) - A(x) \right) \right]$$

$$\times (x^* - x_n).$$

Using the preceding idea we will present the following result.

Theorem 24.3.1. *Let $F : D \subset X \to Y$ be Fréchet-differentiable and let $A(x) \in L(X, Y)$ be an approximation of $F'(x)$. Suppose that there exist an open convex subset D_0 of D, $x^* \in D_0$, $A(x^*)^{-1} \in L(Y, X)$, and constants $K > 0$, $M, L, \mu, l \geq 0$ such that for all $x, y \in D_0$ the following conditions hold:*

- $F(x^*) = 0$,
- $\| A(x^*)^{-1} \left(F'(x) - F'(y) \right) \| \leq K \| x - y \|$,
- $\| A(x^*)^{-1} \left(F'(x) - A(x) \right) \| \leq M \| x - x^* \| + \mu$,
- $\| A(x^*)^{-1} \left(A(x) - A(x^*) \right) \| \leq L \| x - x^* \| + l$,
- $a = \mu + l < 1$, *and*
- $\bar{U}(x^*, r) \subseteq D_0$,

  where

$$r = \frac{1 - (\mu + l)}{\frac{K}{2} + M + L}.$$

Then sequence $\{x_n\}$ generated by Newton-like method (24.2.1) for $x_0 \in U(x^*, r)$ is well defined, remains in $\bar{U}(x^*, r)$, and converges to x^*. Moreover, the following estimates hold for each $n = 0, 1, 2, \ldots$:

$$\| x_{n+1} - x^* \| \leq e_n \| x_n - x^* \| t < r, \tag{24.3.1}$$

where

$$e_n = \frac{\left(\frac{K}{2} + M\right) \|x_n - x^*\| + \mu}{1 - L\|x_n - x^*\| - l}.$$

Proof. By hypothesis, $x_0 \in U(x^*, r)$. Using the fourth hypothesis, we get

$$\|A(x^*)^{-1}\left(A(x_0) - A(x^*)\right)\| \le L\|x_0 - x^*\| + l < Lr + l < 1.$$

It follows from the preceding inequality and the Banach lemma on invertible operators [33] that $A(x_0)^{-1} \in L(Y, X)$ and

$$\|A(x_0)^{-1}A(x^*)\| \le \frac{1}{1 - L\|x_0 - x^*\| - l} < \frac{1}{1 - Lr - l}.$$

Consequently, x_1 exists. Then, we have by the inequality above, Theorem 24.3.1 for $n = 0$, and the preceding hypotheses

$$
\begin{aligned}
\|x_1 - x^*\| &\le \|A(x_0)^{-1}A(x^*)\| \left[\| \int_0^1 A(x^*)^{-1}[F'(x_0 + \theta(x^* - x_0)) \right. \\
&\qquad \left. - F'(x_0)d\theta \|] + \|A(x^*)^{-1}(F'(x_0) - A(x_0))\| \right] \|x_0 - x^*\| \\
&\le \frac{\left(\frac{K}{2} + M\right)\|x_0 - x^*\| + \mu}{1 - L\|x_0 - x^*\| - l} \|x_0 - x^*\| \\
&= e_0 \|x_0 - x^*\| \\
&< \frac{\left(\frac{K}{2} + M\right)r + \mu}{1 - Lr - l} \|x_0 - x^*\| \\
&= \|x_0 - x^*\| \\
&< r,
\end{aligned}
$$

which shows (24.3.1) for $n = 0$ and $x_1 \in U(x^*, r)$. Continuing in an analogous way, by simply replacing x_0, x_1 with x_k, x_{k+1} in the preceding estimates, we arrive at (24.3.1). Furthermore, by the estimate

$$\|x_{k+1} - x^*\| < \|x_k - x^*\|,$$

we deduce that $\lim_{k \to \infty} x_k = x^*$. $\qquad\square$

Now we present the following corollaries and remarks.

Corollary 24.3.2. *Suppose that the hypotheses of Theorem 24.3.1 hold with the mapping Λ replacing A. Then the conclusions of Theorem 24.3.1 for equation (24.1.1) and Newton-like method (24.2.1) hold with mapping Λ replacing A.*

Next the preceding corollary specializes even further in the case of method (24.1.2) when $X = Y = \mathbb{R}$ or $X = Y = \mathbb{C}$.

Corollary 24.3.3. *Suppose that the hypotheses of Corollary 24.3.2 hold and $X = Y = \mathbb{R}$ or $X = Y = \mathbb{C}$. Then the conclusions of Corollary 24.3.2 for equation (24.1.1) and the method (24.1.2) hold.*

Remark 24.3.4. *It is worth noticing that the local convergence analysis presented here is more general and is given under weaker hypotheses on the mappings involved than the corresponding ones given in [13].*

Remark 24.3.5. *The conditions of Theorem 24.3.1 can easily be realized in some interesting cases without knowing the solution x^*. Suppose that mapping F satisfies the autonomous differential equation [33]*

$$F'(x) = P(F(X)) \quad \text{for each } x \in D,$$

where P is a known continuous operator. Then, since $F'(x^*) = P(F(x^*)) = P(0)$ is known, operator $\Lambda(x^*)$ reduces to

$$\Lambda(x^*) = B_*(x^*)^{-1}C_*(x^*),$$

where

$$
\begin{aligned}
B_*(x^*) &= (\beta^2 - 22\beta - 27)P(0) + 3(\beta^2 + 10\beta + 5)P(0) \\
&= 4(\beta^2 + 2\beta - 3)P(0) \\
&= 4(\beta + 3)(\beta - 1)P(0)
\end{aligned}
$$

and

$$
\begin{aligned}
C_*(x) &= 2(\beta P(0) + 3P(0)[3(\beta + 1)P(0) - (\beta + 5)P(0)] \\
&= 4(\beta + 3)(\beta - 1)P^2(0).
\end{aligned}
$$

Hence we get that

$$\Lambda(x^*) = P(0).$$

As an academic example, let us define mapping F on D by $F(x) = e^x - 1$. Then we can set $P(x) = x + 1$ and $\Lambda(x^*) = P(0) = 1$.

In the rest of the chapter we choose $X = Y = \mathbb{C}$.

24.4 DYNAMICAL STUDY

In this Section, some dynamical concepts of complex dynamics that will be used in this chapter are shown (we refer the reader to see [5,6,8–12,15–17,22,24–27, 30]). First of all, in order to simplify the study we consider the conjugacy map given by P. Blanchard, in [14]

$$M(z) = \frac{z - a}{z - b}, \qquad M^{-1}(z) = \frac{zb - a}{z - 1}, \qquad (24.4.1)$$

which obviates the roots and we obtain that the family (24.1.2) is conjugated to

$$G(z, \beta) = \frac{z^4 \left(-11 - 3z - 6\beta + 2z\beta + \beta^2 + z\beta^2\right)}{-3 - 11z + 2\beta - 6z\beta + \beta^2 + z\beta^2} \tag{24.4.2}$$

As the family of iterative methods (24.4.2) applied to polynomials gives a rational function. The focus will be centered on them. Given a rational function $R : \hat{\mathbb{C}} \to \hat{\mathbb{C}}$, where $\hat{\mathbb{C}}$ is the Riemann sphere, the *orbit of a point* $z_0 \in \hat{\mathbb{C}}$ is defined by

$$\{z_0, R(z_0), R^2(z_0), ..., R^n(z_0), ...\}.$$

A point $z_0 \in \bar{C}$, is called a *fixed point* of $R(z)$ if it verifies that $R(z) = z$. Moreover, z_0 is called a *periodic point* of period $p > 1$ if it is a point such that $R^p(z_0) = z_0$ but $R^k(z_0) \neq z_0$, for each $k < p$. Moreover, a point z_0 is called *pre-periodic* if it is not periodic but there exists a $k > 0$ such that $R^k(z_0)$ is periodic.

There exist different types of fixed points depending on its associated multiplier $|R'(z_0)|$. Taking the associated multiplier into account a fixed point z_0 is called:

- *superattractor* if $|R'(z_0)| = 0$
- *attractor* if $|R'(z_0)| < 1$
- *repulsor* if $|R'(z_0)| > 1$
- and *parabolic* if $|R'(z_0)| = 1$.

The fixed points that do not correspond to the roots of the polynomial $p(z)$ are called *strange fixed points*. On the other hand, a *critical point* z_0 is a point which satisfies that, $R'(z_0) = 0$.

The basin of attraction of an attractor α is defined as

$$\mathcal{A}(\alpha) = \{z_0 \in \hat{\mathbb{C}} : R^n(z_0) \to \alpha, \ n \to \infty\}.$$

The *Fatou set* of the rational function R, $\mathcal{F}(R)$, is the set of points $z \in \hat{\mathbb{C}}$ whose orbits tend to an attractor (fixed point, periodic orbit or infinity). Its complement in $\hat{\mathbb{C}}$ is the *Julia set, $\mathcal{J}(R)$*. That means that the basin of attraction of any fixed point belongs to the Fatou set and the boundaries of these basins of attraction belong to the Julia set.

It is clear that $z = 0$ and $z = \infty$ are fixed points (related to the roots a and b, respectively, of the polynomial $p(z)$). On the other hand, we see that $z = 1$ is a strange fixed point, which is associated with the original convergence to infinity. Moreover, there are also another two strange fixed points whose analytical expression, depending on β, are:

$$ex_1(\beta) = \frac{11 + 6\beta - \beta^2 - \sqrt{85 + 180\beta + 22\beta^2 - 28\beta^3 - 3\beta^4}}{2\left(-3 + 2\beta + \beta^2\right)}$$

and

$$ex_2(\beta) = \frac{11 + 6\beta - \beta^2 + \sqrt{85 + 180\beta + 22\beta^2 - 28\beta^3 - 3\beta^4}}{2\left(-3 + 2\beta + \beta^2\right)}.$$

There relations between the strange fixed points are described in the following result.

Lemma 24.4.1. *The number of simple strange fixed points of operator $G(z, \beta)$ is three, except in the following cases:*

i) *If $\beta = -5 - 2\sqrt{5}$ or $\beta = -5 + 2\sqrt{5}$, then $ex_1(\beta) = ex_2(\beta)$ and so there only exist two strange fixed points.*

ii) *If $\beta = \frac{1}{3}\left(1 - 2\sqrt{13}\right)$ or $\beta = \frac{1}{3}\left(1 + 2\sqrt{13}\right)$, then $ex_1(\beta) = ex_2(\beta) = 1$, and as a consequence the family only presents one strange fixed point.*

Moreover, for all values of the parameter β, $ex_1(\beta) = \dfrac{1}{ex_2(\beta)}$.

Related to the stability of that strange fixed points, the first derivative of $G(z, \beta)$ must be calculated.

$$
\begin{aligned}
G'(z, \beta) &= 4z^3 \frac{(33 - 4\beta - 26\beta^2 - 4\beta^3 + \beta^4)z^2}{\left(-3 + \beta(2 - 6z) - 11z + \beta^2(1 + z)\right)^2} \\
&+ 4z^3 \frac{(102 + 84\beta + 8\beta^2 - 4\beta^3 + 2\beta^4)z}{\left(-3 + \beta(2 - 6z) - 11z + \beta^2(1 + z)\right)^2} \\
&+ 4z^3 \frac{33 - 4\beta - 26\beta^2 - 4\beta^3 + \beta^4}{\left(-3 + \beta(2 - 6z) - 11z + \beta^2(1 + z)\right)^2}.
\end{aligned}
$$

Taking into account the form of the derivative it is immediate that the origin and ∞ are superattractive fixed points for every value of β. The stability of the other fixed points is not as trivial as 0 and ∞ and we will study it in a separate way.

First of all, focussing the attention in the strange fixed point $z = 1$, which is related to the convergence of ∞, the following result is shown.

Theorem 24.4.2. *The character of the strange fixed point $z = 1$ is:*

i) *Superattractor if $\beta = 2$ or $\beta = -3$.*

ii) *Attractor if β is inside one of the two ovals that appear in Fig. 24.1.*

iii) *Repulsor if β is outside that ovals.*

Notice that

$$G'(1, \beta) = \frac{4(-3 + \beta)(2 + \beta)}{-7 - 2\beta + \beta^2}.$$

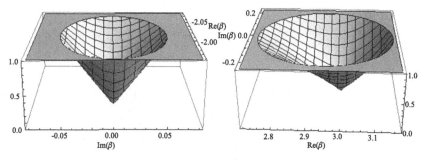

FIGURE 24.1   Stability regions of the strange fixed point $z = 1$.

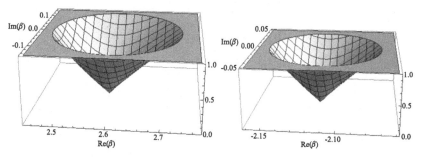

FIGURE 24.2   Stability regions of $ex_1(\beta)$ (and $ex_2(\beta)$).

Then,

$$4(-3 + \beta)(2 + \beta) = 0,$$

if and only if $\beta = -2$ or $\beta = 3$.

In Fig. 24.1 it can be seen the stability regions of $z = 1$, which are two ovals in the complex plane. Notice that the stability of $ex_1(\beta)$ and $ex_2(\beta)$, is the same. The multiplier associated to each one is:

$$G'(ex_1) = G'(ex_2) = \frac{4\left(-19 + \beta(8 + \beta)\left(-6 + \beta^2\right)\right)}{\left(-3 + 2\beta + \beta^2\right)^2}.$$

In Fig. 24.2 the regions of stability of $ex_1(\beta)$ (and $ex_2(\beta)$) is shown. Taking into account this regions the following result summarize the behavior of the strange fixed points.

Theorem 24.4.3. *The stability of $ex_1(\beta)$ and $ex_2(\beta)$ is the same. In particular,*

- *Both points are super-attractors if $\beta = -2 + \sqrt{5} \pm \sqrt{2\left(5 - \sqrt{5}\right)}$ or $\beta = -2 + \sqrt{5} \pm \sqrt{2\left(5 - \sqrt{5}\right)}$.*

- *They are attractor if β is inside one of the two ovals that appear in Fig. 24.1.*

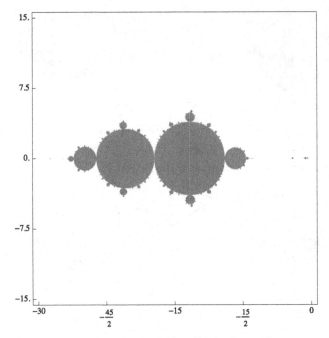

FIGURE 24.3 Parameter space associated to the free critical point $cr_1(\beta)$.

As a conclusion we have noticed that the number and the stability of the fixed points depend on the parameter, β, of the family.

Now, the critical points will be calculated and the parameter spaces associated to the free critical points will be shown. It is well known that there is at least one critical point associated with each invariant Fatou component. The critical points of the family are the solutions of $G'(z, \beta) = 0$. Solving the equation, it is clear that $z = 0$ and $z = \infty$ are critical points, which are related to the roots of the polynomial $p(z)$ and they have associated their own Fatou component. Moreover, there exists critical point no related to the roots, these points are called free critical points. The appearance of free critical points is summarized in the following result.

Lemma 24.4.4. *The free critical points are:*

a) *if $\beta = -2$, $\beta = -1$, $\beta = 3$, $\beta = 1 - 2\sqrt{2}$, $\beta = 1 + 2\sqrt{2}$, $\beta = -7 - 2\sqrt{10}$ or $\beta = -7 + 2\sqrt{10}$,*

$$cr_1(\beta) = \frac{-51 - 42\beta - 4\beta^2 + 2\beta^3 - \beta^4}{33 - 4\beta - 26\beta^2 - 4\beta^3 + \beta^4}$$

$$-2\frac{\sqrt{378 + 1137\beta + 968\beta^2 + 47\beta^3 - 206\beta^4 - 33\beta^5 + 12\beta^6 + \beta^7}}{33 - 4\beta - 26\beta^2 - 4\beta^3 + \beta^4}$$

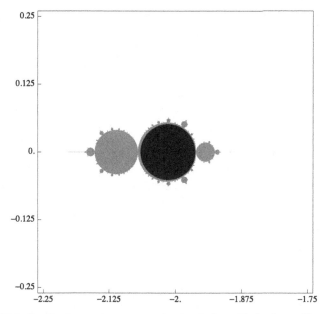

FIGURE 24.4 Details of parameter space associated to the free critical point $cr_1(\beta)$.

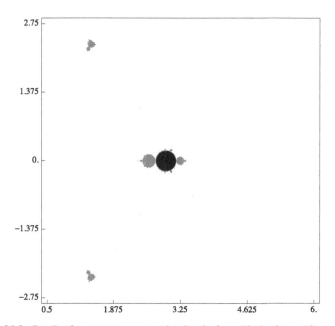

FIGURE 24.5 Details of parameter space associated to the free critical point $cr_1(\beta)$.

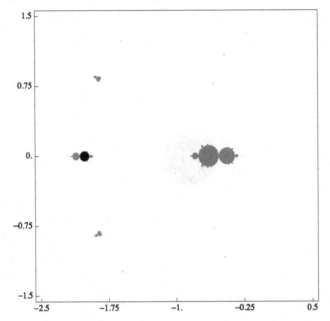

FIGURE 24.6 Details of parameter space associated to the free critical point $cr_1(\beta)$.

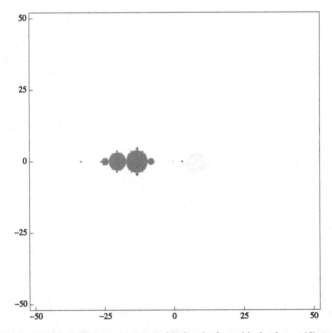

FIGURE 24.7 Details of parameter space associated to the free critical point $cr_1(\beta)$.

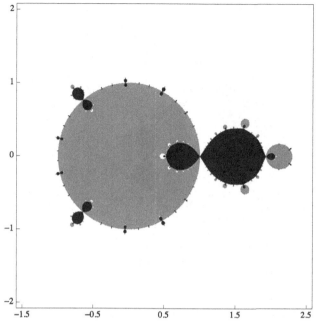

FIGURE 24.8 Dynamical planes for $\beta = -2.1$ in which there exists convergence to the strange fixed points.

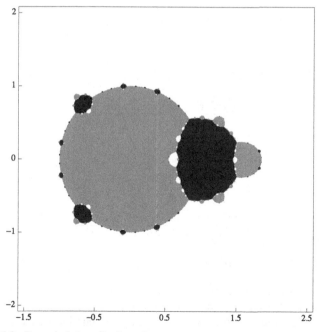

FIGURE 24.9 Dynamical planes for $\beta = -2$.

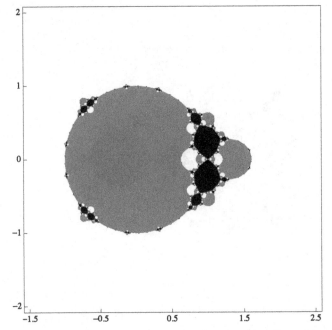

FIGURE 24.10 Dynamical planes for $\beta = -1.94$ in which there exist regions of convergence to 2-cycles.

b) *In other case,*

$$cr_1(\beta) \;=\; \frac{-51-42\beta-4\beta^2+2\beta^3-\beta^4}{33-4\beta-26\beta^2-4\beta^3+\beta^4}$$

$$-2\frac{\sqrt{378+1137\beta+968\beta^2+47\beta^3-206\beta^4-33\beta^5+12\beta^6+\beta^7}}{33-4\beta-26\beta^2-4\beta^3+\beta^4}$$

and

$$cr_2(\beta) \;=\; \frac{-51-42\beta-4\beta^2+2\beta^3-\beta^4}{33-4\beta-26\beta^2-4\beta^3+\beta^4}$$

$$+2\frac{\sqrt{378+1137\beta+968\beta^2+47\beta^3-206\beta^4-33\beta^5+12\beta^6+\beta^7}}{33-4\beta-26\beta^2-4\beta^3+\beta^4}$$

are free critical points. Moreover, $cr_1(\beta) = \frac{1}{cr_2(\beta)}$.

So, there is only one independent free critical point. Without loss of generality, in this chapter the free critical point considered is $cr_1(\beta)$. In order to find the members of the family with anomalies in terms of stability the parameter spaces will be shown.

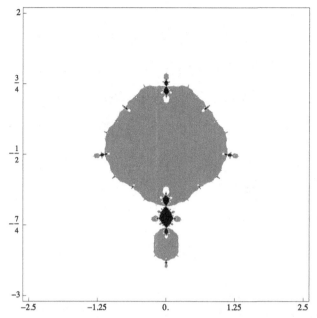

FIGURE 24.11 Dynamical planes for $\beta = 1.35 + 2.23$ in which there exist regions of convergence to 2-cycles.

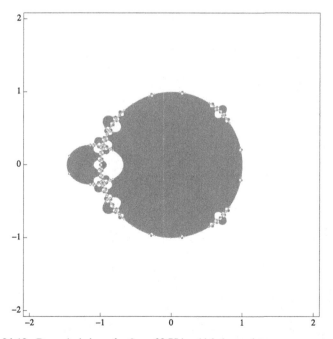

FIGURE 24.12 Dynamical planes for $\beta = -28.75$ in which there exists convergence to cycles.

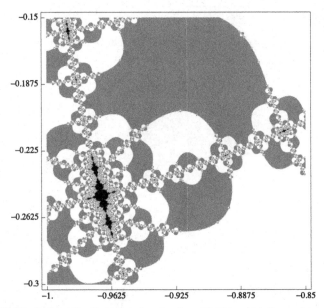

FIGURE 24.13 Zoom of the dynamical plane for $\beta = -28.75$ in which there exists convergence to cycles.

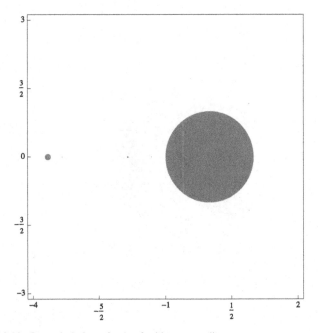

FIGURE 24.14 Dynamical planes for $\beta = 0$ without anomalies.

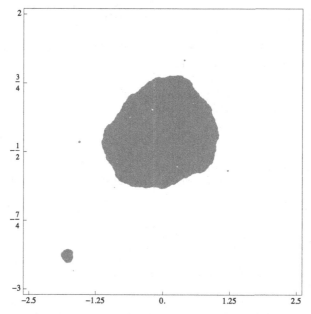

FIGURE 24.15 Dynamical planes for $\beta = i$ without anomalies.

In Fig. 24.3, the parameter space associated to $cr_1(\beta)$ is shown. A point is painted in white if the iteration of the method starting in $z_0 = cr_1(\beta)$ converges to the fixed point 0 (related to root a) or if it converges to ∞ (related to root b) and in black if the iteration converges to 1 (related to ∞). Moreover, the regions in grey correspond to zones of convergence to cycles or other "bad" behavior. As a consequence, every point of the plane which is not white is not a good choice of β in terms of numerical behavior.

In Fig. 24.4 a detail of parameter space near to the region where the main convergence problem occurs is shown. It can be seen that in this region there exist values of β for which the iteration of $cr_1(\beta)$ converges to cycles or even to strange fixed points. On the other hand, in Figs. 24.5–24.7, other regions with no convergence to the roots of the polynomial are shown.

Once the anomalies have been detected the next step consist on finding them in the dynamical planes. In these dynamical planes the convergence to 0 will appear in grey, in white it appears the convergence to ∞ and in black the zones with no convergence to the roots. First of all, in Fig. 24.8 the dynamical planes associated with the values of β for which there exists convergence to any of the strange fixed points are shown.

In Fig. 24.9 appear the dynamical planes associated with the values of β for which there exists convergence to $z = 1$, which is related with the original convergence to infinity.

In Figs. 24.10–24.11 the dynamical planes of some members which has convergence to 2-cycles are shown.

On the other hand, related to the values of β for which other attracting cycles appear can be seen in black in Figs. 24.12–24.13.

Finally, in Figs. 24.14–24.15 some dynamical planes of members with convergence to the roots without any anomaly are shown.

REFERENCES

[1] S. Amat, S. Busquier, S. Plaza, Review of some iterative root-finding methods from a dynamical point of view, Sci. Ser. A Math. Sci. 10 (2004) 3–35.

[2] S. Amat, S. Busquier, S. Plaza, Dynamics of the King and Jarratt iterations, Aequationes Math. 69 (3) (2005) 212–223.

[3] S. Amat, S. Busquier, S. Plaza, A construction of attracting periodic orbits for some classical third-order iterative methods, J. Comput. Appl. Math. 189 (2006) 22–33.

[4] S. Amat, S. Busquier, S. Plaza, Chaotic dynamics of a third-order Newton-type method, J. Math. Anal. Appl. 366 (1) (2010) 24–32.

[5] S. Amat, I.K. Argyros, S. Busquier, Á.A. Magreñán, Local convergence and the dynamics of a two-point four parameter Jarratt-like method under weak conditions, Numer. Algorithms 74 (2) (2017) 371–391.

[6] S. Amat, S. Busquier, C. Bermúdez, Á.A. Magreñán, On the election of the damped parameter of a two-step relaxed Newton-type method, Nonlinear Dynam. 84 (1) (2016) 9–18.

[7] I.K. Argyros, Computational Theory of Iterative Methods, Elsevier Science, San Diego, USA, 2007.

[8] I.K. Argyros, A. Cordero, Á.A. Magreñán, J.R. Torregrosa, On the convergence of a higher order family of methods and its dynamics, J. Comput. Appl. Math. 309 (2017) 542–562.

[9] I.K. Argyros, Á.A. Magreñán, A study on the local convergence and the dynamics of Chebyshev–Halley-type methods free from second derivative, Numer. Algorithms 71 (1) (2016) 1–23.

[10] I.K. Argyros, Á.A. Magreñán, On the convergence of an optimal fourth-order family of methods and its dynamics, Appl. Math. Comput. 252 (2015) 336–346.

[11] I.K. Argyros, Á.A. Magreñán, L. Orcos, Local convergence and a chemical application of derivative free root finding methods with one parameter based on interpolation, J. Math. Chem. 54 (7) (2016) 1404–1416.

[12] I.K. Argyros, Á.A. Magreñán, Local convergence and the dynamics of a two-step Newton-like method, Internat. J. Bifur. Chaos. 26 (5) (2016) 1630012.

[13] R. Behl, Development and Analysis of Some New Iterative methods for Numerical Solutions of Nonlinear Equations, PhD thesis, Punjab University, 2013.

[14] P. Blanchard, The dynamics of Newton's method, in: Proc. Sympos. Appl. Math., vol. 49, 1994, pp. 139–154.

[15] A. Cordero, J.M. Gutiérrez, Á.A. Magreñán, J.R. Torregrosa, Stability analysis of a parametric family of iterative methods for solving nonlinear models, Appl. Math. Comput. 285 (2016) 26–40.

[16] A. Cordero, Á.A. Magreñán, C. Quemada, J.R. Torregrosa, Stability study of eighth-order iterative methods for solving nonlinear equations, J. Comput. Appl. Math. 291 (2016) 348–357.

[17] A. Cordero, L. Feng, Á.A. Magreñán, J.R. Torregrosa, A new fourth-order family for solving nonlinear problems and its dynamics, J. Math. Chem. 53 (3) (2014) 893–910.

[18] F. Chicharro, A. Cordero, J.M. Gutiérrez, J.R. Torregrosa, Complex dynamics of derivative-free methods for nonlinear equations, Appl. Math. Comput. 219 (2013) 7023–7035.

[19] C. Chun, M.Y. Lee, B. Neta, J. Džunić, On optimal fourth-order iterative methods free from second derivative and their dynamics, Appl. Math. Comput. 218 (2012) 6427–6438.

[20] A. Cordero, J. García-Maimó, J.R. Torregrosa, M.P. Vassileva, P. Vindel, Chaos in King's iterative family, Appl. Math. Lett. 26 (2013) 842–848.

[21] A. Cordero, J.R. Torregrosa, P. Vindel, Dynamics of a family of Chebyshev–Halley type methods, Appl. Math. Comput. 219 (2013) 8568–8583.

[22] Y.H. Geum, Y.I. Kim, Á.A. Magreñán, A biparametric extension of King's fourth-order methods and their dynamics, Appl. Math. Comput. 282 (2016) 254–275.

[23] J.M. Gutiérrez, M.A. Hernández, N. Romero, Dynamics of a new family of iterative processes for quadratic polynomials, J. Comput. Appl. Math. 233 (2010) 2688–2695.

[24] J.M. Gutiérrez, Á.A. Magreñán, N. Romero, Dynamic aspects of damped Newton's method, in: Civil-Comp Proceedings, vol. 100, 2012.

[25] J.M. Gutiérrez, Á.A. Magreñán, J.L. Varona, Fractal dimension of the universal Julia sets for the Chebyshev–Halley family of methods, in: AIP Conf. Proc., vol. 1389, 2011, pp. 1061–1064.

[26] J.M. Gutiérrez, Á.A. Magreñán, J.L. Varona, The "Gauss-seidelization" of iterative methods for solving nonlinear equations in the complex plane, Appl. Math. Comput. 218 (6) (2011) 2467–2479.

[27] T. Lotfi, Á.A. Magreñán, K. Mahdiani, J.J. Rainer, A variant of Steffensen–King's type family with accelerated sixth-order convergence and high efficiency index: dynamic study and approach, Appl. Math. Comput. 252 (2015) 347–353.

[28] Á.A. Magreñán, A new tool to study real dynamics: the convergence plane, Appl. Math. Comput. 248 (2014) 215–224.

[29] Á.A. Magreñán, Different anomalies in a Jarratt family of iterative root-finding methods, Appl. Math. Comput. 233 (2014) 29–38.

[30] Á.A. Magreñán, I.K. Argyros, On the local convergence and the dynamics of Chebyshev–Halley methods with six and eight order of convergence, J. Comput. Appl. Math. 298 (2016) 236–251.

[31] B. Neta, M. Scott, C. Chun, Basin attractors for various methods for multiple roots, Appl. Math. Comput. 218 (2012) 5043–5066.

[32] M. Scott, B. Neta, C. Chun, Basin attractors for various methods, Appl. Math. Comput. 218 (2011) 2584–2599.

[33] T. Yamamoto, A convergence theorem for Newton-like methods in Banach spaces, Numer. Math. 51 (1987) 545–557.

Chapter 25

Convergence of iterative methods for multiple zeros

25.1 INTRODUCTION

In this chapter we are concerned with the problem of approximating a locally unique solution x^* of equation

$$F(x) = 0, \qquad (25.1.1)$$

where F is a differentiable function defined on a convex subset D of S with values in S, where S is $\mathbb{R}$ or $\mathbb{C}$.

Many problems from applied sciences, including engineering, can be solved by means of finding the solutions of equations like (25.1.1) using mathematical modeling [5–13,36,40,45]. For example, dynamic systems are mathematically modeled by difference or differential equations, and their solutions usually represent the states of the systems. Except in special cases, the solutions of these equations cannot be found in closed form. This is the main reason why the most commonly used solution methods are iterative.

The dynamical properties related to an iterative method applied to polynomials give important information about its stability and reliability. In recently studies, recognized authors [2,3,14–16,20–35,37–39,44–50] have found interesting dynamical planes, including periodical behavior and others anomalies.

We study the local convergence of the three-step method defined for each $n = 0, 1, 2, \ldots$ by

$$
\begin{aligned}
y_n &= x_n - A_n F(x_n), \\
z_n &= y_n - B_n F(y_n), \qquad (25.1.2) \\
x_{n+1} &= z_n - C_n F(z_n),
\end{aligned}
$$

where x_0 is an initial point, $A_n = A(x_n)$, $B_n = B(x_n)$, $C = C(x_n)$, and $A, B, C : \mathbb{R} \to \mathbb{R}$ are continuous functions. Special choices of functions A, B, and C lead to well known methods such as

- **Two-step Newton's method.**

$$
\begin{aligned}
y_n &= x_n - F'(x_n)^{-1} F(x_n), \\
x_{n+1} &= x_n - F'(y_n)^{-1} F(y_n), \qquad \text{for each } n = 0, 1, 2, \ldots
\end{aligned}
$$

A Contemporary Study of Iterative Methods. DOI: 10.1016/B978-0-12-809214-9.00025-5

- **Corrected Newton's method.**

$$x_{n+1} = x_n - m F'(x_n)^{-1} F(x_n),$$

for $C_n = B_n = 0$ and $A_n = m F'(x_n)^{-1}$, where $m > 1$ is the multiplicity of the solution x^*. Notice that if $m = 1$ we obtain the classical Newton's method

$$x_{n+1} = x_n - F'(x_n)^{-1} F(x_n).$$

- **Method presented by Neta et al. in [41–43].**

$$
\begin{aligned}
y_n &= x_n - \frac{2m}{m+2} F'(x_n)^{-1} F(x_n), \\
x_{n+1} &= x_n - \phi(t_n) F'(x_n)^{-1} F(y_n),
\end{aligned}
\tag{25.1.3}
$$

where $t_n = F'(x_n)^{-1} F(y_n)$ and ϕ is an at least twice differentiable function satisfying the conditions

$$\phi(\lambda) = m, \quad \phi'(\lambda) = -\frac{1}{4} m^3 \left(\frac{m+2}{m} \right)^m,$$

$$\phi''(\lambda) = \frac{1}{4} m^4 \left(\frac{m+2}{m} \right)^{2m} \quad \text{for } \lambda = \left(\frac{m}{m+2} \right)^{m-1}.$$

The weight functions ϕ have been chosen in [35] as follows:

$$\phi(x) = \frac{D + Ex}{1 + Fx},$$

where

$$F = -\frac{m+2}{m\lambda}, \quad D = -\frac{m^2}{2}, \quad E = \frac{(m-2)(m+2)}{2\lambda}.$$

It was shown in [35] that this is a method equivalent to an optimal fourth order method given in [42,43]. In [42,43] the function

$$\psi(x) = \frac{b + cx + dx^2}{1 + ax + \mu x^2},$$

is used (see [17–19,42,43] for the values of b, c, d, a and μ).
- **Laguerre's method.** Choose $\alpha = 1$ and

$$\phi(x) = -1 + \frac{q}{1 + \text{sign}(q - m) \sqrt{\left(\dfrac{q - m}{m} \right) \left[(q - 1) - q \dfrac{F(x) F''(x)}{F'(x)^2} \right]}},$$

where $q \neq 0$, m is a real parameter. In particular, for special choices of q, we obtain $q = 2m$.

- **Osada's method.** Choose $\alpha = 1$ and

$$\phi(x) = -1 + \frac{1}{g}m(m+1) - \frac{(m-1)^2}{2}\frac{F'(x)^2}{F''(x)F(x)}.$$

- **Euler–Chebyshev method.** Choose $\alpha = 1$ and

$$\phi(x) = -1 + \frac{m(3-m)}{2} + \frac{m^2}{2}\frac{F(x)F''(x)}{F'(x)^2}.$$

- **Li et al. [35] fifth order method.**

$$y_n = x_n - \left(\frac{F(x_n + F(x_n)) - F(x_n)}{F(x_n)}\right)^{-1} F(x_n),$$

$$z_n = y_n - \left(\frac{F(x_n + F(x_n)) - F(x_n)}{F(x_n)}\right)^{-1} F(y_n),$$

$$x_{n+1} = z_n - (F[z_n, y_n] + F[z_n, x_n, x_n](z_n - y_n))^{-1} F(z_n),$$

where $F[\cdot, \cdot]$ and $F[\cdot, \cdot, \cdot]$ are the divided differences of F of order one and two, respectively.

- **Sharma et al. [47] fifth order method.**

$$y_n = x_n - \left(\frac{F(x_n + F(x_n)) - F(x_n)}{F(x_n)}\right)^{-1} F(x_n),$$

$$z_n = y_n - \left(\frac{F(x_n + F(x_n)) - F(x_n)}{F(x_n)}\right)^{-1} F(y_n),$$

$$x_{n+1} = z_n - \left(\frac{F[x_n, z_n] + F[y_n, z_n]}{F[x_n, y_n]}\right)^{-1} F(z_n).$$

Many other choices of functions A, B, and C are possible. Although many multistep methods for approximating multiple zeros of nonlinear equations have been given, almost no information is provided for computable radius of convergence and error bounds. That is, we do not know how close to the solution x^* the initial point x_0 should be. In the present paper motivated by the work by Argyros and Ren in [44] and Zhou, Chen, Song in [50] on the corrected Newton's method we present a local convergence analysis for method (25.1.2), where the radius of convergence and the error bounds on the distances $\|x_n - x^*\|$ depend on Lipschitz parameters.

The rest of the chapter is organized as follows: Section 25.2 contains the local convergence analysis of method (25.1.2). Finally, the numerical examples are presented in the concluding Section 25.4.

25.2 AUXILIARY RESULTS

We introduce the definition of the beta function and state a well known result concerning the Taylor expansion with integral form remainder.

Definition 25.2.1. *[1], [49] The incomplete beta function $B(s, p, q)$ is defined by*

$$B(s, p, q) = \int_0^1 t^{p-1}(1-t)^{q-1}dt \quad \text{for each } s \in (0, 1]. \quad (25.2.1)$$

If $s = 1$ then the incomplete beta function reduces to the complete beta function $B(p, q)$ defined by

$$B(p, q) = \int_0^1 t^{p-1}(1-t)^{q-1}dt. \quad (25.2.2)$$

Let

$$\Gamma(p) = \int_0^\infty e^{-t}t^{p-1}dt. \quad (25.2.3)$$

Then the beta function can be given in the form

$$B(p, q) = \frac{\Gamma(p)\Gamma(q)}{\Gamma(p+q)}. \quad (25.2.4)$$

More properties of these functions can be found in [1,49] and references therein.

Let $U(v, \rho)$, $\bar{U}(v, \rho)$ stand for the open and closed balls in S, respectively, with center $v \in S$ and radius $\rho > 0$. Next we state the following Taylor expansion result with integral form remainder.

Theorem 25.2.2. *[1,4,12,44,49,50] Suppose that $f : U(x_0, \rho) \to \mathbb{R}$ is i-times differentiable and $f^{(i)}(x)$ is integrable from x^* to any $x \in U(x^*, r)$. Then the following expansion holds:*

$$f(x) = f(x^*) + f'(x^*)(x - x^*) + \frac{1}{2}f''(x^*)(x - x^*)^2$$

$$+ \cdots + \frac{1}{i!}f^{(i)}(x^*)(x - x^*)^i$$

$$+ \frac{1}{(i-1)!}\int_0^1 \left[f^{(i)}(x^* + t(x - x^*)) - f^{(i)}(x^*) \right](x - x^*)^i(1 - t)^{i-1}dt.$$

$$(25.2.5)$$

Moreover, by differentiating both sides of (25.2.5), we obtain

$$
f'(x) = f'(x^*) + f''(x^*)(x - x^*) + \frac{1}{2} f'''(x^*)(x - x^*)^2
$$

$$
+ \cdots + \frac{1}{(i-1)!} f^{(i)}(x^*)(x - x^*)^{i-1}
$$

$$
+ \frac{1}{(i-2)!} \int_0^1 \left[f^{(i)}(x^* + t(x - x^*)) - f^{(i)}(x^*) \right] (x - x^*)^{i-1}(1-t)^{i-2} dt.
$$

$$(25.2.6)$$

25.3 LOCAL CONVERGENCE

We present the local convergence analysis of method (25.1.2) in this section. It is convenient for the local convergence analysis that follows to introduce some functions and parameters. Let $L_0 > 0$, $p \in (0, 1]$ be given parameters and $\varphi_j :$ $[0, R_0) \to [0, +\infty)$ be continuous and nondecreasing functions for $j = 1, 2, 3$, where

$$
R_0 = \left(\frac{1}{(m-1)B(p+1, m-1)L_0} \right)^{\frac{1}{p}}.
$$

$$(25.3.1)$$

Define functions g_j, h_j on the interval $[0, R_0)$ by

$$
g_1(t) = \frac{(2a + pB(m, p+1))L_0 t^p}{1 - (m-1)B(p+1, m-1)L_0 t^p}
$$

$$
+ \varphi_1(t) \left(\frac{1}{m!} t^{m-1} + \frac{L_0 B(p+1, m)t^{p+m-1}}{(m-1)!} \right),
$$

$$
h_1(t) = g_1(t) - 1,
$$

$$
g_2(t) = g_1(t) + \varphi_2(t) \left[\frac{1}{m!} g_1^m(t) t^{m-1} + \frac{L_0 B(p+1, m)g_1^{m+p}(t)t^{p+m-1}}{(m-1)!} \right],
$$

$$
h_2(t) = g_2(t) - 1,
$$

$$
g_3(t) = g_2(t) + \varphi_3(t) \left[\frac{1}{m!} g_2^m(t) t^{m-1} + \frac{L_0 B(p+1, m)g_2^{m+p}(t)t^{p+m-1}}{(m-1)!} \right],
$$

and

$$
h_3(t) = g_3(t) - 1,
$$

where

$$
a = \int_0^{\frac{1}{m}} t^p (1-t)^{m-2}(1 - mt)dt.
$$

$$(25.3.2)$$

Using the definition of these functions, we can obtain in turn that $h_1(0) = -1$ and $h_1(t) \to +\infty$ as $t \to R_0^-$. It then follows from the intermediate value theorem that function h_1 has a zero in the interval $(0, R_0)$. Denote by r_1 the smallest such zero. Secondly, we have that $h_2(0) = -1 < 0$ and

$$h_2(r_1) = \varphi_2(r_1) \left[\frac{r_1^{m-1}}{m!} + \frac{L_0 B(p+1, m) r_1^{p+m-1}}{(m-1)!} \right] \geq 0,$$

since $g_1(r_1) = 1$. Denote by r_2 the smallest such zero in the interval $(0, r_1]$.
Thirdly, we have that $h_3(0) = -1 < 0$ and

$$h_3(r_2) = \varphi_3(r_2) \left[\frac{r_2^{m-1}}{m!} + \frac{L_0 B(p+1, m) r_2^{p+m-1}}{(m-1)!} \right] \geq 0.$$

Denote by r_3 the smallest such zero in the interval $(0, r_2]$. Then we have that

$$r_3 < r_2 < r_1 < R_0 \tag{25.3.3}$$

and for each $t \in [0, r_3)$

$$0 \leq g_1(t) < 1, \tag{25.3.4}$$

$$0 \leq g_2(t) < 1, \tag{25.3.5}$$

and

$$0 \leq g_3(t) < 1. \tag{25.3.6}$$

Next we present the local convergence analysis of method (25.1.2) using the preceding notation.

Theorem 25.3.1. *Let $F : D \subset \mathbb{R} \to \mathbb{R}$ be an m-times differentiable function. Suppose that there exist $x^* \in D$, $L_0 > 0$, $p \in (0, 1]$, and continuous and nondecreasing functions $\varphi_i : [0, R_0) \to [0, +\infty)$, $i = 1, 2, 3$, such that for each $x \in D$*

$$F(x^*) = F'(x^*) = \cdots = F^{(m-1)}(x^*) = 0, \quad F^{(m)}(x^*) \neq 0, \tag{25.3.7}$$

$$\|F^{(m)}(x^*)^{-1}(F^{(m)}(x) - F^{(m)}(x^*))\| \leq L_0 \|x - x^*\|^p, \text{ for each } x \in D, \tag{25.3.8}$$

$$\|(mF'(x)^{-1} - A(x))F^{(m)}(x^*)\| \leq \varphi_1(\|x - x^*\|), \tag{25.3.9}$$

$$\|B(x)F^{(m)}(x^*)\| \leq \varphi_2(\|x - x^*\|), \tag{25.3.10}$$

$$\|C(x)F^{(m)}(x^*)\| \leq \varphi_3(\|x - x^*\|), \tag{25.3.11}$$

and

$$\bar{U}(x^*, r_3) \subseteq D, \tag{25.3.12}$$

where the radius r_3 is defined previously. Then the sequence $\{x_n\}$ generated for $x_0 \in U(x^*, r_3) \setminus \{x^*\}$ by method (25.1.2) is well defined, remains in $\bar{U}(x^*, r_3)$ for each $n = 0, 1, 2, \ldots$, and converges to x^*. Moreover, the following estimates hold for each $n = 0, 1, 2, \ldots$:

$$\|y_n - x^*\| \leq g_1(\|x_n - x^*\|)\|x_n - x^*\| < \|x_n - x^*\| < r_3, \qquad (25.3.13)$$

$$\|z_n - x^*\| \leq g_2(\|x_n - x^*\|)\|x_n - x^*\| < \|x_n - x^*\|, \qquad (25.3.14)$$

and

$$\|x_{n+1} - x^*\| \leq g_3(\|x_n - x^*\|)\|x_n - x^*\| < \|x_n - x^*\|, \qquad (25.3.15)$$

where the "g" functions are defined above Theorem 25.3.1. Furthermore, if $r_3 < R_1$ and $\bar{U}(x^*, R) \subseteq D$ for $R \in [r_3, R_1)$, then the limit point x^* is the only solution of equation $F(x) = 0$ in $\bar{U}(x^*, R)$. Finally, if $R_1 \leq R_3$, then the solution x^* is unique in $\bar{U}(x^*, R_1)$.

Proof. We shall show estimates (25.3.13)–(25.3.15) using mathematical induction. Using the definition of R_0, r_3, hypothesis $x_0 \in U(x^*, r_3) \setminus \{x^*\}$, (25.3.3), and (25.3.8), we have in turn that

$$\|F^{(m)}(x^*)^{-1}$$
$$\times \left[F^{(m)}(x^*) - (F^{(m)}(x^*) + (m-1)\int_0^1 (F^{(m)}(x^* + t(x_0 - x^*)) \right.$$
$$\left. - F^{(m)}(x^*))(1-t)^{m-2}dt) \right] \|$$
$$= (m-1)\| \int_0^1 F^{(m)}(x^*)^{-1}$$
$$\times \left[\int_0^1 (F^{(m)}(x^* + t(x_0 - x^*)) - F^{(m)}(x^*))(1-t)^{m-2}dt \right] \| \qquad (25.3.16)$$
$$\leq (m-1)L_0 \int_0^1 t^p (1-t)^{m-2}dt \|x_0 - x^*\|^p$$
$$= (m-1)L_0 B(p+1, m-1)\|x_0 - x^*\|^p$$
$$< (m-1)L_0 B(p+1, m-1)r_3^p$$
$$< \frac{(m-1)L_0 B(p+1, m-1)}{(2a + (m+2p)B(p+1, m))L_0}$$
$$= \frac{m+p}{m+2p} < 1,$$

since by (25.3.2)

$$a = B(\frac{1}{m}, p+1, m-1) - mB(\frac{1}{m}, p+2, m-1) > 0. \qquad (25.3.17)$$

It follows from (25.3.16) and the Banach lemma on invertible functions [12, 13,48] that

$$F^{(m)}(x^*) + (m-1) \int_0^1 \left[F^{(m)}(x^* + t(x_0 - x^*)) - F^{(m)}(x^*) \right] (1-t)^{m-2} dt \neq 0$$

and

$$(\| F^{(m)}(x^*)^{-1} (F^{(m)}(x^*) + (m-1) \int_0^1 \left[F^{(m)}(x^* + t(x_0 - x^*)) \right.$$
$$\left. - F^{(m)}(x^*) \right] (1-t)^{m-2} dt \|))^{-1} \qquad (25.3.18)$$
$$< \frac{1}{1 - (m-1)B(p+1, m-1)L_0 \| x_0 - x^* \|^p}.$$

Hence, y_0, z_0, and x_{n+1} are well defined by method (25.1.2) for $n = 0$. We also have by (25.3.8) that

$$\| \int_0^1 F^{(m)}(x^*)^{-1} \left[F^{(m)}(x^* + t(x_0 - x^*)) - F^{(m)}(x^*) \right] \left[(m-1)(t-1)^{m-2} \right.$$
$$\left. - m(1-t)^{m-1} \right] dt \|$$

$$\leq \int_0^1 \| F^{(m)}(x^*)^{-1} \left[F^{(m)}(x^* + t(x_0 - x^*)) - F^{(m)}(x^*) \right] \| \| (m-1)(1-t)^{m-2}$$
$$- m(1-t)^{m-1} dt \|$$

$$\leq L_0 \int_0^1 t^p \| (m-1)(1-t)^{m-2} - m(1-t)^{m-1} dt \| \| x_0 - x^* \|^p$$

$$= L_0 \int_0^1 t^p (1-t)^{m-2} |1 - mt| dt \| \| x_0 - x^* \|^p$$

$$= \left[\int_0^{\frac{1}{m}} t^p (1-t)^{m-2}(1-mt) dt + \int_0^1 t^p (1-t)^{m-2}(1-mt) dt \right]$$
$$\times L_0 \| x_0 - x^* \|^p$$

$$= \left[2 \int_0^{\frac{1}{m}} t^p (1-t)^{m-2}(1-mt) dt + \int_0^1 t^p (1-t)^{m-2}(mt-1) dt \right]$$
$$\times L_0 \| x_0 - x^* \|^p$$

$$= (2a + pB(m, p+1))L_0 \| x_0 - x^* \|^p.$$
$$(25.3.19)$$

Then, using the definition of g_1, r_3, the first substep of method (25.1.2) for $n = 0$, (25.3.4), (25.3.7), (25.3.9), (25.3.18), and (25.3.19), we obtain in turn from

$$
\begin{aligned}
y_0 - x^* &= (x_0 - x^* - mF'(x_0)^{-1}F(x_0)) \\
&\quad + (mF'(x_0)^{-1} - A_0)F^{(m)}(F^{(m)}(x^*)^{-1}F(x_0))
\end{aligned}
$$

that

$$
\| y_0 - x^* \| \leq
$$

$$
\left\| \frac{\int_0^1 \left[F^{(m)}(x^* + t(x_0 - x^*)) - F^{(m)}(x^*) \right] \left[(m-1)(1-t)^{m-2} - m(1-t)^{m-1} \right] dt \, (x_0 - x^*)^p}{F^{(m)}(x^*) + (m-1)\int_0^1 \left[F^{(m)}(x^* + t(x_0 - x^*)) - F^{(m)}(x^*) \right](1-t)^{m-2} dt} \right\|
$$

$$
+ \| (mF'(x_0)^{-1} - A_0)F^{(m)}(x^*) \| \| F^{(m)}(x^*)^{-1}F(x_0) \|
$$

$$
\leq \frac{(2a + pB(m, p+1))L_0 \|x_0 - x^*\|^{p+1}}{1 - (m-1)B(p+1, m-1)L_0 \|x_0 - x^*\|^p}
$$

$$
+ \varphi_1(\|x_0 - x^*\|) \left[\frac{\|x_0 - x^*\|^m}{m!} + \frac{L_0 B(p+1, m)\|x_0 - x^*\|^{m+p}}{(m-1)!} \right]
$$

$$
= g_1(\|x_0 - x^*\|)\|x_0 - x^*\| < \|x_0 - x^*\| < r_3,
$$

$$
(25.3.20)
$$

which shows (25.3.13) for $n = 0$ and $y_0 \in U(x^*, r_3)$ where we also used (25.2.5) to obtain the estimate

$$
\| F^{(m)}(x^*)^{-1}F(x_0) \|
$$

$$
\leq \frac{\|x_0 - x^*\|^m}{m!} + \frac{\|x_0 - x^*\|^m L_0}{(m-1)!} \int_0^1 t^p (1-t)^{m-1} dt \, \|x_0 - x^*\|^p \quad (25.3.21)
$$

$$
= \frac{\|x_0 - x^*\|^m}{m!} + \frac{L_0 \|x_0 - x^*\|^{m+p}}{(m-1)!} B(p+1, m).
$$

Then, using the second substep of method (25.1.2) for $n = 0$, (25.3.3), (25.3.5), (25.3.10), (25.3.20), and (25.3.21) (for $y_0 = x_0$) we get from

$$
z_0 - x^* = (y_0 - x^*) + (B_0 F^{(m)}(x^*))(F^{(m)}(x^*)^{-1}F(y_0))
$$

that

$$
\| z_0 - x^* \| \leq \| y_0 - x^* \| + \| B_0 F^{(m)}(x^*) \| \| F^{(m)}(x^*)^{-1}F(y_0) \|
$$

$$
\leq g_1(\|x_0 - x^*\|)\|x_0 - x^*\| + \varphi_2(\|x_0 - x^*\|) \quad (25.3.22)
$$

$$
\times \left[\frac{\|y_0 - x^*\|^m}{m!} + \frac{L_0 B(p+1, m)\|y_0 - x^*\|^{m+p}}{(m-1)!} \right]
$$

$$
\leq g_2(\|x_0 - x^*\|)\|x_0 - x^*\| < \|x_0 - x^*\| < r_3,
$$

which shows (25.3.14) for $n = 0$ and $z_0 \in U(x^*, r_3)$. Next, using the third substep of method (25.1.2) for $n = 0$, (25.3.3), (25.3.6), (25.3.11), (25.3.21) (for

$z_0 = x_0$) and (25.3.22), we get that

$$
\begin{aligned}
\|x_1 - x^*\| &\le \|z_0 - x^*\| + \|C_0 F^{(m)}(x^*)\| \| F^{(m)}(x^*)^{-1} F(z_0)\| \\
&\le g_2(\|x_0 - x^*\|)\|x_0 - x^*\| + \varphi_3(\|x_0 - x^*\|) \\
&\quad \times \left[\frac{\|z_0 - x^*\|^m}{m!} + \frac{L_0 B(p+1,m)\|z_0 - x^*\|^{m+p}}{(m-1)!} \right] \\
&\le g_3(\|x_0 - x^*\|)\|x_0 - x^*\| < \|x_0 - x^*\| < r_3,
\end{aligned}
\tag{25.3.23}
$$

which shows (25.3.15) for $n = 0$ and $x_1 \in U(x^*, r_3)$. By simply replacing x_0, y_0, z_0, x_1 by x_k, y_k, z_k, x_{k+1} in the preceding estimates, we arrive at estimates (25.3.13)–(25.3.15). Then, from the estimates $\|x_{k+1} - x^*\| < \|x_k - x^*\| < r_3$, we deduce that $\lim_{k \to \infty} x_k = x^*$ and $x_{k+1} \in U(x^*, r_3)$. Finally, to show the uniqueness part for $r_3 < R_1$ and $R \in [r_3, R_1)$, let $y^* \in \bar{U}(x^*, R)$ a solution of equation $F(x) = 0$ in $\bar{U}(x^*, R)$. We can write

$$
0 = F(y^*) - F(x^*) = \frac{1}{(m-1)!} \int_{x^*}^{y^*} F^{(m)}(t)(y^* - t)^{m-1} dt. \tag{25.3.24}
$$

Suppose that $y^* \ne x^*$. Using (25.3.8), we get in turn that

$$
\begin{aligned}
&\left\| \frac{(y^* - x^*)^m}{m} F^{(m)}(x^*)^{-1} \right. \\
&\quad \times \left. \left(\frac{(y^* - x^*)^m}{m} F^{(m)}(x^*) - \int_{x^*}^{y^*} F^{(m)}(t)(y^* - t)^{m-1} dt \right) \right\| \\
&\le \frac{m L_0}{(y^* - x^*)^m} \int_{x^*}^{y^*} (t - x^*)^p (y^* - t)^{m-1} dt \\
&= \frac{m L_0}{(y^* - x^*)^m} (y^* - x^*)^{m+p} B(p+1,m) \\
&\le \frac{m B(p+1,m) L_0}{(m+2p) B(p+1,m) L_0} = \frac{m}{m+2} < 1.
\end{aligned}
\tag{25.3.25}
$$

It follows from (25.3.25) that $\dfrac{(y^* - x^*)^m}{m} F^{(m)}(x^*)^{-1} \int_{x^*}^{y^*} (t - x^*)^p (y^* - t)^{m-1} dt$ is invertible and consequently $\int_{x^*}^{y^*} (t - x^*)^p (y^* - t)^{m-1} dt$ is also invertible, which together with (25.3.24) contradict the hypothesis $y^* \ne x^*$. Hence we conclude that $x^* = y^*$. $\qquad \square$

Remark 25.3.2. 1. *The results obtained here can be used for operators F satisfying the autonomous differential equation [12,13] of the form*

$$
F^{(m)}(x) = Q(F(x)),
$$

where Q is a known continuous operator. Since $F^{(m)}(x^*) = Q(F(x^*)) = Q(0)$, we can apply the results without actually knowing the solution x^*. Let as an example, define $F(x) = x(e^x - 1)$. Notice that in this case $m = 2$. Then we can choose $Q(x) = x + 2e^x + F(x)$.

25.4   NUMERICAL EXAMPLES

Example 25.4.1. *Let*

$$S = \mathbb{R},$$

$$D = [-1, 1],$$

$$x^* = 0,$$

and define function F on D by

$$F(x) = x^5 + x^3. \tag{25.4.1}$$

Then, choosing

$$p = 0.5,$$

$$A(x) = \frac{1000}{x}, \quad B(x) = x, \quad C(x) = 2x,$$

we obtain that

$$R_0 = 1.7777\ldots,$$

$$a = 0.0439886\ldots,$$

and

$$L_0 = 2.$$

Then, by the definition of the "g" functions, we obtain

$$r_3 = 0.00181875\ldots$$

So we can ensure the convergence of the method (25.1.2).

Example 25.4.2. *Let*

$$\mathcal{X} = [-1, 1],$$

$$\mathcal{Y} = \mathbb{R},$$

$$x_0 = 0,$$

and let $F : \mathcal{X} \to \mathcal{Y}$ be given by

$$F(x) = x(e^x - 1).$$

Then, choosing

$$p = 1,$$

$$A(x) = \frac{2}{-1 + e^x + e^x x},$$

$$B(x) = 1,$$

and

$$C(x) = 1,$$

we obtain that

$$R_0 = 1.53871\ldots,$$

$$a = 0.0416666\ldots,$$

and

$$L_0 = \frac{3e - 2}{2}.$$

Then, by the definition of the "g" functions, we obtain

$$r_3 = 0.348646\ldots$$

So we can ensure the convergence of the method (25.1.2).

REFERENCES

[1] M. Abramowitz, I.S. Stegun, Handbook of Mathematical Functions With Formulas, Graphs, and Mathematical Tables, Applied Math. Ser., vol. 55, United States Department of Commerce, National Bureau of Standards, Washington DC, 1964.

[2] S. Amat, S. Busquier, S. Plaza, Dynamics of the King and Jarratt iterations, Aequationes Math. 69 (3) (2005) 212–223.

[3] S. Amat, S. Busquier, S. Plaza, Chaotic dynamics of a third-order Newton-type method, J. Math. Anal. Appl. 366 (1) (2010) 24–32.

[4] S. Amat, M.A. Hernández, N. Romero, A modified Chebyshev's iterative method with at least sixth order of convergence, Appl. Math. Comput. 206 (1) (2008) 164–174.

[5] S. Amat, I.K. Argyros, S. Busquier, Á.A. Magreñán, Local convergence and the dynamics of a two-point four parameter Jarratt-like method under weak conditions, Numer. Algorithms 74 (2) (2017) 371–391.

[6] S. Amat, S. Busquier, C. Bermúdez, Á.A. Magreñán, On the election of the damped parameter of a two-step relaxed Newton-type method, Nonlinear Dynam. 84 (1) (2016) 9–18.

[7] I.K. Argyros, A. Cordero, Á.A. Magreñán, J.R. Torregrosa, On the convergence of a higher order family of methods and its dynamics, J. Comput. Appl. Math. 309 (2017) 542–562.

[8] I.K. Argyros, Á.A. Magreñán, A study on the local convergence and the dynamics of Chebyshev–Halley-type methods free from second derivative, Numer. Algorithms 71 (1) (2016) 1–23.

[9] I.K. Argyros, Á.A. Magreñán, On the convergence of an optimal fourth-order family of methods and its dynamics, Appl. Math. Comput. 252 (2015) 336–346.

[10] I.K. Argyros, Á.A. Magreñán, L. Orcos, Local convergence and a chemical application of derivative free root finding methods with one parameter based on interpolation, J. Math. Chem. 54 (7) (2016) 1404–1416.

[11] I.K. Argyros, Á.A. Magreñán, Local convergence and the dynamics of a two-step Newton-like method, Internat. J. Bifur. Chaos. 26 (5) (2016) 1630012.

[12] I.K. Argyros, Convergence and Application of Newton-Type Iterations, Springer, 2008.

[13] I.K. Argyros, S. Hilout, Numerical Methods in Nonlinear Analysis, World Scientific Publ. Comp., New Jersey, 2013.

[14] R. Behl, Development and Analysis of Some New Iterative Methods for Numerical Solutions of Nonlinear Equations, PhD thesis, Punjab University, 2013.

[15] V. Candela, A. Marquina, Recurrence relations for rational cubic methods I: the Halley method, Computing 44 (1990) 169–184.

[16] V. Candela, A. Marquina, Recurrence relations for rational cubic methods II: the Chebyshev method, Computing 45 (4) (1990) 355–367.

[17] A. Cordero, J.M. Gutiérrez, Á.A. Magreñán, J.R. Torregrosa, Stability analysis of a parametric family of iterative methods for solving nonlinear models, Appl. Math. Comput. 285 (2016) 26–40.

[18] A. Cordero, Á.A. Magreñán, C. Quemada, J.R. Torregrosa, Stability study of eighth-order iterative methods for solving nonlinear equations, J. Comput. Appl. Math. 291 (2016) 348–357.

[19] A. Cordero, L. Feng, Á.A. Magreñán, J.R. Torregrosa, A new fourth-order family for solving nonlinear problems and its dynamics, J. Math. Chem. 53 (3) (2014) 893–910.

[20] F. Chicharro, A. Cordero, J.R. Torregrosa, Drawing dynamical and parameters planes of iterative families and methods, Sci. World J. (2013) 780153.

[21] C. Chun, Some improvements of Jarratt's method with sixth-order convergence, Appl. Math. Comput. 190 (2) (1990) 1432–1437.

[22] A. Cordero, J. García-Maimó, J.R. Torregrosa, M.P. Vassileva, P. Vindel, Chaos in King's iterative family, Appl. Math. Lett. 26 (2013) 842–848.

[23] A. Cordero, J.R. Torregrosa, P. Vindel, Dynamics of a family of Chebyshev–Halley type methods, Appl. Math. Comput. 219 (2013) 8568–8583.

[24] A. Cordero, J.R. Torregrosa, Variants of Newton's method using fifth-order quadrature formulas, Appl. Math. Comput. 190 (2007) 686–698.

[25] J.A. Ezquerro, M.A. Hernández, On the R-order of the Halley method, J. Math. Anal. Appl. 303 (2005) 591–601.

[26] Y.H. Geum, Y.I. Kim, Á.A. Magreñán, A biparametric extension of King's fourth-order methods and their dynamics, Appl. Math. Comput. 282 (2016) 254–275.

[27] J.M. Gutiérrez, Á.A. Magreñán, N. Romero, Dynamic aspects of damped Newton's method, in: Civil-Comp Proceedings, vol. 100, 2012.

[28] J.M. Gutiérrez, Á.A. Magreñán, J.L. Varona, Fractal dimension of the universal Julia sets for the Chebyshev–Halley family of methods, in: AIP Conf. Proc., vol. 1389, 2011, pp. 1061–1064.

[29] J.M. Gutiérrez, Á.A. Magreñán, J.L. Varona, The "Gauss-seidelization" of iterative methods for solving nonlinear equations in the complex plane, Appl. Math. Comput. 218 (6) (2011) 2467–2479.

[30] J.M. Gutiérrez, M.A. Hernández, Recurrence relations for the super-Halley method, Computers Math. Applic. 36 (7) (1998) 1–8.

[31] M.A. Hernández, Chebyshev's approximation algorithms and applications, Computers Math. Applic. 41 (3–4) (2001) 433–455.

[32] M.A. Hernández, M.A. Salanova, Sufficient conditions for semilocal convergence of a fourth order multipoint iterative method for solving equations in Banach spaces, Southwest J. Pure Appl. Math. 1 (1999) 29–40.

[33] P. Jarratt, Some fourth order multipoint methods for solving equations, Math. Comput. 20 (95) (1966) 434–437.

[34] J. Kou, X. Wang, Semilocal convergence of a modified multi-point Jarratt method in Banach spaces under general continuity conditions, Numer. Algorithms 60 (2012) 369–390.

[35] X. Li, C. Mu, J. Ma, L. Hou, Fifth-order iterative method for finding multiple roots if nonlinear equations, Numer. Algorithms 57 (2011) 389–398.

[36] T. Lotfi, Á.A. Magreñán, K. Mahdiani, J.J. Rainer, A variant of Steffensen–King's type family with accelerated sixth-order convergence and high efficiency index: dynamic study and approach, Appl. Math. Comput. 252 (2015) 347–353.

[37] Á.A. Magreñán, Estudio de la dinámica del método de Newton, amortiguado (PhD thesis), Servicio de Publicaciones, Universidad de La Rioja, 2013, http://dialnet.unirioja.es/servlet/tesis?codigo=38821.

[38] Á.A. Magreñán, Different anomalies in a Jarratt family of iterative root-finding methods, Appl. Math. Comput. 233 (2014) 29–38.

[39] Á.A. Magreñán, A new tool to study real dynamics: the convergence plane, Appl. Math. Comput. 248 (2014) 215–224.

[40] Á.A. Magreñán, I.K. Argyros, On the local convergence and the dynamics of Chebyshev–Halley methods with six and eight order of convergence, J. Comput. Appl. Math. 298 (2016) 236–251.

[41] B. Neta, C. Chun, A family of Laguerre methods to find multiple roots of nonlinear equations, Appl. Math. Comput. 219 (2013) 10987–11004.

[42] B. Neta, C. Chun, Erratum "On a family of Laguerre methods to find multiple roots of nonlinear equations", Appl. Math. Comput. 248 (2014) 693–696.

[43] M. Petković, B. Neta, L. Petković, J. Džunić, Multipoint Methods for Solving Nonlinear Equations, Elsevier, 2013.

[44] H. Ren, I.K. Argyros, Convergence radius of the modified Newton method for multiple zeros under Hölder continuous derivative, Appl. Math. Comput. 217 (2) (2010) 612–621.

[45] H. Ren, Q. Wu, W. Bi, New variants of Jarratt method with sixth-order convergence, Numer. Algorithms 52 (4) (2009) 585–603.

[46] W.C. Rheinboldt, An adaptive continuation process for solving systems of nonlinear equations, in: Banach Ctr. Publ., vol. 3, Polish Academy of Science, 1978, pp. 129–142.

[47] R. Sharma, A. Bahl, Sixth order transformation method for finding multiple roots of nonlinear equations and basins for various methods, Appl. Math. Comput. 269 (15 October 2015) 105–117, https://doi.org/10.1016/j.amc.2015.07.056.

[48] J.F. Traub, Iterative Methods for the Solution of Equations, Prentice-Hall Ser. Automat. Comput., Prentice Hall, Englewood Cliffs, NJ, 1964.

[49] S. Wolfram, The Mathematica Book, 5th edition, Wolfram, Media, 2003.

[50] X. Zhou, X. Chen, Y. Song, On the convergence radius of the modified Newton method for multiple roots under the center-Hölder condition, Numer. Algorithms 65 (2) (2014) 221–232.

Index

Printed in the United States
By Bookmasters